Algebraic Geometry

On behalf of the participants, we are pleased to dedicate this special volume of Contemporary Mathematics to Igor Dolgachev on the occasion of his sixtieth birthday. His love of mathematics has inspired everyone around him. We hope that these articles will convey to him the gratitude and admiration of his students, friends and collaborators.

JongHae Keum & Shigeyuki Kondō

CONTEMPORARY MATHEMATICS

422

Algebraic Geometry

Korea-Japan Conference
in Honor of Igor Dolgachev's 60th Birthday
July 5–9, 2004
Korea Institute for Advanced Study
Seoul, Korea

JongHae Keum
Shigeyuki Kondō
Editors

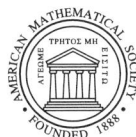

American Mathematical Society
Providence, Rhode Island

2000 *Mathematics Subject Classification.* Primary 11D25, 11F11, 14H10, 14J15, 14J26, 14J28, 14J50, 14J60, 32N15.

Library of Congress Cataloging-in-Publication Data
Korea-Japan Conference on Algebraic Geometry (2004 : Seoul, Korea)
   Algebraic geometry : proceedings of the Korea-Japan Conference on Algebraic Geometry in honor of Igor Dolgachev's 60th birthday, July 5–9, 2004, Korea Institute for Advanced Study, Seoul, Korea / JongHae Keum, Shigeyuki Kondō, editors.
      p. cm. — (Contemporary mathematics, ISSN 0271-4132 ; v. 422)
   Includes bibliographical references.
   ISBN-13: 978-0-8218-4201-0 (alk. paper)
   ISBN-10: 0-8218-4201-3 (alk. paper)
   1. Geometry, Algebraic—Congresses.   I. Dolgachev, I. (Igor V.)   II. Keum, JongHae, 1957–
III. Kondō, Shigeyuki, 1958–   IV. Title.

QA564.K67   2004
516.3′5—dc22
                                                                                    2006049918

Contents

Preface

This volume contains the proceedings of the Korea-Japan Conference on Algebraic Geometry held in honor of Igor Dolgachev on his sixtieth birthday at the Korea Institute for Advanced Study in Seoul, Korea on July 5-9, 2004. Many of the lectures given at the meeting are represented here; in addition, some work that was invited but not presented at the meeting appears.

The papers and the lectures explore a wide variety of problems that illustrate interactions between algebraic geometry and other branches of mathematics. Among the topics covered by this volume are algebraic curve theory, algebraic surface theory, moduli space, automorphic forms, Mordell-Weil lattices, and automorphisms of hyperkähler manifolds.

The conference was organized as an academic collaboration among mathematicians in Korea and Japan. It was not only a place for fruitful mathematical interaction, but also an occasion where old friendships were renewed and many new friendships were made. The atmosphere was very stimulating and enjoyable.

We would like to take this opportunity to thank all the contributors and participants for their efforts in making the conference a big success. We also thank all the referees for their invaluable time spent on these articles. Special thanks should go to the Korea Institute for Advanced Study (KIAS) and Japan Society for the Promotion of Science (JSPS) for their generous financial support.

October 2005
JongHae Keum & Shigeyuki Kondō

List of Participants

Valery Alexeev
U. Georgia, USA, valery@math.uga.edu

Insong Choe
KIAS, Korea, ischoe@kias.re.kr

Youngook Choi
KAIST, Korea, ychoi@math.kaist.ac.kr

Jaeyoo Choy
KIAS, Korea, donvosco@kias.re.kr

Elisabetta Colombo
Milano, Italy,
Elisabetta.Colombo@mat.unimi.it

I. Dolgachev
U Michigan, USA, idolga@umich.edu

B. van Geemen
U Milano, Italy, geemen@mat.unimi.it

M. Gizatullin
Russia, gizmarat@yandex.ru

Byungheup Jun
KIAS, Korea, bhjun@kias.re.kr

T. Katsura
U Tokyo, Japan,
tkatsura@ms.u-tokyo.ac.jp

JongHae Keum
KIAS, Korea, jhkeum@kias.re.kr

Young-Hoon Kiem
Seoul National U, Korea,
kiem@math.snu.ac.kr

Bumsig Kim
KIAS, Korea, bumsig@kias.re.kr

Hoil Kim
Kyungpook National U, Korea,
hikim@knu.ac.kr

Yonggu Kim
Chonnam National U, Korea,
kimm@chonnam.ac.kr

S. Kondō
Nagoya U, Japan,
kondo@math.nagoya-u.ac.jp

Sijong Kwak
KAIST, Korea,
sjkwak@math.kaist.ac.kr

Yongnam Lee
Sogang University, Korea,
ynlee@ccs.sogang.ac.kr

E. Looijenga
U Utrecht, Netherlands,
looijeng@math.uu.nl

Sung Myung
KIAS, Korea, s-myung1@kias.re.kr

V. Nikulin
U Liverpool, UK, vnikulin@liv.ac.uk

H. Ochiai
Nagoya U, Japan,
ochiai@math.nagoya-u.ac.jp

K. Oguiso
Tokyo, Japan, (Current address : Keio
U, Japan), oguiso@hc.cc.keio.ac.jp

Euisung Park
KIAS, Korea, puserdos@kias.re.kr

Jihun Park
Postech, Korea, wlog@postech.ac.kr

Masa-Hiko Saito
Kobe U, Japan,
ym-saito@ya3.so-net.ne.jp

I. Shimada
Hokkaido U, Japan,
shimada@math.sci.hokudai.ac.jp

T. Shioda
Rikkyo U, Japan,
shioda@rkmath.rikkyo.ac.jp

Pho Duc Tai
Hokkaido U, Japan,
tai@math.sci.hokudai.ac.jp

A. Verra
U Rome, Italy,
verra@matrm3.mat.uniroma3.it

C. Werner
Allegheny U, USA,
cwerner@allegheny.edu

D.-Q. Zhang
National University of Singapore,
matzdq@math.nus.edu.sg

List of Talks given at the Conference

Valery Alexeev : Toric degenerations of spherical varieties

B. van Geemen : Some remarks on the Brauer groups of elliptic fibrations

M. Gizatullin : On covariants of a plane quartic associated to its even theta characteristic

T. Katsura : On a stratification of moduli of K3 surfaces

JongHae Keum : Finite groups acting on K3 surfaces in positive characteristic

Young-Hoon Kiem : Desingularizations of moduli space of vector bundles over a curve

Bumsig Kim : Toward mirror symmetry of nonabelian quotients

Hoil Kim : Noncommutative theta functions

S. Kondō : Moduli of hyperelliptic curves of genus 3 and Borcherds products

Yongnam Lee : Singularities of pair, the Grassmannian variety and the Chow form

E. Looijenga : Geometric structures on arrangement complements

V. Nikulin : On correspondences of a K3 surface with itself, II

K. Oguiso : Automorphism groups of hyperkähler manifolds of null entropy

Jihun Park : **Q**-factorial double solids with simple double points

Masa-Hiko Saito : Moduli spaces of parabolic Higgs bundles, their deformations and integrable systems

I. Shimada : Moduli of supersingular K3 surfaces in characteristic 2

T. Shioda : Mordell-Weil lattices and K3 surfaces

A. Verra : Remarks on the uniruledness of M_g for low g

D.-Q. Zhang : The role of the alternating group of degree 6 in the geometry of Leech lattice and K3 surfaces

Contemporary Mathematics
Volume **422**, 2007

Holomorphic Eisenstein Series with Jacobian Twists

Lev A. Borisov

ABSTRACT. For every point on the Jacobian of the modular curve $X_0(l)$ we define and study certain twisted holomorphic Eisenstein series. These are particular cases of a more general notion of twisted modular forms which correspond to sections on the modular curve $X_1(l)$ of the degree zero twists of line bundles of usual modular forms. We conjecture that a point on the Jacobian is rational if and only if the ratios of these twisted Eisenstein series of the same weights have rational coefficients.

1. Introduction

Let $X_0(l)$ denote the modular curve of level l. It is defined as the compactification of the quotient of the upper half plane $\mathcal{H}$ by the group $\Gamma_0(l)$ which consists of matrices $\begin{pmatrix} a & b \\ c & d \end{pmatrix} \in SL_2(\mathbb{Z})$ with $c \in l\mathbb{Z}$. We are interested in the Jacobian of $X_0(l)$. We associate to any point of this Jacobian a collection of holomorphic functions $E_{i,k;h}(\tau)$ on the upper half plane $\mathcal{H}$, which we call h-twisted Eisenstein series. Here $k \geq 3$ is an integer, $i \in \mathbb{Z}/l\mathbb{Z}$, and $h = h(\tau)$ is a modular form of weight two for $\Gamma_0(l)$ which is related to the point of the Jacobian by period integrals.

More specifically, to any weight two form h we associate a unitary character of $\Gamma_0(l)$ by

$$(1.1) \qquad \begin{pmatrix} a & b \\ c & d \end{pmatrix} \mapsto \exp\left(2\pi i \Re \int_{i\infty}^{-\frac{d}{c}} h(s)\, ds\right).$$

Then we use this unitary character to modify a definition of some Eisenstein series. The unitarity of the character assures the convergence. It then turns out that the character and the series depend only on the point of the Jacobian, so one gets well-defined invariants of degree zero invertible sheaves on $X_0(l)$.

We remark that twisted Eisenstein series and modular forms of this paper should not be confused with twisted Eisenstein series of Goldfeld and Gunnells, see [**Go**] and [**GG**], which involve only an additive version of the character (1.1). However, a nonholomorphic analog of the Eisenstein series considered in this paper has already appeared in [**P**].

2000 *Mathematics Subject Classification.* Primary 11F11.
The author was partially supported by NSF grant DMS-0140172.

The paper is organized as follows. In Section 2 we define the Eisenstein series twisted by modular forms $E_{k,i;h}(\tau)$ and prove their convergence. We use the Dirichlet summation trick to get a formula for the Fourier coefficients of $E_{k,i;h}$. In Section 3 we define a more general class of functions, called h-twisted modular forms. The definition is parallel to that of the usual modular forms, except for the extra character (1.1). We give a geometric interpretation of these twisted modular forms as sections of the twists of the line bundles of the usual modular forms on the modular curves by degree zero line bundles, which justifies our terminology. Section 4 extends the Petersson inner product to the twisted case. Section 5 contains the rationality Conjecture 5.1, as well as some highly circumstantial evidence for it. Finally, in Section 6 we list the open problems which currently by far outnumber the results.

Acknowledgements. I thank Paul Gunnells and Gautam Chinta who read the first version of the paper and suggested several useful references.

This paper is dedicated to the 60th birthday of Igor Dolgachev who was my PhD advisor at the University of Michigan in 1993-1996. His love of mathematics inspires everyone around him. It is also a safe bet that he can pinpoint exactly which 19th century mathematician has already proved all the results of this paper.

2. Eisenstein series twisted by modular forms

In this section we introduce the main objects of interest. Let $l > 0$ be a positive integer, which will be fixed throughout the paper. We will denote by τ the coordinate on the upper half plane $\mathcal{H}$ and will use the notation $q = e^{2\pi i \tau}$. Let $h(\tau) = \sum_{n=1}^{\infty} a_n q^n$ be a cusp form of weight two for the congruence subgroup $\Gamma_0(l)$. Let $H(\tau) = \int_{i\infty}^{\tau} h(s)\, ds$ be the antiderivative of h.

We first define the auxiliary series $\hat{E}_{k,\chi;h}(\tau)$.

DEFINITION 2.1. For a positive integer k, a character $\chi : (\mathbb{Z}/l\mathbb{Z})^* \to \mathbb{C}$ and h as above, we define h-twisted Eisenstein series by

$$\hat{E}_{k,\chi;h}(\tau) = \sum_{\Gamma_\infty \backslash \Gamma_0(l)} (c\tau + d)^{-k} \chi(d) e^{2\pi i \Re H(-\frac{d}{c})}.$$

Here Γ_∞ is the infinite cyclic subgroup generated by $\tau \to \tau + 1$.

REMARK 2.2. In Definition 2.1 and throughout the rest of the paper we use the convention $H(-\frac{d}{0}) = 0$. It is motivated by the fact that $-\frac{d}{c}$ is the preimage of $\tau = i\infty$ under $\tau \to \frac{a\tau + b}{c\tau + d}$. So if $c = 0$, this preimage is $i\infty$, and $H(i\infty) = 0$ by the definition.

We can reformulate the above series as follows. The cosets of $\Gamma_0(l)$ over Γ_∞ are given by pairs of integers (c, d) with $c \equiv 0 \bmod l$ and $\gcd(c, d) = 1$. Hence, we have

$$\hat{E}_{k,\chi;h}(\tau) = \sum_{c \in l\mathbb{Z}} \sum_{d \in \mathbb{Z}, \gcd(c,d)=1} (c\tau + d)^{-k} \chi(d) e^{2\pi i \Re H(-\frac{d}{c})}.$$

We will be actually interested in a slight variation of this series, which is defined below.

DEFINITION 2.3. We define

$$E_{k,i;h}(\tau) = \sum_{c \in l\mathbb{Z}} \sum_{d \in \mathbb{Z}, \gcd(l,d)=1} (c\tau + d)^{-k} e^{-2\pi i \frac{di}{l}} e^{2\pi i \Re H(-\frac{d}{c})}.$$

PROPOSITION 2.4. *For $k \geq 3$, the series $E_{k,i;h}(\tau)$ and $\hat{E}_{k,\chi;h}(\tau)$ absolutely converge to holomorphic functions on the upper half plane.*

PROOF. The series of absolute values is $\sum_{(c,d) \in \mathbb{Z}^2 - (0,0)} |c\tau + d|^{-k}$. It has $\leq$ $(const)R$ terms with $|c\tau + d| \in [R, R+1)$. Absolute convergence on compacts is also clear, which implies that the resulting sum is holomorphic. $\qquad\square$

REMARK 2.5. Proposition 2.4 just barely fails for $k = 2$ and fails quite miserably for $k = 1$, which is perhaps the case of most interest.

By splitting the series $E_{k,i;h}$ according to $\gcd(c,d)$ we observe that it is a linear combination of series of type $\hat{E}$ with coefficients that involve the values of Dedekind L-functions at k as well as Gauss sums. To write it out explicitly, let us denote by $\chi_1, \ldots, \chi_{\phi(l)}$ the characters $(\mathbb{Z}/l\mathbb{Z})^* \to \mathbb{C}$.

PROPOSITION 2.6. *For $k \geq 3$ there holds*

$$E_{k,i;h}(\tau) = \frac{1}{\phi(l)} \sum_{j=1}^{\phi(l)} \left(\sum_{t \in (\mathbb{Z}/l\mathbb{Z})^*} e^{-2\pi i \frac{ti}{l}} \chi_j^{-1}(t) \right) L(\chi_j, k) \hat{E}_{k,\chi_j;h}(\tau)$$

where $L(\chi, k)$ denotes the value of the Dedekind L-function at k, and ϕ is the Euler function.

PROOF. For any $i \in \mathbb{Z}/l\mathbb{Z}$ and d coprime to l, we have

$$e^{-2\pi i \frac{di}{l}} = \sum_{j=1}^{\phi(l)} \chi_j(d) r_{i,j}$$

where

$$r_{i,j} = \frac{1}{\phi(l)} \sum_{t \in (\mathbb{Z}/l\mathbb{Z})^*} e^{-2\pi i \frac{ti}{l}} \chi_j^{-1}(t).$$

This implies

$$E_{k,i;h}(\tau) = \sum_{c \in l\mathbb{Z}} \sum_{d \in \mathbb{Z}, \gcd(l,d)=1} \sum_{j=1}^{\phi(l)} (c\tau + d)^{-k} r_{i,j} \chi_j(d) e^{2\pi i \Re H(-\frac{d}{c})}$$

$$= \sum_{n \in \mathbb{Z}_{>0}, \gcd(n,l)=1} \sum_{c_1 \in l\mathbb{Z}} \sum_{d_1 \in \mathbb{Z}, \gcd(c_1,d_1)=1} \sum_{j=1}^{\phi(l)}$$

$$(c_1\tau + d_1)^{-k} n^{-k} r_{i,j} \chi_j(d_1) \chi_j(n) e^{2\pi i \Re H(-\frac{d_1}{c_1})}$$

$$= \sum_{c_1 \in l\mathbb{Z}} \sum_{d_1 \in \mathbb{Z}, \gcd(c_1,d_1)=1} \sum_{j=1}^{\phi(l)} (c_1\tau + d_1)^{-k} r_{i,j} \chi_j(d_1) L(\chi_j, k) e^{2\pi i \Re H(-\frac{d_1}{c_1})}$$

since $L(\chi_j, k) = \sum_{n>0, \gcd(n,l)=1} n^{-k} \chi_j(n)$.

So we have

$$E_{k,i;h}(\tau) = \sum_{j=1}^{\phi(l)} L(\chi_j, k) r_{i,j} \hat{E}_{k,\chi_j;h}(\tau)$$

$$= \frac{1}{\phi(l)} \sum_{j=1}^{\phi(l)} \sum_{t \in (\mathbb{Z}/l\mathbb{Z})^*} L(\chi_j, k) e^{-2\pi i \frac{ti}{l}} \chi_j^{-1}(t) \hat{E}_{k,\chi_j;h}(\tau).$$

The absolute convergence that allows us to change the order of summation follows from the proof of Proposition 2.4. $\qquad \square$

We will now use the Dirichlet summation formula to find Fourier expansions of $E_{k,i;h}(\tau)$. Namely, we have for $k \geq 2$ and $\Im x > 0$

$$(2.1) \qquad \sum_{n \in \mathbb{Z}} (x+n)^{-k} = \sum_{m>0} \frac{(-2\pi i)^k m^{k-1}}{(k-1)!} e^{2\pi i m x}.$$

Equation (2.1) allows us to rewrite the formula for $E_{k,i;h}(\tau)$.

PROPOSITION 2.7. For $k \geq 3$ we have

$$E_{k,i;h}(\tau) = \sum_{d \in \mathbb{Z}, gcd(d,l)=1} d^{-k} e^{-2\pi i \frac{di}{l}} + \frac{(-2\pi i)^k}{(k-1)!} \sum_{c \in l\mathbb{Z}_{>0}} \sum_{d_0 \in (\mathbb{Z}/c\mathbb{Z}), gcd(d_0,l)=1}$$
$$c^{-k} \sum_{m>0} m^{k-1} q^m \left(e^{-2\pi i \frac{d_0 i}{l}} + (-1)^k e^{2\pi i \frac{d_0 i}{l}} \right) e^{2\pi i (m \frac{d_0}{c} + \Re H(-\frac{d_0}{c}))}$$

where $q = e^{2\pi i \tau}$.

PROOF. The terms with $c = 0$ correspond to the q^0 term. Here we use the convention $H(-\frac{d}{0}) = 0$ of Remark 2.2.

For a given value of $c > 0$, the possible values of d are given by $d = d_0 + nc$ with $0 < d_0 < c$, $gcd(d_0, l) = 1$ and $n \in \mathbb{Z}$. For each d_0 the value of $\Re H(-\frac{d}{c})$ is independent of n. Indeed, one easily sees that $H(\tau + 1) = H(\tau)$, since $h(\tau)$ was a cusp form. Then one uses (2.1) to sum $(c\tau + d_0 + nc)^{-k} = c^{-k}(\tau + \frac{d_0}{c} + n)^{-k}$ over all integer n.

Finally, we notice that terms for $c < 0$ can be obtained for those for $|c| > 0$ by changing d_0 to $-d_0$. Consequently, we sum for $c > 0$ only by taking into account terms with $(-c)$. $\qquad \square$

As a corollary, we observe that $E_{k,i;h}(\tau)$ has a Fourier expansion.

COROLLARY 2.8. For $k \geq 3$ one has

$$E_{k,i;h}(\tau) = R_0 + \sum_{m>0} R_m q^m$$

where

$$R_m = \frac{(-2\pi i)^k m^{k-1}}{(k-1)!} \sum_{c \in l\mathbb{Z}_{>0}} \sum_{d_0 \in (\mathbb{Z}/c\mathbb{Z}), gcd(d_0,l)=1}$$
$$c^{-k} \left(e^{-2\pi i \frac{d_0 i}{l}} + (-1)^k e^{2\pi i \frac{d_0 i}{l}} \right) e^{2\pi i (m \frac{d_0}{c} + \Re H(-\frac{d_0}{c}))}, \quad m > 0$$
$$R_0 = \sum_{d \in \mathbb{Z}, gcd(d,l)=1} d^{-k} e^{-2\pi i \frac{di}{l}}.$$

PROOF. Follows from Proposition 2.7. $\qquad \square$

REMARK 2.9. It is instructive to see what happens for $h = H = 0$. Then for a fixed $c = c_0 l$ the sum over d_0 that are the same $\mod l$ is zero unless m is divisible by c_0. Indeed, for a given $d_0 \mod l$ we are summing over $d_0 = s + rl \mod c_0 l$ with $0 \leq r < c_0$. The resulting sum of $e^{2\pi i m \frac{r}{c_0}}$ is zero unless $c_0 | m$. This gives

$$R_m = \frac{(-2\pi i)^k m^{k-1}}{l^k (k-1)!} \sum_{c_0 | m} c_0^{-k} \sum_{\substack{d_0 \in (\mathbb{Z}/c_0 l\mathbb{Z}), \\ gcd(d_0,l)=1}} \left(e^{-2\pi i d_0 (\frac{i}{l} - \frac{m}{lc_0})} + (-1)^k e^{2\pi i d_0 (\frac{i}{l} + \frac{m}{lc_0})} \right)$$

$$= \frac{(-2\pi i)^k m^{k-1}}{l^k(k-1)!} \sum_{c_0|m} c_0^{-k} c_0 \sum_{j\in(\mathbb{Z}/l\mathbb{Z})^*} \left(e^{-2\pi i j(\frac{i-\frac{m}{c_0}}{l})} + (-1)^k e^{2\pi i j(\frac{i+\frac{m}{c_0}}{l})} \right)$$

$$= \frac{(-2\pi i)^k}{l^k(k-1)!} \sum_{r|m} r^{k-1} \sum_{j\in(\mathbb{Z}/l\mathbb{Z})^*} \left(e^{-2\pi i \frac{j(i-r)}{l}} + (-1)^k e^{2\pi i \frac{j(i+r)}{l}} \right).$$

This can be recognized as the coefficient by q^m of a linear combination of the Eisenstein series from [**BG2**]. Indeed, it is the sum over the divisors r of m of an odd ($\mathrm{mod}\, l$)-polynomial function of r of degree $(k-1)$, see [**BG2**]. In the particular case of prime l we get

$$\frac{(-2\pi i)^k}{l^{k-1}(k-1)!} \sum_{r|m} r^{k-1} \sum_{j\in(\mathbb{Z}/l\mathbb{Z})^*} \left(\delta_r^{i\,\mathrm{mod}\,l} + (-1)^k \delta_r^{-i\,\mathrm{mod}\,l} - \frac{2}{l}\delta_k^{0\,\mathrm{mod}2} \right)$$

where δ is the Kronecker symbol.

REMARK 2.10. It is easy to see that the coefficients by q^m of $\frac{1}{(2\pi i)^k} E_{k,i;0}$ are rational for $m > 0$. Indeed, $\sum_{j\in(\mathbb{Z}/l\mathbb{Z})^*} \left(e^{-2\pi i \frac{j(i-r)}{l}} + (-1)^k e^{2\pi i \frac{j(i+r)}{l}} \right)$ is invariant under the Galois group of $\mathbb{Q}[e^{2\pi i/l}] \supset \mathbb{Q}$. Since the space of Eisenstein series has a basis whose elements have all Fourier coefficients in $\mathbb{Q}$, the q^0 coefficients of $\frac{1}{(2\pi i)^k} E_{k,i;0}$ are rational as well. This is the main reason we prefer $\frac{1}{(2\pi i)^k} E_{k,i;h}$ to their close relatives $\frac{1}{(2\pi i)^k} \hat{E}_{k,\chi;h}$ that do not have rational Fourier coefficients even in the untwisted case.

Armed with the formulas for its Fourier coefficients, we can now try to define $E_{k,i;h}$ formally by their Fourier expansions in the interesting cases $k = 1$ and $k = 2$. This approach works in the untwisted case, but fails in general. Indeed, let us analyze the convergence of the series in Corollary 2.8. We rewrite it as

$$\sum_{c_0\in\mathbb{Z}_{>0}} \sum_{d_0\in(\mathbb{Z}/c_0l\mathbb{Z}),gcd(d_0,l)=1} c_0^{-k} \left(e^{-2\pi i \frac{d_0 i}{l}} + (-1)^k e^{2\pi i \frac{d_0 i}{l}} \right) e^{2\pi i(m\frac{d_0}{c} + \Re H(-\frac{d_0}{c_0 l}))}.$$

For a given $d_0 \,\mathrm{mod}\, l = j \,\mathrm{mod}\, l$, we are dealing with convergence of

$$\sum_{c_0\in\mathbb{Z}_{>0}} \sum_{d_0\in(\mathbb{Z}/c_0l\mathbb{Z}),d_0\equiv j\,\mathrm{mod}\,l} c_0^{-k} e^{2\pi i(m\frac{d_0}{c} + \Re H(-\frac{d_0}{c_0 l}))}.$$

There is no absolute convergence of the double sum, even in the $k = 2$ case. However, as the nontwisted case $H = 0$ suggests, one can try to look at the finite sums for fixed c_0 first and then look at the resulting series. It is conceivable that this series will converge absolutely, at least for $k = 2$. Indeed, for small h the values of the sums may be close to those for $h = 0$ where all but a finite number are zero. But this is definitely not a proof. For instance, the values of $\Re H(-\frac{d_0}{c_0 l})$ are not bounded, although one can show that they grow at most logarithmically in c_0.

3. Twisted modular forms

In this section we put the definition of the h-twisted Eisenstein series into a more general context of the h-twisted modular forms. We denote by $X_0(l)$ and $X_1(l)$ the modular curves for the congruence subgroups $\Gamma_0(l)$ and $\Gamma_1(l)$. The group $\Gamma_1(l)$ is the subgroup of $\Gamma_0(l)$ whose diagonal elements are $1 \,\mathrm{mod}\, l$.

DEFINITION 3.1. An h-twisted modular form of weight k with respect to $\Gamma_1(l)$ is a holomorphic function f on the upper half plane, which satisfies

(1) For any $\begin{pmatrix} a & b \\ c & d \end{pmatrix} \in \Gamma_1(l)$ there holds

$$f(\frac{a\tau+b}{c\tau+d}) = (c\tau+d)^k f(\tau) e^{-2\pi i \Re H(-\frac{d}{c})}.$$

(2) For any $\begin{pmatrix} a & b \\ c & d \end{pmatrix} \in SL_2(\mathbb{Z})$ the function

$$f(\frac{a\tau+b}{c\tau+d})(c\tau+d)^{-k}$$

is bounded near $i\infty$.

We immediately see that $E_{k,i;h}(\tau)$ and $\hat{E}_{k,\chi;h}(\tau)$ are h-twisted modular forms of level l.

PROPOSITION 3.2. For $k \geq 3$, the functions $E_{k,i;h}(\tau)$ and $\hat{E}_{k,\chi;h}(\tau)$ are h-twisted modular form of weight k for $\Gamma_1(l)$.

PROOF. We will first check the transformation properties of $E_{k,i;h}(\tau)$ under $\tau \to \frac{a\tau+b}{c\tau+d}$. If the sum in the definition of $E_{k,i;h}$ is taken over (c_1, d_1), then we can rewrite the sum for $E_{k,i;h}(\frac{a\tau+b}{c\tau+d})$ in terms of the sum over $(c_2, d_2) = (c_1 a + d_1 c, c_1 b + d_1 d)$ as

$$E_{k,i;h}(\frac{a\tau+b}{c\tau+d}) = \sum_{c_2 \in l\mathbb{Z}} \sum_{\substack{d_2 \in \mathbb{Z}, \\ \gcd(d_2,l)=1}} (c_2\tau+d_2)^{-k}(c\tau+d)^k e^{-2\pi i \frac{d_1 i}{l}} e^{2\pi i \Re H(-\frac{d_1}{c_1})}.$$

We notice that $d_1 \equiv d_2 \bmod l$, so the first exponential term equals $e^{-2\pi i \frac{d_2 i}{l}}$. It then suffices to show that $H(-\frac{d_1}{c_1}) = H(-\frac{d_2}{c_2}) - H(-\frac{d}{c})$. Since $H(\tau)$ is the antiderivative of a weight two $\Gamma_0(l)$ modular form h, the transformation properties of $h(\tau)$ imply

$$H(\frac{-d\tau+b}{c\tau-a}) = H(\tau) + C$$

where C is independent of τ. By plugging in $\tau = i\infty$, we calculate C to get $H(\frac{d_1\tau+b_1}{c_1\tau-a_1}) = H(\tau)+H(-\frac{d}{c})$. Then we plug in $\tau = -\frac{d_1}{c_1}$ to get $H(-\frac{d_2}{c_2}) = H(-\frac{d_1}{c_1}) + H(-\frac{d}{c})$.

We now check that for any $\begin{pmatrix} a & b \\ c & d \end{pmatrix} \in SL_2(\mathbb{Z})$ the function

$$E_{i,k;h}(\frac{a\tau+b}{c\tau+d})(c\tau+d)^{-k}$$

is bounded near $i\infty$. We have

$$E_{k,i;h}(\frac{a\tau+b}{c\tau+d})(c\tau+d)^{-k} = \sum_{c_1 \in l\mathbb{Z}} \sum_{\substack{d_1 \in \mathbb{Z}, \\ \gcd(d_1,l)=1}} (c_2\tau+d_2)^{-k} e^{-2\pi i \frac{d_1 i}{l}} e^{2\pi i \Re H(-\frac{d_1}{c_1})}.$$

Each term is bounded by $|c_2\tau+d_2|^{-k}$. We extend the summation set to sum over $(c_2, d_2) \in \mathbb{Z}^2 - \{(0,0)\}$ and assume $\Im\tau \geq 1$. Then $|c_2\tau+d_2| \geq 1$, and we consider the subsums for $|c_2\tau+d_2| \in [m, m+1)$ for all positive integer m. If we slice the annulus $|z| \in [m, m+1)$ into m equal pieces according to the polar angle, we see that each slice contains at most a constant number of elements of $\mathbb{Z}\tau + \mathbb{Z}$, since this lattice has no elements of length less than one. Importantly, this constant C is independent of τ. This gives

$$\sum_{(c_2,d_2)} |c_2\tau + d_2|^{-k} \leq C \sum_{m>0} m^{-k+1},$$

which converges for $k \geq 3$.

The argument for $\hat{E}_{k,\chi;h}$ is similar and is left to the reader.   □

We can extend the Proposition 3.2 to describe the transformation properties of $E_{k,i;h}$ and $\hat{E}_{k,\chi;h}$ under $\Gamma_0(l)$.

PROPOSITION 3.3. *For $k \geq 3$, for every $\tau \to \frac{a\tau+b}{c\tau+d} \in \Gamma_0(l)$ one has*

$$E_{k,i;h}(\frac{a\tau+b}{c\tau+d}) = E_{k,ai;h}(\tau)(c\tau+d)^k e^{-2\pi i \Re H(-\frac{d}{c})}$$

$$E_{k,\chi;h}(\frac{a\tau+b}{c\tau+d}) = E_{k,\chi;h}(\tau)(c\tau+d)^k \chi(a) e^{-2\pi i \Re H(-\frac{d}{c})}.$$

PROOF. The argument for $E_{k,i;h}$ follows that of Proposition 3.2. The only difference is what happens to the term $e^{-2\pi i \frac{d_1 i}{l}}$, where one now observes that $d_2 \equiv d_1 d \bmod l$, so $d_1 \equiv a d_2 \bmod l$.

The argument for $\hat{E}_{k,\chi;h}$ is similar and is left to the reader.   □

REMARK 3.4. The space of h-twisted modular forms of weight k and with respect to $\Gamma_1(l)$ admits a natural action of $\Gamma_0(l)/\Gamma_1(l) \cong (\mathbb{Z}/l\mathbb{Z})^*$. It sends $f(\tau)$ to

$$f(\frac{a\tau+b}{c\tau+d})(c\tau+d)^{-k} e^{2\pi i \Re H(-\frac{d}{c})}.$$

The eigenvectors of this action will be called h-twisted modular forms of nebentypus χ. They satisfy

$$f(\frac{a\tau+b}{c\tau+d}) = f(\tau)(c\tau+d)^k \chi(a) e^{-2\pi i \Re H(-\frac{d}{c})}$$

for $\begin{pmatrix} a & b \\ c & d \end{pmatrix}$ in $\Gamma_0(l)$. We notice that $\hat{E}_{k,\chi;h}$ have nebentypus χ.

REMARK 3.5. We will call the nebentypus $\chi = 1$ the h-twisted $\Gamma_0(l)$ modular forms. For example, $E_{k,0;h}(\tau)$ is one such form for $k \geq 3$. As in the usual case, the action of $-\mathbf{1} \in \Gamma_0(l)$ causes the spaces of odd weight to be zero.

It is clear that for a given h, the space of h-twisted modular forms for $\Gamma_1(l)$ is a graded module over the ring of modular forms for $\Gamma_1(l)$. In what follows we give a geometric interpretation of this space. We can safely assume $l \geq 3$ so the action of $\Gamma_1(l)$ on the upper half plane is free. We denote by $X_1(l)$ the corresponding modular curve which is obtained by the compactification of its open subset $\mathcal{H}/\Gamma_1(l)$. There is a line bundle $\mathcal{L}$ of weight one modular forms on $X_1(l)$, such that the space $M_k(\Gamma_1(l))$ of usual modular forms of weight k is naturally identified with the space of global sections $H^0(X_1(l), \mathcal{L}^{\otimes k})$.

PROPOSITION 3.6. *For each h there is a degree zero invertible $\mathcal{F} = \mathcal{F}(h)$ such that the space of weight k h-twisted modular forms of level n is naturally identified with $H^0(X_1(l), \mathcal{F} \otimes \mathcal{L}^{\otimes k})$.*

PROOF. For an h-twisted modular form f we define the order of zero of f at a cusp to be the exponent in the corresponding Fourier expansion. Combined with the zeroes in the open set $\mathcal{H}/\Gamma_1(l)$, they give a divisor on $X_1(l)$ which we denote by $Div(f)$.

We observe that the divisor class of $Div(f)$ depends only on h and k. Indeed, the ratio $f(\tau) = \frac{f_1(\tau)}{f_2(\tau)}$ of two weight k h-twisted modular forms is a $\Gamma_1(l)$ modular function. It remains to observe that $Div(f) = Div(f_1) - Div(f_2)$. More generally, for h-twisted forms f_i of weights k_i one has $Div(f_1) - Div(f_2) \sim (k_1 - k_2)Div(L)$ where $Div(\mathcal{L})$ is the divisor of any (meromorphic) modular form of weight 1. We define $\mathcal{F}$ as the invertible sheaf that corresponds to $Div(E_{4,0;h}) - Div(E_{4,0;0})$.

Every h-twisted modular form of weight k gives a global section of $\mathcal{F} \otimes \mathcal{L}^{\otimes k}$. Conversely, given a global section of $\mathcal{F} \otimes \mathcal{L}^{\otimes k}$, one can write it as a product of $E_{4,0;h}(\tau)$ and a modular function on $X_1(l)$. This product is easily seen to be a holomorphic function on $\mathcal{H}$ that satisfies the definition of the h-twisted modular form of weight k.

It remains to see that the degree of $\mathcal{F}$ is zero. Consider a ratio of an h-twisted and a usual form of the same weight k, which we can assume to be nonzero at the cusps. It is a meromorphic function $g(\tau)$ on the upper half plane, which is bounded at the cusps. To find the number of its zeroes and poles in the fundamental domain, one needs to integrate to find the sum of residues of $d \log g(\tau)$ in it. Transformation properties of g imply that $d \log g(\tau)$ is a meromorphic differential form on $X_1(l)$. Consequently, the sum of its residues is zero, which leads to $\deg \mathcal{F} = 0$, as desired. $\square$

PROPOSITION 3.7. *In the notations above, the invertible sheaf $\mathcal{F}(h)$ is invariant under the action of $\Gamma_0(l)/\Gamma_1(l)$. Moreover, it admits a natural linearization for this action.*

PROOF. The key point here is that $h(\tau)$ was a $\Gamma_0(l)$ modular form. The linearization is provided by the action of $\Gamma_0(l)/\Gamma_1(l)$ from Remark 3.4. $\square$

REMARK 3.8. We may as well talk about the invertible sheaf $\mathcal{F}_0 = \mathcal{F}_0(h)$ of degree zero on the modular curve $X_0(l)$. It can be either defined as the $\Gamma_0(l)/\Gamma_1(l)$-invariant part of the pushforward $\pi_*\mathcal{F}$ for the quotient map $\pi : X_1(l) \to X_0(l)$, or as a sheaf that corresponds to the divisor $Div(E_{4,0;h}) - Div(E_{4,0;0})$ on $X_0(l)$. We have $\pi^*\mathcal{F}_0 \cong \mathcal{F}$.

REMARK 3.9. The operation of tensoring by an invertible sheaf is commonly referred to as twisting, which justifies our terminology. Unfortunately, this term is already used in [**GG**] in the context of Eisenstein series. We hope that this does not lead to a confusion, since the series we construct in this paper are quite different from those in [**GG**].

THEOREM 3.10. *Any degree zero invertible sheaf $\mathcal{F}_0$ on $X_0(l)$ is isomorphic to $\mathcal{F}_0(h)$ of Remark 3.8 for some weight two $\Gamma_0(l)$ cusp form h.*

PROOF. The map $h \mapsto \mathcal{F}_0(h)$ can be described in terms of the periods. Let r be the genus of $X_0(l)$, which we assume to be positive. Denote by $K_{X_0(l)}$ the sheaf of holomorphic 1-forms on $X_0(l)$. The Jacobian of $X_0(l)$ is isomorphic to the quotient of $\mathbb{C}^r = H^0(X_0(l), K_{X_0(l)})^{\vee}$ by the lattice of periods, defined as the image of $H_1(X_0(l), \mathbb{Z})$ via integration, see [**GH**]. We would like to calculate the point of the Jacobian that corresponds to $\mathcal{F}_0(h)$ in these terms. This means that for any $g \in H^0(X_0(l), K_{X_0(l)})$ we would like to find the integral of g from zeroes of some h-twisted weight k form f_h to zeroes of some usual weight k form f, up to the lattice of periods.

By an easy Riemann-Roch calculation, we can safely assume that f_h and f do not vanish at the extra symmetry points or cusps, where by extra symmetry points we mean those that are equivalent to i or $e^{2\pi i/3}$ modulo $SL_2(\mathbb{Z})$. Let $G(\tau) = \int_{i\infty}^{\tau} g(s)\,ds$ be an antiderivative of g. In order to find the integral of g from the zeroes of f to the zeroes of f_h, we need to find the integral

$$I = \frac{1}{2\pi i} \int_{\partial \mathcal{H}/\Gamma_0(l)} G(\tau) \left(\frac{f_h'(\tau)}{f_h(\tau)} - \frac{f'(\tau)}{f(\tau)} \right) d\tau$$

over the boundary of a fundamental domain of $\Gamma_0(l)$ action on $\mathcal{H}$. Different choices of fundamental domains will give answers that differ by periods of g.

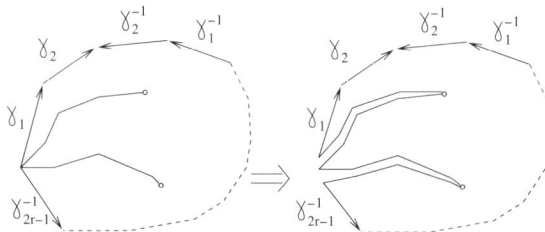

FIGURE 1.

Let us pick a fundamental domain as follows. Pick a point $p \in X_0(l)$ and cut $X_0(l)$ to represent it by a $4r$-gon with pairs of sides identified in the usual manner, such that all vertices map to p. We will assume that the cusps, the points of extra symmetry, as well as the zeroes of f_h and f lie in the interior of this polygon. We will draw cuts from a vertex of the $4r$-gon to the cusps and extra symmetry points, see Figure 1. We can represent the resulting $4r$-gon without these extra cuts by a region in the upper half plane by picking a preimage of a point in the middle and extending all the paths from it by continuity. This will have an effect of opening up the cuts to the cusps and to those extra symmetry points whose isotropy subgroup in $\Gamma_0(l)$ is bigger than just $\{\pm 1\}$, see Figure 1.

The boundary arcs of the resulting fundamental domain $\mathcal{D}$ are paired up according to the gluing required to make $X_0(l)$. These arcs will then be related by a coordinate transformation $\tau \to \frac{a\tau+b}{c\tau+d}$ from $\Gamma_0(l)$. We first investigate pairs of paths that go to the cusps or symmetry points of $X_0(l)$. Whether or not these cuts open up, the values of $G(\tau)\left(\frac{f_h'(\tau)}{f_h(\tau)} - \frac{f'(\tau)}{f(\tau)}\right)$ at the opposite sides of the cut are the same. Indeed, $G(\frac{a\tau+b}{c\tau+d}) = G(\tau)$ for any $\tau \to \frac{a\tau+b}{c\tau+d}$ from $\Gamma_0(l)$ that has a fixed point. Similar statement holds for $H(\tau)$, and the transformation properties of f_h and f show that $d\log f_h - d\log f$ is unchanged under such transformations.

So we are left to integrate over the arcs $p_j \to q_j$ of the boundary that correspond to the sides of the original $4r$-gon. Pair them up according to the gluings. The two arcs γ_j and γ_j^{-1} (see Figure 1) differ by $\tau \to \frac{a_j\tau+b_j}{c_j\tau+d_j}$ from $\Gamma_0(l)$, and we are looking at

$$I = \frac{1}{2\pi i} \sum_{j=1}^{2r} \int_{p_j}^{q_j} \left(G(\frac{a_j\tau+b_j}{c_j\tau+d_j}) \left(\frac{f_h'(\frac{a_j\tau+b_j}{c_j\tau+d_j})}{f_h(\frac{a_j\tau+b_j}{c_j\tau+d_j})} - \frac{f'(\frac{a_j\tau+b_j}{c_j\tau+d_j})}{f(\frac{a_j\tau+b_j}{c_j\tau+d_j})} \right)(c_j\tau+d_j)^{-2} \right.$$

$$\left. - G(\tau)\left(\frac{f_h'(\tau)}{f_h(\tau)} - \frac{f'(\tau)}{f(\tau)} \right) \right) d\tau$$

Transformation properties of G, f_h and f allow us to simplify the integrand to be

$$\left((G(\tau)+\alpha_j)\left(\frac{f_h'(\tau)}{f_h(\tau)} - \frac{f'(\tau)}{f(\tau)} \right) - G(\tau)\left(\frac{f_h'(\tau)}{f_h(\tau)} - \frac{f'(\tau)}{f(\tau)} \right) \right)$$

$$= \alpha_j \left(\frac{f_h'(\tau)}{f_h(\tau)} - \frac{f'(\tau)}{f(\tau)} \right)$$

where $\alpha_j = G(\frac{a_j\tau+b_j}{c_j\tau+d_j}) - G(\tau)$. Hence we get

$$I = \frac{1}{2\pi i} \sum_{j=1}^{2r} \int_{p_j}^{q_j} \alpha_j \left(\left(\frac{f_h'(\tau)}{f_h(\tau)} - \frac{f'(\tau)}{f(\tau)} \right) d\tau = \frac{1}{2\pi i} \sum_{j=1}^{2r} \alpha_j \log\left(\frac{f_h(\tau)}{f(\tau)} \right) \Big|_{p_j}^{q_j} \right.$$

$$= -\sum_{j=1}^{2r} \alpha_j \Re H(-\frac{\hat{d}_j}{\hat{c}_j}) \bmod periods,$$

where $\tau \to \frac{\hat{a}_j\tau+\hat{b}_j}{\hat{c}_j\tau+\hat{d}_j}$ is an element of $\Gamma_0(l)$ that sends $p_j \mapsto q_j$. Indeed, the transformation properties of f_h and f lead to the transformation properties of the logarithms, up to integer constants that are independent of g, and α_i are the period integrals.

We observe that α_j are the period integrals of $g(\tau)\,d\tau$ over the standard symplectic basis of $H_1(X_0(l), \mathbb{Z})$, as are $H(-\frac{\hat{d}_j}{\hat{c}_j})$ are the integrals of $h(\tau)\,d\tau$ over the same basis. Indeed, $H(\frac{\hat{a}_j\tau+\hat{b}_j}{\hat{c}_j\tau+\hat{d}_j}) = H(\tau) - H(-\frac{\hat{d}_j}{\hat{c}_j})$ in view of $H(i\infty) = 0$, so $-H(-\frac{\hat{d}_j}{\hat{c}_j}) = H(q_j) - H(p_j) = \int_{p_j}^{q_j} h(\tau)\,d\tau$. To show that any point of the Jacobian can be given by $\mathcal{F}_h$, it is enough to show that the pairing (of real vector spaces)

$$H_1(X_0(l), \mathbb{R}) \times H^0(X_0(l), K_{X_0(l)}) \to \mathbb{R}, \quad (\gamma, w) \mapsto \Re \int_\gamma w$$

is nondegenerate. This is a standard general fact, which is true for any projective curve X, but we provide the argument below for the benefit of the reader.

By picking a basis $(w_1, \ldots, w_r)$ dual to a half of a symplectic basis $(\gamma_1, \ldots, \gamma_{2r})$, the period matrix $\Omega = (\int_{\gamma_j} w_i)$ is given by $(\mathbf{I}_r, Z)$, where $\Im Z$ is positive definite (see [**GH**]). If $\int_{\gamma_j} w$ is purely imaginary for all j, then we first see that $w = \sum_{i=1}^r \lambda_i w_i$ with purely imaginary λ_i. Then (λ_i) lies in the kernel of Z, so $\Im Z > 0$ implies $(\lambda_i) = \mathbf{0}$. $\qquad\square$

COROLLARY 3.11. *Two weight two cusp forms h_1 and h_2 have $\mathcal{F}_1 \cong \mathcal{F}_2$ if and only if $\Re(H_1(-\frac{d}{c}) - H_2(-\frac{d}{c})) \in \mathbb{Z}$ for all $c \in l\mathbb{Z}$ and $d \in \mathbb{Z}, \gcd(d, l) = 1$. In particular, the Eisenstein series $E_{k,i;h}$ depend only on the point of the Jacobian.*

PROOF. The *if* part is clear, since in this case the Eisenstein series are identical. To see the *only if* part, we observe that in the proof of Theorem 3.10, the sheaf $\mathcal{F}$ is given by a linear combination of the periods with coefficients $\Re H(-\frac{\hat{d}_j}{\hat{c}_j})$, $j = 1, \ldots, 2r$. If $\mathcal{F}_1 \cong \mathcal{F}_2$, then

$$(3.1) \qquad \Re(H_1(-\frac{\hat{d}_j}{\hat{c}_j}) - H_2(-\frac{\hat{d}_j}{\hat{c}_j})) \in \mathbb{Z}$$

for $j = 1, \ldots, 2r$. For any $\begin{pmatrix} a & b \\ c & d \end{pmatrix}$ in $\Gamma_0(l)$, the values of $\Re H_1(-\frac{d}{c}) - \Re H_2(-\frac{d}{c})$ are integer linear combinations of the periods in (3.1). Indeed, this follows from the fact that (3.1) gives integrals of $(h_1(\tau) - h_2(\tau))\,d\tau$ over a basis of $H_1(X_0(l), \mathbb{Z})$. Finally, any ratio $-\frac{d}{c}$ with $c \in l\mathbb{Z}$ and $gcd(d, l) = 1$ can be obtained from a $\Gamma_0(l)$ matrix, after one cancels off common factors of c and d.   □

REMARK 3.12. Corollary 3.11 shows that each Fourier coefficient of $E_{k,i;h}$ gives a function on the Jacobian of $X_0(l)$. However, these functions are clearly not holomorphic in general, else they would have to be constant.

REMARK 3.13. An easy application of the Riemann-Roch theorem shows that the dimension of the space of h-twisted modular forms of weight $k \geq 2$ coincides with that of the space of usual modular forms of weight k. This holds for either $\Gamma_0(l)$ or $\Gamma_1(l)$, as well as for any nebentypus.

4. Petersson inner product for twisted modular forms and twisted Eisenstein series

In this section we define the Petersson inner product on the space of h-twisted modular forms of given weight k and investigate its basic properties.

DEFINITION 4.1. A $\Gamma_1(l)$-modular h-twisted form is called a *cusp form*, if in the second condition of Definition 3.1 the function

$$f(\frac{a\tau + b}{c\tau + d})(c\tau + d)^{-k}$$

has limit 0 as $\tau \to i\infty$ for all $\begin{pmatrix} a & b \\ c & d \end{pmatrix}$.

DEFINITION 4.2. Let f and g be two h-twisted modular forms for $\Gamma_1(l)$ of weight k. Assume that at least one of then is a cusp form. We define Petersson inner product $\langle f, g \rangle$ by

$$\langle f, g \rangle = \int_{\Gamma_1(l)\backslash\mathcal{H}} f(\tau)\overline{g(\tau)}y^{k-2}\,dxdy$$

where $\tau = x + iy$. The convergence is assured by the cusp condition.

REMARK 4.3. We have to check that the above definition is independent of the choice of the fundamental domain. If $\tau = \frac{a\tau_1 + b}{c\tau_1 + d}$, then

$$2iy = \frac{a\tau_1 + b}{c\tau_1 + d} - \frac{a\bar{\tau}_1 + b}{c\bar{\tau}_1 + d} = \frac{(a\tau_1 + b)(c\bar{\tau}_1 + d) - (c\tau_1 + d)(a\bar{\tau}_1 + b)}{(c\tau_1 + d)(c\bar{\tau}_1 + d)},$$

$$\frac{\tau_1 - \bar{\tau}_1}{(c\tau_1 + d)(c\bar{\tau}_1 + d)} = \frac{2iy_1}{(c\tau_1 + d)(c\bar{\tau}_1 + d)}$$

and

$$dxdy = \frac{1}{(c\tau_1 + d)^2(c\bar{\tau}_1 + d)^2}\, dx_1 dy_1$$

so

$$f(\tau)\overline{g(\tau)}y^{k-2}\, dxdy = f(\frac{a\tau_1 + b}{c\tau_1 + d})\overline{g(\frac{a\tau_1 + b}{c\tau_1 + d})}\frac{y_1^{k-2}}{(c\tau_1 + d)^k(c\bar{\tau}_1 + d)^k}\, dx_1 dy_1$$

$$= f(\tau_1)e^{-2\pi i\Re H(-\frac{d}{c})}\overline{g(\tau_1)}e^{2\pi i\Re H(-\frac{d}{c})}y_1^{k-2}\, dx_1 dy_1 = f(\tau_1)\overline{g(\tau_1)}y_1^{k-2}\, dx_1 dy_1.$$

REMARK 4.4. It is clear that the above defined Petersson pairing restricts to a nondegenerate Hermitean form on the space of h-twisted cusp forms.

The following proposition extends the orthogonality of Eisenstein series and cusp forms (see [**L**]) to the h-twisted case, for the particular class of Eisenstein series considered in this paper.

PROPOSITION 4.5. *For $k \geq 3$ we have*

$$\langle E_{k,i;h}, g \rangle = 0$$

for any weight k h-twisted cusp form g for $\Gamma_1(l)$.

PROOF. It is enough to check that $\langle \hat{E}_{k,\chi;h}, g \rangle = 0$ for all characters $\chi : (\mathbb{Z}/l\mathbb{Z})^* \to \mathbb{C}^*$. We can also assume that g is an eigenform for Γ_0-action, as in Remark 3.4. Denote the character of g by ψ. We have

$$\langle \hat{E}_{k,\chi;h}, g \rangle = \int_{\Gamma_1(l)\backslash\mathcal{H}} \sum_{\Gamma_\infty\backslash\Gamma_0(l)} (c\tau + d)^{-k}\chi(d)e^{2\pi i\Re H(-\frac{d}{c})}\overline{g(\tau)}y^{k-2}\, dxdy.$$

We switch the order of integration and summation and then switch to a different domain for each term. Namely, we make a substitution $\tau = \frac{d\tau_1 - b}{-c\tau_1 + a}$ for each term of the $\hat{E}$-series. This gives

$$\langle \hat{E}_{k,\chi;h}, g \rangle = \sum_{\Gamma_\infty\backslash\Gamma_0(l)} \int_{\Gamma_1(l)\backslash\mathcal{H}} (-c\tau_1 + a)^k\chi(d)e^{2\pi i\Re H(-\frac{d}{c})}.$$

$$\cdot(-c\bar{\tau}_1 + a)^k e^{-2\pi i\Re H(-\frac{d}{c})}\overline{\psi(d)}\,\overline{g(\tau_1)}\frac{y_1^{k-2}dx_1 dy_1}{(-c\tau_1 + a)^k(-c\bar{\tau}_1 + a)^k}$$

$$= \sum_{\Gamma_\infty\backslash\Gamma_0(l)} \int_{\Gamma_1(l)\backslash\mathcal{H}} \chi(d)\overline{\psi(d)}\,\overline{g(\tau_1)}y_1^{k-2}\, dx_1 dy_1.$$

This allows us to rewrite the pairing as

$$\sum_{\Gamma_1(l)\backslash\Gamma_0(l)} \chi(d)\overline{\psi(d)}\int_{\Gamma_\infty\backslash\mathcal{H}} \overline{g(\tau_1)}y_1^{k-2}\, dx_1 dy_1.$$

We then observe that $\Gamma_\infty\backslash\mathcal{H}$ can be thought of as $0 \leq x \leq 1, y \geq 0$ and for a fixed y

$$\int_{iy}^{1+iy} \overline{g(\tau)}\, dx = \sum_{m>0} \bar{r}_m e^{-2\pi imy}\int_{x=0}^{x=1} e^{-2\pi imx}\, dx = 0.$$

At the last step we used that g has a Fourier expansion with a zero leading term. Indeed, we have $H(i\infty) = 0$, which implies that $g(\tau + 1) = g(\tau)$, and the $g(i\infty) = 0$ follows from g being a cusp form. $\square$

5. Rationality conjecture

The motivation behind our investigation of twisted Eisenstein series is that they might provide a tool for studying the Jacobian of the modular curve $X_0(l)$. We now recall that $X_0(l)$ comes equipped with the natural $\mathbb{Q}$-structure, which then gives a $\mathbb{Q}$-structure on its Jacobian. The following conjecture relates the rationality of the coefficients of twisted Eisensten series and the rationality of the point on the Jacobian.

CONJECTURE 5.1. *The sheaf $\mathcal{F}_0(h)$ on $X_0(l)$ gives a rational point on the Jacobian if and only if for each k there exists a nonzero formal power series $R_k(q)$ such that all coefficients of the q-series $E_{k,i;h} R_k(q)$ are* rational *for all i. Equivalently, the ratios of $E_{k,i;h}/E_{k,j;h}$ have rational Fourier coefficients for all k, i and j, provided $E_{k,j;h} \not\equiv 0$.*

REMARK 5.2. A stronger version of Conjecture 5.1 would require $R_k = R^k$ for some series $R = R(q)$.

REMARK 5.3. It is reasonable to generalize Conjecture 5.1 to all number fields. Namely, let K be a number field. Then the K-rational points of the Jacobian of $X_0(l)$ are conjecturally characterized by the condition that Fourier coefficients of $E_{k,i;h}/E_{k,j;h}$ are in K for all k, i and j with $E_{k,j;h} \not\equiv 0$. We thank the referee who suggested to include this generalization.

We devote the remainder of the section to the evidence in favor of Conjecture 5.1. Since the evidence is mostly circumstantial anyway, we will be somewhat sketchy and leave the details to the reader.

First of all, the *if* part of the conjecture holds up to a finite index subgroup as long as $E_{k,i;h}$ do not have common zeroes for some $k > 0$. Indeed, the existence of R_k implies that the module over the ring of $\Gamma_1(l)$ modular forms which is generated by $E_{k,i;h}$ is defined over $\mathbb{Q}$. By the usual correspondence between graded modules and sheaves we see that the invertible sheaf $\mathcal{F}(h) \otimes \mathcal{L}^{\otimes k}$ is defined over $\mathbb{Q}$. Since $\mathcal{L}$ is defined over $\mathbb{Q}$, we see that the invertible sheaf $\mathcal{F}(h)$ on $X_1(l)$ is defined over $\mathbb{Q}$. Denote $\pi : X_1(l) \to X_0(l)$ and recall that $\mathcal{F}(h) = \pi^* \mathcal{F}_0(h)$. Then

$$\mathcal{F}_0^{\phi(l)/2} \cong (\Lambda^{\phi(l)/2} \pi_* \mathcal{F}(h)) \otimes (\Lambda^{\phi(l)/2} \pi_* \pi^* \mathcal{O})^{-1}$$

is defined over $\mathbb{Q}$.

For the *only if* part of the conjecture, observe that it holds for $h = 0$, in view of Remark 2.10. In general, for every rational point on the Jacobian of $X_1(l)$ that corresponds to a sheaf $\mathcal{F} = \pi^* \mathcal{F}_0$, the space of global sections $H^0(\mathcal{F} \otimes L^{\otimes k})$ has a basis that is defined over $\mathbb{Q}$. The essence of the Conjecture 5.1 is that the h-twisted Eisenstein series $E_{k,i;h}$ are rational linear combinations of the basis elements. Since q^0 coefficients and the $\Gamma_0(l)/\Gamma_1(l)$ transformation properties of $E_{k,i;h}$ are independent of the twisting, if the linear span of $E_{k,i;h}$ is defined over rationals, then so are the individual elements, at least in the case of prime l. Indeed, one can show that in this case the matrix of q^0 coefficients of $\Gamma_0(l)/\Gamma_1(l)$-translates of $E_{k,i}$ is nondegenerate and rational. Consequently, if $Span(E_{k,i;h})$ has a rational basis, $E_{k,i;h}$ have rational coefficients in it.

Unfortunately, it is not at all obvious that $Span(E_{k,i;h})$ is defined over $\mathbb{Q}$. The space of all h-twisted forms of weight k splits up as a direct sum of the space of cusp forms and its orthogonal complement under the Petersson pairing. One can hope that this splitting respects the rational structures. In the case of prime l, the series $E_{k,i;h}(\tau)$ roughly account for half the dimension of this orthogonal complement. However, if one could define $E_{k,i;h}(\tau)$ for weight $k = 1$, then one would expect that all weight one h-twisted modular forms lie in $Span(E_{k,i;h})$, by an application of Riemann-Roch formula.

On the plus side, it is quite easy to give a sufficient condition that assures that the Fourier coefficients of $\frac{1}{(2\pi i)^k} E_{k,i;h}$ are real. Recall that an $\mathbb{R}$-structure on the modular curve is given by the involution that sends $h(\tau)$ to $\overline{h(-\bar{\tau})}$. Let us calculate $\overline{E_{k,i;h}(-\bar{\tau})}$.

PROPOSITION 5.4. *Let* $g = \overline{h(-\tau)}$. *For* $k \geq 3$ *one has*

$$\overline{E_{k,i;h}(-\bar{\tau})} = (-1)^k E_{k,i;g}(\tau)$$

PROOF. and let G be the antiderivative $\int_{i\infty}^{\tau} g(s)\, ds$ of g. Observe that $\overline{H(-\bar{\tau})} = -G(\tau)$. We have

$$\overline{E_{k,i;h}(-\bar{\tau})} = \sum_{c \in l\mathbb{Z}} \sum_{d \in \mathbb{Z}, \gcd(l,d)=1} (-c\tau + d)^{-k} e^{2\pi i \frac{di}{l}} e^{-2\pi i \Re H(-\frac{d}{c})}$$

$$= \sum_{c \in l\mathbb{Z}} \sum_{d \in \mathbb{Z}, \gcd(l,d)=1} (-c\tau + d)^{-k} e^{2\pi i \frac{di}{l}} e^{2\pi i \Re G(\frac{d}{c})} = E_{k,-i;g}(\tau)$$

where we have switched from c to $-c$ in the summation. It remains to observe $E_{k,-i;g} = (-1)^k E_{k,i;g}$, in view of the change from (c,d) to $(-c,-d)$ in the summation. $\square$

COROLLARY 5.5. *If all Fourier coefficients of* $h(\tau)$ *are real, then all Fourier coefficients of* $\frac{1}{(2\pi i)^k} E_{k,i;h}(\tau)$ *are real.*

PROOF. Notice that If $F(\tau) = \sum_m r_m q^m$ then $\overline{F(-\bar{\tau})} = \sum_m \overline{r_m} q^m$. Then use the above proposition with $h = g$. $\square$

6. Open problems

There are numerous open problems, whose solution would greatly enhance the understanding of the h-twisted modular forms. We list several of them.

- Can one efficiently calculate Fourier coefficients of an h-twisted Eisenstein series? The problem is that the definition only provides a relatively slowly convergent series. Lack of explicit coefficients makes it impossible to do any kind of computer experiments to uncover relations among the series or to test Conjecture 5.1.
- Can one define h-twisted Eisenstein series for weights $k = 1$ or 2? What will happen to their modularity properties? The similar question in the untwisted case is answered by looking at the Fourier expansions, but this approach seems to fail in general.
- In the untwisted case, these Eisenstein series can be constructed from logarithmic derivatives of the theta function, see [**BG1**]. Is there a twisted analog of the theta function, with the same properties? In a related comment, the notion of h-twisting can be extended to the Jacobi forms.
- Do any twisted forms have nice infinite product expansions?

- What is the meaning of Hecke operators in this setting? It is easy to see that the usual definition of the Hecke operators does not produce operators on the spaces of h-twisted forms. Also, if h is a Hecke eigenform, does this lead to any properties of h-twisted forms? A related question is what should be the analog of Dirichlet series and Euler expansions in the h-twisted case.
- What is the meaning of the Fricke involution in this setting?
- What would be an analog of Rankin-Selberg method for the Petersson inner product of a usual cusp form and product of two twisted Eisenstein series of opposite twists? The argument of Proposition 4.5 shows that the unfolding trick generally works, but the lack of Euler expansions is a major hindrance.
- Is it possible to see what the twisted Eisenstein series are in the case of the divisors coming from the Heegner points?
- What is the relation between the Eisenstein series of opposite twists?
- In the untwisted case, there are quadratic relations on the Eisenstein series that mimic the relations on modular symbols. Is there any analog of such relations in the twisted case? For instance, dimension counts show that there are a lot of linear relations between products of h-twisted and untwisted Eisenstein series. Can one write at least some of them explicitly? Together with the rationality conjecture, this might provide an approach to the Birch-Swinnerton-Dyer conjecture, see [**T**], although at this point in time this is, at best, a very long shot.

References

[BG1] L. Borisov, P. Gunnells, *Toric varieties and modular forms.* Invent. Math. **144** (2001), no. 2, 297–325.

[BG2] L. Borisov, P. Gunnells, *Toric modular forms of higher weight.* J. Reine Angew. Math. 560 (2003), 43–64.

[Go]   D. Goldfeld, *Zeta functions formed with modular symbols.* Automorphic forms, automorphic representations, and arithmetic (Fort Worth, TX, 1996), 111–121, Proc. Sympos. Pure Math., 66, Part 1, Amer. Math. Soc., Providence, RI, 1999.

[GG]   D. Goldfeld, P. Gunnells, *Eisenstein series twisted by modular symbols for the group* SL_n. Math. Res. Lett. **7** (2000), no. 5-6, 747–756.

[GH]   P. Griffiths, J. Harris, *Principles of algebraic geometry.* Reprint of the 1978 original. Wiley Classics Library. John Wiley & Sons, Inc., New York, 1994.

[L]   S. Lang, *Introduction to modular forms.* Grundlehren der mathematischen Wissenschaften, No. 222. Springer-Verlag, Berlin-New York, 1976.

[P]   Y. Petridis, *Spectral deformations and Eisenstein series associated with modular symbols.* Int. Math. Res. Not. 2002, no. 19, 991–1006.

[T]   J. Tate, *On the conjectures of Birch and Swinnerton-Dyer and a geometric analog.* Séminaire Bourbaki, Vol. 9, Exp. No. 306, 415–440, Soc. Math. France, Paris, 1995. 11G40 (14G10)

DEPARTMENT OF MATHEMATICS, UNIVERSITY OF WISCONSIN, MADISON, WI, 53706, USA, borisov@math.wisc.edu

Contemporary Mathematics
Volume **422**, 2007

Hessians and the moduli space of cubic surfaces

Elisa Dardanelli and Bert van Geemen

This paper is dedicated to Igor Dolgachev on his 60th birthday.

ABSTRACT. The Hessian of a general cubic surface is a nodal quartic surface, hence its desingularisation is a K3 surface. We determine the transcendental lattice of the Hessian K3 surface for various cubic surfaces (with nodes and/or Eckardt points for example). Classical invariant theory shows that the moduli space of cubic surfaces is a weighted projective space. We describe the singular locus and some other subvarieties of the moduli space.

Moduli spaces of cubic surfaces have been intensively studied recently. The main reason was the discovery by Allcock, Carlson and Toledo [**ACT**] that the moduli space of cubic surfaces is a ball quotient. Another proof of this fact, using K3 surfaces rather than cubic threefolds, is given in [**DGK**]. There is a classical, quite different, way to associate a K3 surface to a general cubic surface, it is simply by taking the Hessian of the cubic polynomial defining the surface. We refer to the desingularization of the corresponding nodal quartic surface as the Hessian K3 surface of the cubic surface.

In this paper we study these Hessian K3 surfaces. The Hessian, with its polarization given by the natural map to $\mathbb{P}^3$, determines the cubic surface. This gives a birational isomorphism between the moduli space of M-lattice polarized K3 surfaces, where M is the Néron-Severi group of the general Hessian, and the moduli space of cubic surfaces.

As a first step in the study of the moduli space of these K3 surfaces we determine the transcendental lattices of various types of Hessians. We find that certain naturally defined divisors in the moduli space (Hessians of cubics with a node, an Eckardt point, without a Sylvester form, Kummer surfaces) are Heegner divisors, that is, their transcendental lattices are sublattices of a lower rank of the transcendental lattice of a general Hessian. Various very special cubic surfaces (with many nodes and/or Eckardt points) have transcendental lattices of rank two, these are usually referred to as singular (or, nowadays, as attractive) K3 surfaces. As there is no general method to determine the transcendental lattice of a K3 surface, we have to use ad-hoc methods. To study the Hessians, we use the Sylvester form of a cubic surface which is also very useful in describing the moduli space.

1991 *Mathematics Subject Classification.* 14J28, 14J15.
Key words and phrases. Cubic surfaces, K3 surfaces, Moduli.

In the final section of this paper we recall the classical description of the moduli space of cubic surfaces using 19th century invariant theory. The moduli space is a weighted projective space. We describe various divisors and subvarieties in the moduli space. We also describe its singular locus.

This paper is based on the PhD thesis of the first author under the direction of the second author. We are indebted to the referee for helpful suggestions.

1. The Sylvester form and the Hessian

1.1. The Sylvester form. Given a general homogeneous polynomial F (a form) of degree three in four variables, there are five linear forms x_i in four variables such that any four of the five x_i are linearly independent and there are five $\lambda_i \in \mathbb{C} - \{0\}$ such that

$$\sum_{i=0}^{4} x_i = 0, \qquad F = \sum_{i=0}^{4} \lambda_i x_i^3.$$

We refer to these equations as the Sylvester form of F.

The x_i are uniquely determined by F (up to permutation and multiplication by a common non-zero scalar) and the λ_i are uniquely determined by F and the x_i ([**S**], pp. 125-137, [**Rod**], pp. 72ff. and also [**C**] pp. 295 ff.).

1.2. The pentahedron. In case F has a Sylvester form, the union of the five planes in $\mathbb{P}^3$ defined by $x_i = 0$ is called the pentahedron of F. Each plane is called a face, the intersection of two distinct faces is called an edge and the intersection of three distinct faces is a vertex of the pentahedron. The ten edges and ten vertices are denoted by

$$L_{ij}: \quad x_i = x_j = 0, \qquad\qquad P_{ijk}: \quad x_i = x_j = x_k = 0.$$

1.3. The discriminant divisor. It is not hard to verify that the cubic surface defined by a Sylvester form is smooth iff, for all choices of signs, we have:

$$\sum_{i=0}^{4} \pm\sqrt{\lambda_0 \ldots \hat{\lambda}_i \ldots \lambda_4} \neq 0.$$

Assuming the first sign to be $+1$, this condition is equivalent to the product of the 2^4 such expressions, with all possible signs, to be non-zero. This product is homogeneous of degree 32 in the λ_i (in fact the 'discriminant' divisor which parametrizes singular cubic surfaces in $\mathbb{P}^{19} = \mathbb{P}H^0(\mathbb{P}^3, \mathcal{O}(3))$ has degree 32, cf. [**S**], App. III p. 179, [**GKZ**], p. 38).

For a general cubic form F we can use the linear forms $x_0, \ldots, x_3$ as coordinates on $\mathbb{P}^3$ and $x_4 = -(x_0 + \ldots + x_3)$. The general cubic surface defined by $F = 0$ is now determined by $p = (\lambda_0 : \ldots : \lambda_4) \in \mathbb{P}^4$, and p is unique up to permutation of the λ_i.

1.4. The Hessian. The Hessian H_G of a cubic form G in $n + 1$ variables u_i is the homogeneous polynomial of degree $n + 1$ defined by

$$H_G = \det\left(\frac{\partial^2 G}{\partial u_i \partial u_j}\right).$$

1.5. The Hessian K3 surface. The Hessian surface of a cubic surface S defined by $F = 0$ is the quartic surface Y defined by $H_F = 0$ (in case $H_F \equiv 0$ we do not define the Hessian surface). We will call $Y = Y_S$ the Hessian of S.

In case S is defined by a Sylvester form, its Hessian Y is defined by

$$\left\{ \begin{array}{l} H_F = \sum_{i=0}^{4} \lambda_0 \ldots \hat{\lambda}_i \ldots \lambda_4 x_0 \ldots \hat{x}_i \ldots x_4 = (\lambda_0 \ldots \lambda_4)(x_0 \ldots x_4) \sum_{i=0}^{4} \frac{1}{\lambda_i x_i} = 0, \\[2ex] \hphantom{H_F = \sum_{i=0}^{4} \lambda_0} \sum_{i=0}^{4} x_i = 0. \end{array} \right.$$

The ten edges L_{ij} of the pentahedron lie on the Hessian and the ten vertices P_{ijk} are singular points of the Hessian. One verifies that S is smooth iff the singular locus of Y consists of these 10 points.

The desingularization X of Y is a K3 surface, which we refer to as the Hessian K3 surface of S (or simply Hessian if no confusion is likely). The natural map

$$\pi : X \longrightarrow Y, \qquad \pi(N_{ijk}) = P_{ijk}$$

contracts 10 smooth rational curves N_{ijk} to the singular points P_{ijk} of Y and is an isomorphism on the complement. The strict transforms in X of the lines L_{ij} in Y are denoted by N_{ij}.

1.6. Enriques surfaces. The Hessian Y of a cubic surface S defined by a Sylvester form has a birational involution

$$\iota_Y : Y \longrightarrow Y, \qquad (x_0 : \ldots : x_4) \longmapsto \left(\frac{1}{\lambda_0 x_0} : \ldots : \frac{1}{\lambda_4 x_4} \right).$$

This involution lifts to a fixed point free involution, denoted by ι_X, on the K3 surface X which interchanges the (-2)-curves N_{ij} and N_{klm} where $\{i,j,k,l,m\} = \{0, \ldots, 4\}$. The quotient of X by ι_X is thus an Enriques surface with at least 10 nodal curves, the images of the N_α.

The Hessian K3 surface X can be obtained as a Reye congruence. Identify the hyperplane $\sum x_i = 0$ in $\mathbb{P}^4$ with a $\mathbb{P}^3$ with homogeneous coordinates $x_0, \ldots, x_3$ (so $x_4 = -(x_0 + \ldots + x_3)$). The graph of ι_Y, that is the image of the rational map

$$Y \longrightarrow \mathbb{P}^3 \times \mathbb{P}^3, \qquad x \longmapsto (x, \iota_Y(x)),$$

lies in the intersection of four hypersurfaces of bidegree $(1,1)$. In fact if $(z_0 : \ldots : z_3)$ are homogeneous coordinates on the second factor and $z_4 = -(z_0 + \ldots + z_3)$, then we have the obvious equations

$$\lambda_0 x_0 z_0 - \lambda_j x_j z_j = 0, \qquad j = 1, 2, 3, \qquad \lambda_0 x_0 z_0 - \lambda_4 \left(\sum_{i=0}^{3} x_i \right) \left(\sum_{i=0}^{3} z_i \right) = 0$$

for the graph of ι_Y.

1.7. The Néron-Severi group of the general Hessian K3 surface. The ten (-2)-curves N_{ijk} are disjoint, and so are the ten (-2)-curves N_{ij}. Moreover, $N_{ij}N_{klm} = 1$ iff $\sharp\{i,j,k,l,m\} = 3$ and is zero otherwise. A machine computation shows that the rank of the 20×20 matrix of intersection numbers of the N_α is equal to 16. A $\mathbb{Z}$-basis of the sublattice NS_{gen} of $NS(X)$ generated by these curves is given by all twenty curves except N_{234}, N_{14}, N_{23} and N_{24}. The discriminant of NS_{gen} is equal to $2^4 \cdot 3$. From [**DK**] it follows that NS_{gen} is the Néron-Severi group of a general Hessian K3 surface (i.e. one with Picard number 16). The perpendicular of NS_{gen} in $H^2(X, \mathbb{Z})$ is the transcendental lattice T_{gen} which was determined in [**DK**] (for an alternative method, see Lemma 2.2):

$$T_{gen} = U \oplus U(2) \oplus A_2(-2).$$

The following lemma is very useful for determining the transcendental lattices of 'singular' Hessian K3 surfaces, that is of those with Picard number 20.

LEMMA 1.1. *Let T be an even lattice of rank two:*

$$T = \begin{pmatrix} 2n & a \\ a & 2m \end{pmatrix}.$$

Then there is a primitive embedding $T \hookrightarrow T_{gen}$ if and only if at least one among a, n and m is even. In this case T embeds into $U \oplus U(2)$.

PROOF. An embedding $T \hookrightarrow T_{gen}$, is equivalent to giving $x = (x_1, \ldots, x_6)$, $y = (y_1, \ldots, y_6) \in \mathbb{Z}^6 = T_{gen}$, such that

$$
\begin{aligned}
2x_1 x_2 + 4x_3 x_4 - 4(x_5^2 + x_6^2 - x_5 x_6) & = 2n, \\
2y_1 y_2 + 4y_3 y_4 - 4(y_5^2 + y_6^2 - y_5 y_6) & = 2m, \\
x_1 y_2 + x_2 y_1 + 2(x_3 y_4 + x_4 y_3) - 2(2x_5 y_5 + 2x_6 y_6 - x_5 y_6 - x_6 y_5) & = a.
\end{aligned}
$$

If $n \equiv m \equiv a \equiv 1(2)$, this system has no solution, since from the first two equations we get $x_1 x_2 \equiv 1(2)$ and $y_1 y_2 \equiv 1(2)$, so $x_1 \equiv x_2 \equiv y_1 \equiv y_2 \equiv 1(2)$ and hence $x_1 y_2 + x_2 y_1 \equiv 0(2)$. Thus in the third equation the left hand side is even, while the right hand side is odd.

Conversely, let $x := (n, 1, 0, 0, 0, 0)$, then y must satisfy

$$
\begin{aligned}
y_1 y_2 + 2y_3 y_4 - 2(y_5^2 + y_6^2 - y_5 y_6) & = m, \\
n y_2 + y_1 & = a.
\end{aligned}
$$

Substituting $y_1 = a - n y_2$ in the first equation, we get:

$$2y_3 y_4 - 2(y_5^2 + y_6^2 - y_5 y_6) = n y_2^2 - a y_2 + m.$$

Hence we can choose:

$$y = (a - n y_2, y_2, 1, (n y_2^2 - a y_2 + m)/2, 0, 0),$$

unless $n \equiv a(2)$ and $m \equiv 1(2)$, the only case in which $n y_2^2 - a y_2 + m$ is odd for all $y_2 \in \mathbb{Z}$. Since, by assumption, one of n, m and a is even, in this case n (and a) must be even. Now we begin with $y = (1, m, 0, 0, 0, 0)$ and the same argument provides an x. It is easy to see that with these choices of x and y the embedding of T is primitive. □

2. Eckardt points

DEFINITION 2.1. A smooth cubic surface contains 27 lines and has 45 plane sections which are unions of three lines. In case three coplanar lines meet in a single point, that point is called an Eckardt point.

2.1. The Sylvester form and Eckardt points. A smooth cubic surface S defined by a Sylvester form $\sum \lambda_i x_i^3$ as in 1.1 has an Eckardt point iff $\lambda_i = \lambda_j$ for some $i \neq j$ ([**S**], pp. 145 ff.)

If $\lambda_i = \lambda_j$, the corresponding Eckardt point P of S is the vertex P_{klm} of the pentahedron (with $\{i, \ldots, m\} = \{0, \ldots, 4\}$). The plane $x_i + x_j = 0$ (which is the tangent plane $T_P S$ to S in P_{klm}) cuts out the three lines meeting in P_{klm} and S has an involution (induced by permuting x_i and x_j). In particular, S has an Eckardt point iff a vertex of the pentahedron lies on S.

A Sylvester form with $\lambda_0 = \lambda_1 = \lambda_2$ defines a cubic surface with three Eckardt points in general, there are 4 Eckardt points if also $\lambda_3 = \lambda_4$ and there are 10 Eckardt points if all λ_i are equal (this surface is known as the Clebsch diagonal surface). In case $\lambda_0 = \lambda_1$ and $\lambda_2 = \lambda_3$, but otherwise the λ_i's are distinct, the surface has 2 Eckardt points. In case $\lambda_0 = \lambda_1 = \lambda_2 = \lambda_3$ but $\lambda_0 \neq \lambda_4$, the surface has six Eckardt points.

2.2. New curves on the Hessian. For an Eckardt point P the intersection of the tangent plane $T_P S$ and the Hessian Y_S consists of the line L_{ij} (with multiplicity two) and of the pair of lines defined by

$$\lambda_k \lambda_l x_k x_l + \lambda_k \lambda_m x_k x_m + \lambda_l \lambda_m x_l x_m = 0, \qquad x_k + x_l + x_m = 0,$$

which meet in P_{klm}, we refer to these as new lines. The strict transform of these lines in the Hessian K3 surface are two new disjoint (-2)-curves which both meet N_{klm}, but none of the other N_{abc}'s.

2.3. The Clebsch diagonal cubic surface. The diagonal cubic surface is the smooth cubic surface S_{10} defined by the Sylvester form:

$$S_{10}: \qquad \sum_{i=0}^{4} x_i^3 = 0, \qquad \sum_{i=0}^{4} x_i = 0,$$

so all $\lambda_i = 1$. This surface has 10 Eckardt points and thus its Hessian Y_{10} has 10 new pairs of lines. Let ω be a primitive cube root of unity and let

$$C_{234} = \{(s : -s : t : t\omega : t\omega^2) : (s : t) \in \mathbb{P}^1\} \qquad \subset Y_{10} \qquad (\omega^2 + \omega + 1 = 0)$$

be one of the new lines on the Eckardt point $P_{234} = (1 : -1 : 0 : 0 : 0)$. For $\sigma \in S_5$, the symmetric group on $\{0, \ldots, 4\}$, we denote by $C_{\sigma(2), \sigma(3), \sigma(4)}$ the line obtained by permuting the coordinates. In particular,

$$C_{234} = C_{342} = C_{423}, \quad C_{243} = C_{324} = C_{432} = \{(s : -s : t : t\omega^2 : t\omega) : (s : t) \in \mathbb{P}^1\}$$

is the other new line on P_{234}. We use the same names to denote the corresponding (-2)-curves on the Hessian K3 surface X_{10} of S_{10}.

The intersection numbers of N_{ab} and N_{abc} with C_{klm} are zero unless one has $\sharp\{a, b, k, l, m\} = 5$ resp. $\sharp\{a, b, c, k, l, m\} = 3$, in which case the intersection number is 1. Moreover, C_{abc} and C_{klm} intersect, with intersection number 1, iff there are two equal indices in the same order (up to cyclic permutations, e.g. C_{124} intersects C_{041}), C_{abc} and C_{abd} meet in a point on X_{10} whose image in Y_{10} has coordinates $x_a = \omega$, $x_b = \omega^2$, $x_c = x_d = 1$ and $x_e = -1$ where $\{a, b, c, d, e\} = \{0, 1, 2, 3, 4\}$. The following classes in the Néron-Severi group NS_{10} of X_{10}

$$c_{ij} = C_{abc} - C_{acb} \quad \in NS_{10} := NS(X_{10}), \qquad \{i, j, a, b, c\} = \{0, \ldots, 4\},$$

are easily seen to be perpendicular to NS_{gen}. They satisfy:

$$c_{ij}^2 = -4, \qquad c_{ik} \cdot c_{il} = 2, \qquad c_{ij} \cdot c_{kl} = 0$$

for distinct indices i, j, k, l, so we have a copy of $A_4(-2)$ in T_{gen} (cf. Remark 2.3). The symmetric group S_5 is a subgroup of $Aut(X_{10})$ (acting via permutations of the coordinates) and acts via permutation of the indices on the c_{ij}, so it acts as $W(A_4) \cong S_5$ on the copy of $A_4(-2)$.

LEMMA 2.2. *Let X_k be the Hessian K3 surface of a smooth cubic surface with Sylvester form having k Eckardt points (in particular, $k \in \{0, 1, 2, 3, 4, 6, 10\}$). Let T_k be the transcendental lattice of X_k.*

$$
\begin{array}{ll}
T_0 = T_{gen} = U \oplus U(2) \oplus A_2(-2) & \mathrm{discr}(T_{gen}) = 48, \\
T_1 = U \oplus U(2) \oplus <-12> & \mathrm{discr}(T_1) = -48, \\
T_2 = U \oplus <4> \oplus <-12> & \mathrm{discr}(T_2) = 48, \\
T_3 = U \oplus U(6) & \mathrm{discr}(T_3) = 36, \\
T_4 = U(3) \oplus <4> & \mathrm{discr}(T_4) = -36, \\
T_6 = U \oplus <24> & \mathrm{discr}(T_6) = -24. \\
T_{10} = \left(\mathbb{Z}^2, \begin{pmatrix} 4 & 1 \\ 1 & 4 \end{pmatrix} \right) & \mathrm{discr}(T_{10}) = 15.
\end{array}
$$

PROOF. We start with the computation of NS_{10} and T_{10}. The 40×40 matrix of intersection products of the curves N_α and C_β has rank 20, hence the Néron-Severi group of X_{10} has maximal rank. A $\mathbb{Z}$-basis of the lattice generated by these 40 curves is given by the basis of NS_{gen} (given in 1.7) and the 4 curves C_{234}, C_{134}, C_{124} and C_{032}. A machine verified that the discriminant of this lattice is -15. Since 15 is square free, we can conclude that NS_{10} is generated by these 40 curves. It is easy to find two orthogonal $E_8(-1)$'s in this lattice, for example:

$$
N_{034} \; -\!- \; N_{04} \; -\!- \; N_{024} \; -\!- \; C_{024} \; -\!- \; C_{124} \; -\!- \; C_{041} \; -\!- \; N_{23}
$$
$$
| \\
N_{24}
$$

$$
N_{12} \; -\!- \; N_{012} \; -\!- \; N_{01} \; -\!- \; C_{234} \; -\!- \; C_{142} \; -\!- \; C_{143} \; -\!- \; N_{134}
$$
$$
| \\
N_{013}
$$

Using a machine, it is not hard to find the rank 4 lattice L' such that $NS_{10} \cong E_8(-1)^2 \oplus L'$, we found a copy of U in L' (a standard basis is: $-2N_{134} - 5N_{124} - N_{123} + 4N_{034} + 5N_{024} + N_{023} + 7N_{014} + 6N_{013} + 8N_{012} + 13N_{01} + 2N_{02} + 8N_{04} + N_{12} - N_{13} + 5C_{234} - 3C_{134} - 4C_{124} + C_{032}$ and $-2N_{134} - 5N_{124} - N_{123} + 4N_{034} + 5N_{024} + N_{023} + 7N_{014} + 6N_{013} + 8N_{012} + 13N_{01} + 2N_{02} + 8N_{04} + N_{12} - N_{13} + 5C_{234} - 3C_{134} + 4C_{124})$. The orthogonal complement of U in L' is isomorphic to the lattice denoted by $T_{10}(-1)$ in the theorem (a basis for it is: $2N_{124} + 2N_{123} - 5N_{034} - 2N_{024} - 2N_{023} - N_{014} + 2N_{012} + N_{01} - 3N_{03} - 4N_{04} + 3N_{12} + 2N_{13} - 3N_{34} + C_{234} + C_{134} - C_{032}$ and $(-4N_{134} - 9N_{124} - 4N_{123} + 8N_{034} + 9N_{024} + 4N_{023} + 11N_{014} + 9N_{013} + 11N_{012} + 19N_{01} + 4N_{02} + 2N_{03} + 14N_{04} - N_{12} - 3N_{13} + 7C_{234} - 5C_{134} - 6C_{124} + 2C_{032})$. In general, the embedding of a rank 20 lattice like NS_{10} into the K3 lattice $E_8(-1)^2 \oplus U^3$ is not unique up to isomorphism. However, in this case we know that the orthogonal complement of NS_{10} in $H^2(X_{10}, \mathbb{Z})$ is an even rank two lattice with discriminant 15. The classification of positive definite binary quadratic forms shows that there are two such lattices, one is T_{10}, the other has matrix with rows $2, 1$ and $1, 8$.

The discriminant form of the Néron-Severi lattice is the opposite of the one of the transcendental lattice (after identifying the discriminant groups). The discriminant form NS_{10} is equal to minus the one of T_{10}, but the discriminant form of the other lattice is different, hence we found T_{10}.

A result of Nikulin implies that the embedding of T_{10} in the K3 lattice is unique up to isometry (cf. [**M**], cor. 2.10). Choosing a convenient embedding, it is then easy to find an explicit isomorphism of NS_{10} with a sublattice of the K3 lattice.

To find the other lattices, we use that the number of moduli d_k of K3 surfaces with a fixed number k of Eckardt points is easy to find using the Sylvester form. By the Torelli theorem the rank of the transcendental lattice of a general K3 in this family is at least $d_k + 2$, hence NS_k has rank at most $20 - d_k$. Computations show that in each case the lattice spanned by NS_{gen} and the C_α's on the general member of the family has rank $20 - d_k$. It is not hard to check that this lattice is perpendicular to certain elements in NS_{10}, choosing the Sylvester forms defining the X_k such that $\lambda_0 = \lambda_1 = \lambda_2 = \lambda_3$ for X_6, $\lambda_0 = \lambda_1 = \lambda_2$ and $\lambda_3 = \lambda_4$ for X_4, $\lambda_0 = \lambda_1 = \lambda_2$ for X_3, $\lambda_0 = \lambda_1$ and $\lambda_2 = \lambda_3$ for X_2 and $\lambda_0 = \lambda_1$ for X_1 we have:

(1) $NS_6 = < c_{04} - c_{14} + c_{24} - c_{34} >^\perp \hookrightarrow NS_{10}$.

(2) $NS_4 = < c_{14} - c_{24} + c_{03} - c_{13} + c_{23} >^\perp \hookrightarrow NS_{10}$;

(3) $NS_3 = < c_{34} >^\perp \hookrightarrow NS_4$ and $NS_3 = < c_{23} - c_{13} + c_{03} >^\perp \hookrightarrow NS_6$;
   Obviously, to embed NS_2 in NS_6 we need to make a permutation of the indices and consider X_2', i.e. the Hessian of the cubic with two Eckardt points and Sylvester representation with coefficients $\lambda_0 = \lambda_1$ and $\lambda_2 = \lambda_3$. Thus the embedding is: $NS_2' = < c_{02} - c_{03} - c_{12} + c_{13} >^\perp \hookrightarrow NS_6$;

(4) $NS_2 = < c_{02} - c_{12} >^\perp \hookrightarrow NS_4$;

(5) $NS_1 = < c_{34} >^\perp \hookrightarrow NS_2$ and $NS_1 = < c_{02} - c_{12} >^\perp \hookrightarrow NS_3$;

(6) $NS_{gen} = < c_{01} >^\perp \hookrightarrow NS_1$.

Next we determined the transcendental lattice T_i as $NS_i^\perp$ in the K3 lattice. In particular, in this way we determined NS_{gen} and its perpendicular T_{gen} in the K3 lattice. $\square$

REMARK 2.3. We give an explicit primitive embedding of $A_4(-2)$ in T_{gen}. Recall that $T_{gen} = U \oplus U(2) \oplus A_2(-2)$, an element $x \in T_{gen}$ will be written as $x = (x_1, \ldots, x_6) \in \mathbb{Z}^6$, the quadratic form on T_{gen} is $2x_1x_2 + 4x_3x_4 - 2(2x_5^2 - x_5x_6 + 2x_6^2)$.

A basis of $A_4(-2)$ is given by the c_{ij} (cf. section 2.3) and the embedding of $A_4(-2)$ in T_{gen} is:

$$c_{01} = (0,0,0,0,0,1), \qquad c_{12} = (0,0,0,-2,1,0)$$

$$c_{23} = (0,0,1,2,-2,-1), \qquad c_{34} = (4,-2,-3,-5,4,2).$$

The orthogonal complement of this sublattice in T_{gen} is isomorphic to T_{10} and is spanned by

$$t_1 = (2,1,0,0,0,0), \qquad t_2 = (5,-2,-3,-6,4,2), \qquad \text{with} \quad (t_i \cdot t_j) = \begin{pmatrix} 4 & 1 \\ 1 & 4 \end{pmatrix}.$$

Moreover, the lattice

$$T_{10} \oplus A_4(-2) \hookrightarrow T_{gen} = U \oplus U(2) \oplus A_2(-2)$$

has index 5 and T_{gen} is generated by $T_{10} \oplus A_4(-2)$ and

$$\tfrac{1}{5}(2(t_1 + t_2) + c_{01} + 2c_{12} + 3c_{23} + 4c_{34}) = (6,-2,-3,-6,4,2).$$

REMARK 2.4. The quotient of the K3 surface X_{10} by its fixed point free involution $\iota_{X_{10}}$ (cf. 1.6) is an Enriques surface with a finite automorphism group, see [**K**], type IV.

3. Nodal cubic surfaces

3.1. Cayley's four nodal cubic surface. A cubic surface can have up to four nodes. There is a unique surface S_{4n} having four nodes, it is Cayley's cubic surface defined by

$$\frac{1}{z_0} + \frac{1}{z_1} + \frac{1}{z_2} + \frac{1}{z_3} = 0,$$

it has the Sylvester form:

$$S_{4n}: \qquad \sum x_i = 0, \qquad x_0^3 + x_1^3 + x_2^3 + x_3^3 + \frac{1}{4}x_4^3 = 0.$$

Its nodes (in the Sylvester form) are $p_0 = (-1 : 1 : 1 : 1 : -2)$ and the other three, p_1, p_2, p_3, are obtained by permuting the first four coordinates.

These points are also singular on the Hessian Y_{4n}, and they give rise to (-2)-curves M_i on the Hessian K3 surface X_{4n}. An explicit computation shows that the lines M_{ij} connecting two nodes p_i, p_j of the Hessian are on the Hessian. Thus we find another six (-2)-curves on X_{4n} which we also denote by M_{ij}.

In case the cubic surface has k nodes, we denote the (-2)-curves on the Hessian K3 by $M_0, \ldots, M_{k-1}$ and M_{ij} ($0 \le i < j \le k-1$).

LEMMA 3.1. *Let T_{kn} be the transcendental lattice of the Hessian K3 surface of a general cubic surface with k nodes. Then we have:*

$$T_{4n} = <2> \oplus <6>$$

and:

$$T_{3n} = T_{4n} \oplus <-2>, \qquad T_{2n} = T_{4n} \oplus <-2>^2, \qquad T_{1n} = T_{4n} \oplus <-2>^3.$$

PROOF. The proof is similar to the one of Lemma 2.2. The non-zero intersection numbers involving the ten new (-2)-curves on the Hessian K3 surface X_{4n} of the Cayley cubic are:

$$
\begin{array}{llll}
(M_k, M_{ij}) = 1 & \text{if} & \#\{i,j,k\} = 2, \\
(N_{ij4}, M_{kl}) = 1 & \text{if} & \#\{i,j,k,l\} = 4, \\
(N_{ij}, M_{kl}) = 1 & \text{if} & \#\{i,j,k,l\} = 2,
\end{array}
$$

and of course $M_i^2 = M_{ij}^2 = -2$. A machine computation confirms that the rank of the lattice spanned by the 30 (-2)-curves in X_{4n} is indeed 20 and that a basis of this lattice is given by the 16 curves in the basis of $NS(X_{gen})$ (cf. 1.7) and the 4 curves $M_{01}, M_{02}, M_{03}, M_{12}$. The discriminant of this lattice is -12. Thus the discriminant of the Néron-Severi lattice is either -12 or -3, but if it were -3 the transcendental lattice had to be A_2, which is impossible by Lemma 1.1. Thus these 20 curves span the Néron-Severi lattice.

There are two perpendicular $E_8(-1)$'s in the Néron-Severi lattice:

$$
N_{012} \quad \text{---} \quad N_{02} \quad \text{---} \quad N_{024} \quad \text{---} \quad N_{24} \quad \text{---} \quad N_{234} \quad \text{---} \quad N_{34} \quad \text{---} \quad N_{134}
$$
$$
\qquad\qquad\qquad\qquad\qquad |
$$
$$
\qquad\qquad\qquad\quad N_{04}
$$

$$N_{013} \quad -- \quad N_{03} \quad -- \quad M_{03} \quad -- \quad M_3 \quad -- \quad M_{23} \quad -- \quad M_2 \quad -- \quad M_{12}$$
$$\mid$$
$$M_0$$

A further computation then shows that

$$NS(X_{4n}) \cong E_8(-1)^2 \oplus U \oplus < -2 > \oplus < -6 > .$$

Thus T_{4n} has rank two, discriminant 12 and its discriminant form is the opposite of the one on $NS(X_{4n})$ and there is a unique such lattice, which is $< 2 > \oplus < 6 >$.

Another way to obtain this result is by realizing the Hessian as a double cover of the plane by projecting from the node M_0. This double cover branches over the union of three lines and a cubic with a node. The pencil of lines on the node of the cubic curve defines an elliptic fibration on X_{4n} which turns out to have six singular fibers, three of type I_6 and three of type I_2. According to the table in [**SZ**] there is only one such K3 surface (case 4) and its transcendental lattice is indeed $< 2 > \oplus < 6 >$. This surface also appears as a double cover of the plane branched over three conics in [**P**], Example 3, p.304.

There is a unique, up to isomorphism, embedding of T_{4n} into the K3 lattice. It is not hard to find an explicit isomorphism of the Néron-Severi lattice with the orthogonal complement of the image of T_{4n}. The Néron Severi lattice NS_{kn} of X_{kn} is now easily seen to be:

$$NS_{3n} = (T_{4n} \oplus < M_3 >)^\perp, \quad \ldots \quad , NS_{1n} = (T_{4n} \oplus < M_1, M_2, M_3 >)^\perp,$$

next one finds T_{kn} as the orthogonal complement of NS_{kn}. $\qquad \square$

4. Cubics with Eckardt points and nodes

4.1.   In this section we consider the Hessians X_{anb} of three cubic surfaces with a nodes and b Eckardt points with $(a, b) = (1, 6), (1, 4)$ and $(3, 4)$. These K3 surfaces are 'singular', i.e. have Néron-Severi groups of rank 20. In two of the three cases we succeeded in determining the transcendental lattice completely.

LEMMA 4.1.   (1) *The Hessian K3 surface X_{1n6} of the cubic surface S_{1n6} with one node and 6 Eckardt points with Sylvester form*

$$\sum x_i = 0, \qquad x_0^3 + x_1^3 + x_2^3 + x_3^3 + \frac{1}{16}x_4^3 = 0$$

has transcendental lattice

$$T_{1n6} = < 2 > \oplus < 24 > .$$

(2) *The Hessian K3 surface X_{1n4} of the cubic surface S_{1n4} with one node and 4 Eckardt points with Sylvester form*

$$\sum x_i = 0, \qquad x_0^3 + x_1^3 + x_2^3 + \frac{4}{9}x_3^3 + \frac{4}{9}x_4^3 = 0$$

has transcendental lattice

$$< 6 > \oplus < 12 > .$$

(3) *The Hessian K3 surface X_{3n4} of the cubic surface S_{3n4} with three nodes and 4 Eckardt points with Sylvester form*

$$\sum x_i = 0, \qquad x_0^3 + x_1^3 + x_2^3 + 4x_3^3 + 4x_4^3 = 0$$

has transcendental lattice

$$T_{3n4} = <4> \oplus <6>.$$

PROOF. The only singular point of S_{1n6} is the point $p_0 = (1 : \ldots : 1 : -4)$ which is also singular on the Hessian, which has eleven singular points. The point p_0 is not a vertex nor does it not lie on any of the edges L_{ij} of the pentahedron. It is easy to verify that p_0 does not lie on any line C_{abc} from a pair in $T_P S_{1n6}$ where P is an Eckardt point of S_{1n6} (cf. 2.2). Therefore the Néron-Severi group of X_{1n6} contains the direct sum of the Néron-Severi group of the general Hessian K3 with 6 Eckardt points and the lattice $< -2 >$, spanned by the (-2)-curve over p_0:

$$NS_6 \oplus < -2 > = E_8(-1)^2 \oplus U \oplus < -24 > \oplus < -2 > \hookrightarrow NS_{1n6}.$$

As the discriminant of this sublattice is -48, the discriminant of NS_{1n6} is either -48 or there is a vector v, primitive in $NS_6 \oplus < -2 >$, such that $v/2 \in NS_{1n6}$. As $E_8(-1)^2 \oplus U$ is a direct summand, we can write $v = ae + bf$ with $a, b \in \{0, 1\}$ and $e^2 = -24, ef = 0, f^2 = -2$. As $(v/2)^2 = -6a^2 + b^2/2 \in 2\mathbb{Z}$ (since NS_{1n6} is even) we get $b = 0$, as $< -24 >$ is primitive in NS_6 and hence, after specialization, in NS_{1n6}, we get $a = 0$. Therefore $NS_6 \oplus < -2 > \cong NS_{1n6}$. The transcendental lattice T_{1n6} is a positive definite even rank two lattice whose discriminant form is the opposite of the one of NS_{1n6}, and there is only one such lattice.

The unique singular point of S_{1n4} is $p_0 = (-2 : -2 : -2 : 3 : 3)$ which does not lie on any of the lines in the Hessian. Thus one finds as above an embedding of lattices:

$$NS_4 \oplus < -2 > = E_8(-1)^2 \oplus U(3) \oplus < -4 > \oplus < -2 > \hookrightarrow NS_{1n4}.$$

The discriminant of $NS_4 \oplus < -2 >$ is 72, so the index of $NS_4 \oplus < -2 >$ in NS_{1n4} divides 6. Let $v \in NS_4 \oplus < -2 >$ such that $v/6 \in NS_{1n4}$. Again we can restrict ourselves to

$$v = ((u_1, u_2), a, b) \in U(3) \oplus < -4 > \oplus < -2 >, \qquad v^2 = 6u_1u_2 - 4a^2 - 2b^2.$$

If $v/6 \in NS_{1n4}$, an even overlattice of $NS_4 \oplus < -2 >$, we must have $(v/6) \cdot ((1,0),0,0) \in \mathbb{Z}$, so $u_2/2 \in \mathbb{Z}$, similarly $u_1 \in 2\mathbb{Z}$. Next $(v/6) \cdot ((0,0),1,0) \in \mathbb{Z}$, so $2a/3 \in \mathbb{Z}$ and similarly $b \in 3\mathbb{Z}$. Hence $v/6 = (u_1'/3, u_2'/3, a'/2, b'/2)$ with $u_1', \ldots, b' \in \mathbb{Z}$. Therefore $2\mathbb{Z} \ni (v/6)^2 = (2/3)u_1'u_2' - (a')^2 - (b')^2/2$, from which we deduce u_1' or $u_2' \in 3\mathbb{Z}$, $b' \in 2\mathbb{Z}$ and $a' \in 2\mathbb{Z}$, whence, assuming $u_2' \in 3\mathbb{Z}$ and going mod $NS_4 \oplus < -2 >$, we can only have $v/6 = ((u_1'/3,0),0,0)$. Thus $v/6 \in NS_4$, a primitive sublattice, so $u_1' \in 3\mathbb{Z}$. Hence $NS_4 \oplus < -2 > = NS_{1n4}$. From this one computes the discriminant form on $NS_{1n4}^{\perp}$ and one finds that it determines a unique even positive definite lattice of rank two.

The three singular points of S_{3n4} are:

$$p_0 = (-2 : 2 : 2 : -1 : -1), \; p_1 = (2 : -2 : 2 : -1 : -1), \; p_2 = (2 : 2 : -2 : -1 : -1).$$

The lines M_{ij} spanned by p_i and p_j lie on the Hessian. The p_i's do not lie on any of the 10 lines L_{ij} but the lines M_{ij} and L_{ij} intersect, finally the strict transform of M_{ij} in the Hessian K3 surface and N_{klm} intersect if $\sharp\{i, j, k, l, m\} = 5$. Of the lines C_{abc} on the Hessian of a cubic with Eckardt points, only the lines C_{012} and C_{021} remain, the others collapse onto the M_{ij}. A machine computation shows that the lattice spanned by these (-2)-curves has rank 20 and discriminant -24. A basis is given by the 16 curves in the basis of $NS(X_{gen})$ (cf. 1.7) and $C_{012}, M_{01}, M_{02}, M_{12}$. Since the determinant of an even rank two lattice cannot be 6, we conclude that this lattice is NS_{gen}. To find its discriminant form we used the following two copies of $E_8(-1)$:

$$N_{23} \;\;--\;\; N_{123} \;\;--\;\; N_{13} \;\;--\;\; N_{013} \;\;--\;\; N_{01} \;\;--\;\; M_{01} \;\;--\;\; M_0$$
$$\qquad\qquad\qquad\qquad | $$
$$N_{134}$$

$$N_{124} \;\;--\;\; N_{24} \;\;--\;\; N_{024} \;\;--\;\; N_{04} \;\;--\;\; N_{034} \;\;--\;\; M_{12} \;\;--\;\; M_2$$
$$\qquad\qquad\qquad\qquad | $$
$$N_{02}$$

and we found an isomorphism

$$NS_{3n4} \cong E_8(-1)^2 \oplus U \oplus <-4> \oplus <-6>.$$

There is a unique even, rank 2, positive definite lattice with opposite discriminant group. $\qquad\square$

5. Cubic forms without a Sylvester form

5.1. No Sylvester form. We defined the Sylvester form of a cubic form F in 4 variables using 5 linear forms $x_0, \ldots, x_4$ in 4 variables such that any four of the five are linearly independent, cf. 1.1, normalized such that $\sum x_i = 0$ and satisfying $F = \sum \lambda_i x_i^3$ for some $\lambda_i \in \mathbb{C}$.

There are basically two types of cubic forms defining smooth cubic surfaces which do not admit a Sylvester form (we do not know precise results on forms defining singular surfaces). In the first case one still has $F = \sum \lambda_i x_i^3$, but at least four of the five x_i are linearly dependent, say $x_4 = a_1 x_1 + a_2 x_2 + a_3 x_3$. Then one has $F = \lambda_0 x_0^3 + G(x_1, x_2, x_3)$ with a homogeneous polynomial G in three variables. The corresponding cubic surfaces are called cyclic surfaces. Note that the Hessian of a cyclic cubic surface is reducible, which proves that these surfaces do not have a Sylvester form.

The second type of cubic forms F which do not admit a Sylvester forms are those which are not a sum of five cubes of linear forms. These are obtained as a limit of Sylvester forms where at least two of the five x_i coincide, we will refer to these as non-Sylvester forms. The corresponding surfaces were described in [**S**] and [**Rod**]. They are of two types, denoted here by $ns1$ and $ns2$, the dimensions of these families are three and two respectively. We give explicit one parameter families of Sylvester forms specializing to these surfaces in the proof of Theorem 6.1.

5.2. The first case. A cubic surface of type $ns1$ is defined by a form F of the following type:

$$S_{ns1}: \qquad F = x_1^3 + x_2^3 + x_3^3 - x_0^2(a_0 x_0 + 3a_1 x_1 + 3a_2 x_2 + 3a_3 x_3) = 0.$$

This form defines a smooth cubic surface iff, for all choices of sign,

$$a_0 + 2(\pm a_1^{3/2} \pm a_2^{3/2} \pm a_3^{3/2}) \neq 0.$$

In case one of the $a_i = 0$ for $1 \leq i \leq 3$, the form F defines a cyclic surface. We will assume that S_{ns1} is smooth and that $a_1 a_2 a_3 \neq 0$ from now on.

The Hessian of F is, up to a scalar multiple,

$$H_F = x_1 x_2 x_3 (a_0 x_0 + \ldots + a_3 x_3) + x_0^2 (a_1^2 x_2 x_3 + a_2^2 x_1 x_3 + a_3^2 x_1 x_2).$$

Obviously, $p_0 = (1 : 0 : 0 : 0)$ is singular on the Hessian surface Y. Projecting from p_0 realizes the Hessian surface as a double cover of the plane (with coordinates x_1, x_2, x_3) branched over the curve defined by:

$$x_1 x_2 x_3 [a_0^2 x_1 x_2 x_3 - 4(a_1 x_1 + a_2 x_2 + a_3 x_3)(a_1^2 x_2 x_3 + a_2^2 x_1 x_3 + a_3^2 x_1 x_2)] = 0,$$

that is, the union of three lines and a nonsingular cubic curve. As a singular point of the branch curve is the image of a singular point or the image of a line on the Hessian surface it is now easy to determine the singular locus of Y. One finds that Y has exactly 7 singular points, p_0 and the six points:

$$p_1 = (0 : 1 : 0 : 0), \quad p_2 = (0 : 0 : 1 : 0), \quad p_3 = (0 : 0 : 0 : 1),$$

(these points lie on lines $x_i = x_j = 0$ $(1 \leq i < j \leq 3)$ in Y which project to points),

$$q_1 = (0 : 0 : a_3 : -a_2), \quad q_2 = (0 : -a_3 : 0 : a_1), \quad q_3 = (0 : a_2 : -a_1 : 0).$$

In particular, the cubic surface S_0 defined by F cannot have a Sylvester form (in that case the Hessian has exactly 10 singular points) nor is it cyclic (in that case the Hessian surface is reducible).

5.3. The second case. A cubic surface of type $ns2$ is defined as follows.

$$S_{ns2}: \qquad G = x_1^3 + x_2^3 + 2\lambda x_3^3 - 3x_3(\mu x_1 x_3 + x_2 x_3 + x_0^2) = 0.$$

If $\mu = 0$ this is a cyclic cubic surface and the surface S_{ns2} is smooth iff for all choices of sign:

$$\lambda \pm \mu^{3/2} \pm 1 \neq 0.$$

The Hessian of G is (up to scalar multiple):

$$H_G = x_1 x_2 x_3(-2\lambda x_3 + \mu x_1 + x_2) + x_3^3(x_1 + \mu^2 x_2) - x_0^2 x_1 x_2.$$

Again $p_0 = (1 : 0 : 0 : 0)$ is a singular point. Projecting from p_0 gives the branch curve defined by:

$$x_1 x_2 x_3 [x_1 x_2(-2\lambda x_3 + \mu x_1 + x_2) + x_3^3(x_1 + \mu^2 x_2)] = 0,$$

which is the union of three lines and a smooth cubic curve. The Hessian surface has only four singular points, p_0, and

$$p_1 = (0 : 1 : 0 : 0), \quad p_2 = (0 : 0 : 1 : 0), \quad q = (0 : 1 : -\mu : 0).$$

As before, we conclude that these surfaces do not admit a Sylvester form nor are they cyclic.

LEMMA 5.1. *Let X_{nsk}, $k = 1, 2$, be the Hessian K3 surface of a general cubic surface S_{nsk} which does not admit a Sylvester form. Let T_{nsk} be the transcendental lattice of X_{nsk}. Then we have:*

$$T_{ns1} = U \oplus U(2) \oplus < -4 >, \qquad T_{ns2} = U \oplus U(2).$$

PROOF. First we determine the Néron-Severi group of X_{ns2} using the elliptic fibration defined by the pencil of lines on the image $\bar{q}$ of q in $\mathbb{P}^2$. From a study of the branch curve one finds that there are 6 singular fibers, of type $I_0^*, I_8^*, I_1, I_1, I_1, I_1$. The lines $x_i = x_3 = 0$ $(i = 1, 2)$ define sections of the fibration. The rank of the Néron-Severi group of the general X_{ns2} is 18 (the surfaces X_{ns2} have two moduli), which is two plus the number of components of the singular fibers not meeting one of the sections. Hence the Mordell-Weil group of the fibration is finite. The torsion subgroup of a fiber of type I_0^* is $(\mathbb{Z}/2\mathbb{Z})^2$ and the two-torsion of an I_1-fiber is $\mathbb{Z}/2\mathbb{Z}$. As the torsion of the Mordell-Weil group injects in the torsion subgroup of the fibers it must be either trivial or be isomorphic to $\mathbb{Z}/2\mathbb{Z}$. As there are two sections, we conclude that the Mordell-Weil group of the elliptic fibration is $\mathbb{Z}/2\mathbb{Z}$. The Shioda-Tate formula ([**SI**], Lemma 1.3) now shows that the discriminant of $NS(X_{ns2})$ is 4.

The 13 (-2)-curves in the I_8^* fiber (which map to the line $x_3 = 0$ in $\mathbb{P}^2$, note the configuration of curves is two copies of D_6, mapping to $x_1 = x_3 = 0$ and $x_2 = x_3 = 0$, linked with a vertex which is a (-2)-curve mapping to the line $x_3 = 0$), the (-2)-curve over $\bar{q}$, the two sections and the four of the five (-2)-curves in the I_0^* fiber (these 4 curves map to the point $x_1 = x_2 = 0$ in $\mathbb{P}^2$) give a diagram with 20 vertices. It is a square with 3 points on each edge and each of the four vertices of the square is linked to a further point, which is not linked to anything else. In particular, it is very easy to find two perpendicular E_8's in the lattice generated by these 20 curves and to compute that the orthogonal complement to this sublattice is $U(2)$. Hence $NS(X_{ns2}) \cong E_8(-1)^2 \oplus U(2)$. By [**Ni**], Thm. 1.14.4, there is, up to isometry, a unique embedding of $U(2)$ into U^3. From this one finds that $T_{ns2} = NS(X_{ns2})^\perp \cong U(2) \oplus U$.

Similar to the case of X_{ns2}, the family of lines on the image of q_3 in $\mathbb{P}^2$ defines an elliptic fibration X_{ns1} with two sections, $x_i = x_3 = 0$, $i = 1, 2$. There are 7 singular fibers of type $I_4^*, I_4, I_0^*, I_1, I_1, I_1, I_1$. There are 15 components of singular fibers which do not meet one of the sections, so the Mordell-Weil group of this fibration finite. As before we conclude that the Mordell-Weil group of the elliptic fibration is $\mathbb{Z}/2\mathbb{Z}$. The Shioda-Tate formula shows that the discriminant of $NS(X_{ns1})$ is 2^4.

It is amusing to observe that the following 20 (-2)-curves have an intersection diagram which is a cube with a point on the middle of each edge: the 9 components of the I_4^* fiber; the following 3 components of the I_4-fiber: the line $l : x_0 = a_1 x_1 + a_2 x_2 + a_3 x_3 = 0$ in Y_{ns1} and the (-2)-curves over q_1 and q_2 in X_{ns2}; the (-2)-curves over p_0, q_3; the two sections; and 4 of the 5 components of the I_0^* fiber.

It is clear that X_{ns2} is a limit of X_{ns1} (two of the lines in the branch curve become tangent to the cubic component, equivalently, the fibers of type I_4 and I_4^* coalesce to a fiber of type I_8^*). Thus we get an inclusion $NS(X_{ns1}) \subset NS(X_{ns2})$ and it is not hard to see that $NS(X_{ns1}) = (n_1 - n_2)^\perp$ where n_1, n_2 are two disjoint (-2)-curves in X_{ns2}, they are the inverse image of the strict transform of the exceptional divisor in the first blow up of $\mathbb{P}^2$ in the image of p_1 and p_2. Thus the lattice $T(X_{ns2}) \oplus < -4 >$ is a sublattice of $T(X_{ns1})$, but since these lattices have the same discriminant, they are equal. $\qquad\square$

REMARK 5.2. Specializing a general Hessian to one of type $ns1$ induces an inclusion $NS_{gen} \hookrightarrow NS_{ns1}$. One can verify that $NS_{gen}^\perp$ in NS_{ns1} is generated by a class with selfintersection -12. Note that $t = (1, 2) \in A_2(-2)$ has $t^2 = -12$ and that its orthogonal complement is $< (1, 0) >$ with $(1, 0)^2 = -4$, so

$$(0, 0, (1, 2))^\perp \cong U \oplus U(2) \oplus < -4 > \cong T_{ns1} \qquad (\subset U \oplus U(2) \oplus A_2(-2) = T_{gen}).$$

5.4. Eckardt points. As the Eckardt points of a cubic surface S are the singular points of its Hessian Y which are on S ([**S**], p.146), it is easy to find the possible configurations of Eckardt points on a surface of type $ns1$. We determine the transcendental lattices of these surfaces below.

LEMMA 5.3. *Let S_{ns1} be a cubic surface without a Sylvester form as in section 5.2, from which we also adopt the notation and conventions:*

$$S_{ns1}:\qquad a_0 x_0^3 + x_1^3 + x_2^3 + x_3^3 - 3x_0^2(a_1 x_1 + a_2 x_2 + a_3 x_3) = 0.$$

(1) *In case $a_0 = 0$ and all a_i^3's are distinct, the point p_0 is the unique Eckardt point on S_{ns1}. The general Hessian K3 surface has Picard number 18 and transcendental lattice*

$$U(2)^2.$$

(2) *In case $a_0 \neq 0$ but $a_i^3 = a_j^3 \neq a_k^3$ the point q_k is the unique Eckardt point on S_{ns1}. The general Hessian K3 surface has Picard number 18 and transcendental lattice*

$$U \oplus < -4 > \oplus < 4 > .$$

(3) *In case $a_0 = 0$ and $a_i^3 = a_j^3 \neq a_k^3$, the points p_0, q_k are the only Eckardt points on S_{ns1}. The general Hessian K3 surface has Picard number 19 and transcendental lattice*

$$U(2) \oplus < 4 > .$$

(4) *In case $a_0 \neq 0$ and $a_i^3 = a_j^3 = a_k^3$, the points q_1, q_2, q_3 are the only Eckardt points on S_{ns1}. The general Hessian K3 surface has Picard number 19 and transcendental lattice*

$$U \oplus < 12 > .$$

(5) *In case $a_0 = 0$ and $a_i^3 = a_j^3 = a_k^3$, the points p_0, q_1, q_2, q_3 are the only Eckardt points on S_{ns1}. The general Hessian K3 surface has Picard number 20 and transcendental lattice*

$$A_2(2) = \begin{pmatrix} 4 & -2 \\ -2 & 4 \end{pmatrix}.$$

PROOF. We start with the case $a_0 = 0$ and $a_i^3 = a_j^3 = a_k^3$, we denote the cubic surface by S. (Note that the transformation $x_j \mapsto \omega^{n_j} x_j$, with $j = 1, 2, 3$ and with a cube root of unity ω, also gives an equation of type $ns1$ but with $a_j \mapsto \omega^{-n_j} a_j$). For every Eckardt point p the tangent plane to the cubic surface in p intersects the Hessian in a double line and two new lines. In particular, p_0 gives the pair of lines lines $d^\pm$ on S defined by $\sum_{i=1}^3 x_i = 0, \sum_{1 \leq i < j \leq 3} x_i x_j = 0$, and similarly the q_i give the pairs of lines $x_k + x_l = x_0^2 + x_i^2 = 0$. The lattice L generated by a basis of $NS(X_{ns1})$, d^+ and the lines $c_1^+ : x_2 + x_3 = x_0 + ix_1 = 0$, $c_2^- : x_1 + x_3 = x_0 - ix_2 = 0$ has rank 20 and discriminant 12. The only rank two, even, positive definite lattice with discriminant 3 is A_2, but this cannot be the transcendental lattice of a Hessian by Lemma 1.1. Hence L must be the Néron-Severi lattice of the Hessian K3 surface X of S and a computation of the discriminant group shows that the transcendental lattice is $A_2(2)$. It is also not hard to find two orthogonal E_8's in the Néron-Severi group of X and thus to find an embedding of $NS(X)$ into the K3 lattice.

For the other cases it is easy to find the rank r of the transcendental lattice from a count of the moduli and to find curves on the surface which span a lattice of rank $22 - r$. Then one can verify that these curves all lie in the following sublattices of $L = NS(X)$, and using the embedding of $NS(X)$ into the K3 lattice, one finds the transcendental lattices.

(1) The Néron-Severi group is $< (c_1^+ - c_1^-), (c_2^+ - c_2^-) - (c_3^+ - c_3^-) >^\perp$ in $NS(X)$.

(2) The Néron-Severi group is $< (c_i^+ - c_i^-) - (c_j^+ - c_j^-), d^+ - d^- >^\perp$ in $NS(X)$.

(3) The Néron-Severi group is $< (c_i^+ - c_i^-) - (c_j^+ - c_j^-) >^\perp$ in $NS(X)$.

(4) The Néron-Severi group is $< d^+ - d^- >^\perp$ in $NS(X)$.

$\square$

REMARK 5.4. The one dimensional family of Hessians in Lemma 5.3.4 with transcendental lattice $U \oplus < 12 >$ was studied by Peters and Stienstra in [**PS**]. They also observed the 'cube' formed by 20 of the curves in the Néron-Severi lattice (see the proof of Lemma 5.1). The total space of this family is a Calabi-Yau threefold (the Fermi threefold) and was studied by Verrill [**V**], section 4.

6. The moduli space of cubic surfaces

6.1.   Classical invariant theory shows that the moduli space of cubic surfaces $\mathcal{M}$ is isomorphic to the weighted projective space $\mathbb{P}(1, 2, 3, 4, 5)$. Each point in $\mathcal{M}$ corresponds to the isomorphism class of a cubic surface with at most 4 nodes, except for one point. That point corresponds to the semi stable, non-stable, cubic surfaces, the unique closed orbit in this set is the orbit of $l^3 = xyz$. In Theorem 6.1 we show that it maps to $(8 : 1 : 0 : 0 : 0) \in \mathcal{M}$ where we use Salmon's generating invariants (cf. section 6.3).

We recall the classical description of this moduli space using the Sylvester forms. We discuss some divisors, and their classes in the Chow group of the moduli space. We determine the divisor parametrizing cubic surfaces without a Sylvester form in Theorem 6.1. Finally we make some comments on the singular locus of the moduli space.

6.2. Invariants. The ring of invariants of the action of $SL(4, \mathbb{C})$ on the space of cubic forms in 4 variables is generated by the invariant polynomials I_n of degree n for $n = 8, 16, 24, 32, 40, 100$. Since 100 is not divisible by 8 and I_{100}^2 is polynomial in the other generators, the moduli space of cubic surfaces (the *Proj* of the ring of invariants) is the weighted projective space $\mathbb{P}(1, 2, 3, 4, 5)$. Note that I_8 and I_{100} are, up to a scalar multiple, unique whereas $I_{16}, \ldots, I_{40}$ are only unique up to the addition of homogeneous weighted polynomials of lower degree (for example, for any $a, b, c \in \mathbb{C}$ $a \neq 0$, the invariant $aI_{24} + bI_8^3 + cI_8I_{16}$ can also be used as a generator of degree 24). We will use Salmon's convention to choose the generators.

6.3. Generating invariants and the Sylvester form. The generating invariants are easily computed for a Sylvester form

$$\sum_{i=0}^{4} x_i = 0, \qquad \sum_{i=0}^{4} \lambda_i x_i^3 = 0.$$

Let σ_i be the i-th symmetric function in $\lambda_0, \ldots, \lambda_4$. Then ([**Sal**] p. 197):

$$I_8 = \sigma_4^2 - 4\sigma_3\sigma_5, \quad I_{16} = \sigma_5^3\sigma_1, \quad I_{24} = \sigma_5^4\sigma_4, \quad I_{32} = \sigma_5^6\sigma_2, \quad I_{40} = \sigma_5^8.$$

The quotient of $\mathbb{P}_\lambda^4$, the parameter space of the Sylvester forms, by the group S_5 is also a weighted projective space $\mathbb{P}(1, 2, 3, 4, 5)_\sigma$ with coordinates the elementary symmetric functions. The formulas above define a birational isomorphism:

$$\phi : \mathbb{P}(1, 2, 3, 4, 5)_\sigma \dashrightarrow \mathbb{P}(1, 2, 3, 4, 5)_I, \qquad (\sigma_1 : \ldots : \sigma_5) \longmapsto (I_8 : \ldots : I_{40}).$$

The base locus of ϕ is given by $\sigma_4 = \sigma_5 = 0$. The birational inverse of ϕ is:

$$\psi : \mathbb{P}(1, 2, 3, 4, 5)_I \dashrightarrow \mathbb{P}(1, 2, 3, 4, 5)_\sigma,$$

$$(I_8 : \ldots : I_{40}) \longmapsto (I_{16} : I_{32} : (I_{24}^2 - I_8 I_{40})/4 : I_{24} I_{40} : I_{40}^2),$$

the base locus of ψ is the point $(1 : 0 : 0 : 0 : 0)$ which corresponds to the Fermat cubic surface, see Theorem 6.1.

6.4. The boundary divisor. The locus of singular cubic surfaces, the boundary of the moduli space, is the divisor in $\mathcal{M} \cong \mathbb{P}(1, 2, 3, 4, 5)$ defined by:

$$(I_8^2 - 2^6 I_{16})^2 = 2^{14}(I_{32} + 2^{-3} I_8 I_{24}).$$

The degree of this invariant is 32 (cf. [**S**], App. III p. 179, [**GKZ**], p. 38), the formula in [**Sal**], p. 198, omits the exponent -3, but we verified that upon substituting the σ_i in the I_n in our formula one finds the expression given in section 1.3. As I_{40} does not appear in this formula, the boundary divisor is a cone over a $\mathbb{P}(1, 2, 3, 4)$.

6.5. The tritangent divisor. The tritangent divisor parametrizes cubic surfaces with an Eckardt point. The general such surface has a Sylvester form in which two of the λ_i coincide. Thus this divisor is defined by the discriminant of the polynomial $\prod_{i=0}^{4}(x - \lambda_i)$, which is the invariant I_{100} ([**Sal**], p. 197). The discriminant is a polynomial of weight 20 in the symmetric functions in the λ_i and gives, upon pull-back along the map ψ from section 6.3, a polynomial of weight $320 = 20 \cdot 16$ in the basic invariants $I_8, \ldots, I_{40}$. This polynomial has a factor I_{40}^3, the other factor is $I_{100}^2 = f(I_8, \ldots, I_{40})$ which defines the tritangent divisor in $\mathcal{M} \cong \mathbb{P}(1, 2, 3, 4, 5)$.

THEOREM 6.1. *The divisor in the moduli space of cubic surfaces $\mathcal{M}$ which parametrizes cubic surfaces which do not admit a Sylvester form is defined by $I_{40} = 0$.*

The subvariety parametrizing the surfaces of type ns2 is defined by $I_{24} = I_{40} = 0$.
The subvariety parametrizing the cyclic surfaces is defined by $I_{24} = I_{32} = I_{40} = 0$.
The point $I_{16} = I_{24} = I_{32} = I_{40} = 0$ parametrizes the Fermat cubic surface.
The non-stable, semi-stable cubic surface defined by $t^3 = xyz$ maps to the point $(8 : 1 : 0 : 0 : 0) \in \mathcal{M}$.

PROOF. A family of five linear forms where $x_0, \ldots, x_3$ are coordinates on a $\mathbb{P}^3$ and where two of the linear forms coincide in the limit $t \to 0$ is:

$$x_0(t) = x_0, \quad x_i(t) = a_i t x_i \quad (i = 2, 3, 4), \quad x_4(t) = -x_0 - t(a_1 x_1 + a_2 x_2 + a_3 x_3).$$

Note that any four of the five linear forms are independent when the $a_i \neq 0$ and $t \neq 0$ and that $\sum x_i(t) = 0$. The family of cubic surfaces S_t, defined by Sylvester forms:

$$\begin{cases} (a_0 + t^{-1})x_0(t)^3 + (a_1 t)^{-3} x_1(t)^3 + (a_2 t)^{-3} x_2(t)^3 + (a_3 t)^{-3} x_3(t)^3 + t^{-1} x_4(t)^3 = 0, \\ \sum_{i=0}^{4} x_i(t) = 0 \end{cases}$$

has as limit the cubic surface S_{ns1} defined by:

$$S_{ns1} := S_0 : \qquad a_0 x_0^3 + x_1^3 + x_2^3 + x_3^3 - 3x_0^2(a_1 x_1 + a_2 x_2 + a_3 x_3) = 0.$$

Replacing x_0 by $-x_0$ we get the cubic form:

$$F = x_1^3 + x_2^3 + x_3^3 - x_0^2(a_0 x_0 + 3a_1 x_1 + 3a_2 x_2 + 3a_3 x_3)$$

considered in 5.2.

To find the invariants of the general surface of type $ns1$ we compute the invariants of the surface S_t in the one parameter family S_t and take the limit $t \to 0$. The result is:

$$[S_{ns1}] = [\lim_{t \to 0} S_t] = (-4\rho_1 + a_0^2 : \rho_2 : 2\rho_3 : \rho_1\rho_3 : 0) \qquad (\in \mathbb{P}(1,2,3,4,5)_I),$$

where ρ_i is the i-th elementary symmetric function in a_1^3, a_2^3, a_3^3. From the equation of S_{ns1} it is clear that its isomorphism class depends only on the symmetric functions of a_1^3, a_2^3, a_3^3. Thus the three dimensional family $ns1$ maps to the divisor $I_{40} = 0$ which shows that this divisor is the closure of the locus of surfaces of type $ns1$. Note that for a general point of $I_{40} = 0$, in particular one for which the other coordinates are non-zero, one can recover ρ_2 and ρ_3 from the second and third coordinate, the fourth coordinate then determines ρ_1 and finally a_0 is recovered from the first coordinate.

A one parameter family where three of the linear forms coincide in the limit $t \to 0$ is:

$$x_0(t) = -x_3 - t^2 x_0, \quad x_1(t) = \mu t^2 x_1, \quad x_2 = t^2 x_2, \quad x_3 = 2x_3 + tx_0$$

and $x_4(t) = -(x_0(t) + \ldots + x_3(t))$. Again any four of the 5 $x_i(t)$ are independent for $t \neq 0$. The family of cubic surfaces, defined by Sylvester forms:

$$S_t : \begin{cases} t^{-2}x_0(t)^3 + (\mu t^2)^{-3}x_1(t)^3 + t^{-6}x_2(t)^3 + (\frac{\lambda}{4} + \frac{1}{4t^2})x_3(t)^3 + t^{-2}x_4(t)^3 = 0, \\ \sum_{i=0}^{4} x_i(t) = 0 \end{cases}$$

has as limit the cubic surface S_{ns2} defined by (after substituting $\frac{1}{\sqrt{2}}x_0$ for x_0):

$$S_{ns2} := S_0 : \qquad G = x_1^3 + x_2^3 + 2\lambda x_3^3 - 3x_3(\mu x_1 x_3 + x_2 x_3 + x_0^2) = 0.$$

The same procedure as above gives:

$$[S_{ns2}] = (-8\lambda : 1 + \mu^3 : 0 : \mu^3 : 0) \qquad (\in \mathbb{P}(1,2,3,4,5)_I),$$

where λ, μ are the coefficients of the equation of S_{ns2}.

Finally we consider the cyclic surfaces. These can be obtained from S_{ns1} with one of the $a_i = 0$, but the following family has a particularly nice limit, a sum of cubes of linear forms. Consider the one parameter family of Sylvester forms:

$$x_i(t) = (1+t)x_i, \qquad x_4(t) = tx_4, \qquad x_3(t) = -(x_0(t) + x_1(t) + x_2(t) + x_4(t)),$$

where $i = 0, 1, 2$. Any four of the five $x_i(t)$ are independent for $t \neq 0$. The family of cubic surfaces, defined by Sylvester forms:

$$S_t : \quad \lambda_0 x_0(t)^3 + \ldots + \lambda_3 x_3(t)^3 + (\frac{\lambda_4}{t^3})x_4(t)^3 = 0, \quad \sum_{i=0}^{4} x_i(t) = 0,$$

has as limit the cyclic cubic surface S_{cyc} defined by

$$S_{cyc} := S_0 : \qquad G = \lambda_4 x_4^3 - \lambda_3(x_0 + x_1 + x_2)^3 + \lambda_0 x_0^3 + \lambda_1 x_1^3 + \lambda_2 x_2^3 = 0.$$

Computing the invariants as before gives:

$$[S_{cyc}] = (\tau_3^2 - 4\tau_2\tau_4 : \tau_4^3 : 0 : 0 : 0) \qquad (\in \mathbb{P}(1,2,3,4,5)_I),$$

where τ_i is the i-th elementary symmetric function in $\lambda_0, \ldots, \lambda_3$ (note that λ_4 does not appear as is obvious from the equation of S_{cyc}). The Fermat cubic is the surface S_{cyc} with $\lambda_3 = 0$, hence $\tau_4 = \prod_{i=0}^{3} \lambda_i = 0$.

Putting $\lambda_i = 1$, $0 \leq i \leq 3$, we obtain the cubic surface $x_4^3 = 3(x_0 + x_1)(x_0 + x_2)(x_1 + x_2)$, which is isomorphic to the strictly semi-stable surface $t^3 = xyz$, and it defines the point $(-8 : 1 : 0 : 0 : 0) = (8 : 1 : 0 : 0 : 0) \in \mathcal{M} = \mathbb{P}(1, 2, 3, 4, 5)$. Note that this is the unique point in the intersection of the boundary divisor, defined by the equation from 6.4, with the curve $I_{24} = I_{32} = I_{40} = 0$ parametrizing cyclic surfaces. $\qquad\square$

6.6. The Kummer divisor. The Kummer K3 surface of a principally polarized abelian surface is the Hessian K3 surface of a cubic surface. The divisor in $\mathcal{M}$ which we obtain in this way will be called the Kummer divisor. Rosenberg ([**Ros**], Cor. 1.2) proved that the Kummer divisor in $\mathcal{M}$ is defined by:

$$I_8 I_{24} + 8 I_{32}.$$

The Néron-Severi group of the Kummer surface has rank 17 and has transcendental lattice isomorphic to $T_{kum} = U(2) \oplus U(2) \oplus < -4 >$. The lattice T_{kum} is the orthogonal complement of an element t with $t^2 = -12$ in T_{gen}.

6.7. The locus $I_{40} = I_{100}^2 = 0$. The intersection of the tritangent divisor $I_{100}^2 = 0$ and the non-Sylvester divisor $I_{40} = 0$ consists of three irreducible components (of dimension two). In fact, putting $I_{40} = 0$ in $I_{100}^2 = f(I_8, \ldots, I_{40})$ one finds (up to scalar multiple):

$$f(I_8, \ldots, I_{32}, 0) = I_{24}^3 (I_8 I_{24} + 8 I_{32}) g(I_8, \ldots, I_{32})$$

where g is given by:

$$g = 16 I_{16}^3 I_{24}^2 + 27 I_{24}^4 - 72 I_{16} I_{24}^2 I_{32} - 16 I_{16}^2 I_{32}^2 + 64 I_{32}^3.$$

The first factor, $I_{24} = 0$, corresponds to the non-Sylvester surfaces of type $ns2$, cf. Theorem 6.1. These surfaces do indeed have an Eckardt point: $p_0 = (1 : 0 : 0 : 0)$ (notation as in section 5.3) is singular on the Hessian and lies on S_{ns2}, hence it is an Eckardt point.

The second factor is the intersection of $I_{40} = 0$ with the Kummer divisor. Parametrizing the non-Sylvester divisor as in the proof of Theorem 6.1, this component is given by $a_0 = 0$ and these surfaces in fact have an Eckardt point (cf. Lemma 5.3.1).

The last factor pulls-back to the discriminant of the polynomial $(x - a_1^3)(x - a_2^3)(x - a_3^3)$, hence gives the surfaces with Eckardt point considered in Lemma 5.3.2.

6.8. The singular points of the moduli space. The singular locus of a weighted projective space was determined in [**DD**], Prop. 7. For $\mathcal{M} = \mathbb{P}(1, 2, 3, 4, 5)$ we get:

$$\mathbb{P}(1, 2, 3, 4, 5)_{sing} = \{(0 : 0 : 1 : 0 : 0)\} \cup \{0 : 0 : 0 : 0 : 1)\} \cup \{(0 : a : 0 : b : 0)\},$$

with $a, b \in \mathbb{C}$ not both zero.

Let $\mathbb{P}(q_1, \ldots, q_n)$ be a weighted projective space where we (may) assume that any $n - 1$ of the q_i are relatively prime. This algebraic variety is covered by affine open subsets U_i which are cyclic quotients of vector spaces ([**Do**], Proposition 1.3.3 and its proof):

$$U_i = \{(x_1 : \ldots : x_n) \in \mathbb{P}(q_1, \ldots, q_n) : x_i \neq 0\} \cong \mathbb{C}^{n-1}/\mu_{q_i},$$

where the action of a q_i-th root of unity ζ on $\mathbb{C}^{n-1}$ is given by

$$(x_1, \ldots, x_{i-1}, x_{i+1}, \ldots, x_n) \longmapsto (\zeta^{q_1} x_1, \ldots, \zeta^{q_{i-1}} x_{i-1}, \zeta^{q_{i+1}} x_{i+1}, \ldots, \zeta^{q_n} x_n).$$

In particular, $U_1 \subset \mathbb{P}(1, 2, 3, 4, 5)$ is isomorphic to $\mathbb{C}^4$, U_3, U_5 are the quotient of $\mathbb{C}^4$ by an automorphisms of order three and five respectively and both have an isolated singularity.

6.9. The cubic surfaces in the singular locus. The singular locus of $\mathcal{M}$ parametrizes, as expected, certain cubic surfaces with non-trivial automorphism groups.

The cubic surface corresponding to $(0 : 0 : 1 : 0 : 0)$ has $I_{40} = 0, I_{24} \neq 0$, hence it is of non-Sylvester type $ns1$ (Lemma 6.1), moreover, it is a singular point (section 6.4). The parametrization given in the proof of Theorem 6.1 shows that $\rho_1 = \rho_2 = a_0 = 0$, from which one finds that $a_i^3 = \omega^i a_3^3$ for a primitive cube of unity ω and the equation of the surface is: $x_1^3 + x_2^3 + x_3^3 - 3x_0^2(x_1 + \epsilon x_2 + \epsilon^2 x_3) = 0$ where ϵ is a primitive 9th root of unity. A change of variables gives the equation $(\omega = \epsilon^6)$:

$$x_1^3 + \omega x_2^3 + \omega^2 x_3^3 - 3x_0^2(x_1 + x_2 + x_3) = 0.$$

This surface has two singular points, $p_{\pm} = (\pm 1 : 1 : \omega : \omega^2)$. It has an automorphism of order three given by

$$(x_0 : x_1 : x_2 : x_3) \longmapsto (\omega x_0 : x_2 : x_3 : x_1).$$

The cubic surface corresponding to $(0 : 0 : 0 : 0 : 1)$ is singular and has the Sylvester form:

$$\sum_{i=0}^{4} \eta^i x_i^3 = 0, \qquad \sum_{i=0}^{4} x_i = 0,$$

where η is a primitive 5-th root of unity. This surface has the unique singular point $(1 : \eta^2 : \eta^4 : \eta : \eta^3)$, it is an ordinary double point. The map:

$$(x_0 : \ldots : x_3 : x_4) \longmapsto (x_1 : \ldots : x_4 : x_0)$$

is an automorphism of order five of this surface.

The rational curve defined by $I_8 = I_{24} = I_{40} = 0$ parametrizes cubic surfaces of non-Sylvester type $ns2$ (Theorem 6.1):

$$S_\mu : \quad x_1^3 + x_2^3 - 3x_3(\mu x_1 x_3 + x_2 x_3 + x_0^2) = 0$$

and $\mu^3/(1 + \mu^3)^2 = b/a^2$ (or, if $a = 0$, $\mu^3 = -1$). The point $(1 : 0 : 0 : 0)$ is an Eckardt point. Each of these surfaces has an automorphism of order 4 given by:

$$(x_0 : x_1 : x_2 : x_3) \longmapsto (ix_0 : x_1 : x_2 : -x_3).$$

This family was studied for example in [**S**], p. 151, [**Na**], Thm. 2, p. 3, [**Ho**], Thm. 5.3. Any surface in the family has the property that its automorphism group modulo the subgroup generated by the involutions defined by the Eckardt points is non-trivial. This quotient group has two elements except if $\mu^6 = 1$. When $\mu = \omega$, a primitive cube root of unity, the surface has two singular points $(1 : 0 : \omega^2 : \pm 1)$ and it has another Eckardt point: $(0 : 1 : -\omega : 0)$. In case $\mu^3 = -1$ one obtains a surface with quotient group isomorphic to $\mathbb{Z}/4\mathbb{Z}$. This surface defines the point $(0 : 0 : 0 : 1 : 0) \in \mathcal{M}$.

The surface with $\mu = 0$, which corresponds to $(0 : 1 : 0 : 0 : 0) \in \mathcal{M}$, has affine equation (putting $x_0 = y, x_1 = t, x_2 = x, x_3 = 1$ in the equation above):

$$S_0 : \quad t^3 + 3y^2 + x^3 - 3x = 0$$

which shows that it is a cyclic surface (as expected, see Theorem 6.1) which branches over the smooth elliptic curve $E : 3y^2 + x^3 - 3x = 0$ in $\mathbb{P}^2$, it has $j(E) = 1728$, in fact the automorphism of S_0 of order 4 above induces one on E with a fixed point.

References

[ACT] D. Allcock, J. A. Carlson and D. Toledo, *The Complex Hyperbolic Geometry of the Moduli Space of Cubic Surfaces*, J. Algebraic Geom. **11** (2002) 659–724.

[C] O. Chisini, *La superficie cubica II*, Period. Mat. **35** (1957) 286–300.

[DD] A. Dimca, S. Dimiev, *On analytic coverings of weighted projective spaces*, Bull. London Math. Soc. **17** (1985) 234–238.

[DGK] I. Dolgachev, B. van Geemen, S. Kondo, *A Complex Ball Uniformization of the moduli space of cubic surfaces via periods of K3 surfaces*, J. Reine Angew. Math. **588** (2005) 99–148.

[DK] I. V. Dolgachev and J. Keum, *Birational Automorphisms of Quartic Hessian Surfaces*, Trans. Amer. Math. Soc. **354** (2002) 3031–3057.

[Do] I. Dolgachev, *Weighted projective varieties*, in: Group actions and vector fields (Vancouver, B.C., 1981), 34–71, LNM 956, Springer, Berlin, 1982.

[GKZ] I. M. Gelfand, M. M. Kapranov and A. V. Zelevinsky, *Discriminants, Resultants and Multidimensional Determinants*, Birkhäuser 1994.

[H] J. I. Hutchinson, *The Hessian of the Cubic Surface*, Bull. Amer. Math. Soc. **5** (1899) 282–292.

[Ho] T. Hosoh, *Automorphism groups of Cubic surfaces*, J. Algebra **192** (1997) 651–677.

[K] S. Kondō, *Enriques surfaces with finite automorphism groups*, Japan. J. Math. (N.S.) **12** (1986), 191–282.

[M] D. R. Morrison, *On K3 surfaces with large Picard number*, Invent. Math. **75** (1984) 105–121.

[Na] I. Naruki, *Cross Ratio Variety as a Moduli Space of Cubic Surfaces*, Proc. London Math. Soc. **42** (1982) 1–30.

[Ni] V. Nikulin, *Integral Quadratic Bilinear Forms and some of their Applications*, Math. USSR Izvestija **14** (1980) 103–167.

[P] U. Persson, *Double Sextics and Singular K3 Surfaces*, in: Algebraic geometry (Sitges 1983),262–328, LNM 1124, Springer, Berlin 1985.

[PS] C. Peters and J. Stienstra, *A pencil of K3-surfaces related to Apéry's recurrence for $\zeta(3)$ and Fermi surfaces for potential zero*, Arithmetic of complex manifolds (Erlangen, 1988), 110–127, LNM 1399, Springer, Berlin, 1989.

[Rod] C. Rodenberg, *Zur Classification der Flaechen dritter Ordnung*, Math. Ann. **14** (1879) 46–110.

[Ros] J. E. Rosenberg, *Hessian Quartic Surfaces that are Kummer Surfaces*, eprint math.AG/9903037.

[S] B. Segre, *The non-singular cubic surfaces*, Oxford 1942.

[Sal] G. Salmon, *A Treatise on the Analytic Geometry of Three Dimensions*, Chelsea Publishing Company, New York 1965.

[SI] T. Shioda and H. Inose, *On Singular K3 Surfaces*, Complex Analysis and Algebraic Geometry, A collection of Papers Dedicated to K. Kodaira, Cambridge University Press 1977.

[SZ] I. Shimada, D.-Q. Zhang, *Classification of extremal elliptic K3 surfaces and fundamental groups of open K3 surfaces*, Nagoya Math. J. **161** (2001) 23–54.

[V] H. A. Verrill, *The L-series of certain Rigid Calabi-Yau Threefolds*, J. Number Theory **81** (2000) 310–334.

[Vi] È. B. Vinberg, *The Two Most Algebraic K3 Surfaces*, Math. Ann. **265** (1983) 1–21.

PIAZZA BIANCANI 17, 12100 CUNEO, ITALIA
E-mail address: eli.gian@tiscali.it

DIPARTIMENTO DI MATEMATICA, UNIVERSITÀ DI MILANO, VIA SALDINI 50, I-20133 MILANO, ITALIA
E-mail address: geemen@mat.unimi.it

Contemporary Mathematics
Volume **422**, 2007

On covariants of plane quartic associated to its even theta characteristic

Marat Gizatullin

To Igor

ABSTRACT. We describe explicitly some invariants, covariants, Cremona transformations, surfaces and linear complexes associated with a plane quartic curve equipped with an even theta characteristic. For the invariants, new identities are obtained.

1. Introduction

Our goal is to describe explicitly some invariants and covariants of a plane quartic curve C equipped with an even theta characteristic θ. In general, such a covariant is irrational with respect to the coefficients of the equation of C. The problem of finding an explicit description of the invariants was motivated by the problem of obtaining formulas for the action of the standard quadratic Cremona transformation on a general plane quartic by a birational representation of the Cremona group in the space of quartics. For the case of plane cubics, the formulas are presented in [**Gi1**]. For the case of plane quartic, the existence of the action of the Cremona group was established also, but the desired formulas for the standard action have not been written down so far.

Our new results are the following.

In section 2 we discuss Toeplitz' irrational invariant Λ of nets of quadrics. All the distinct but mutually equivalent definitions of the invariant and the details of the definitions are used in the subsequent sections.

In section 3, for any plane quartic together with an even theta (C, θ) we construct certain quartic surfaces.

In section 4, we construct certain transformations, one of them is a Cremona cubo-cubic space transformations, two other are projective transformations. The matrices of the projective transformations are helpful for producing invariants. A new quartic surface arises in the section, but this surface is almost useless for the constructions of new invariants.

1991 *Mathematics Subject Classification.* Primary 14H50; Secondary 14L24, 14Q05.

Key words and phrases. Plane quartic curve, Even theta characteristic, Invariant, Covariant, Syzygy, Cremona transformation, Thomae's formula.

In sections 5, 6, 7, 8, 9 we describe some irrational invariants. I hope that one of them is indeed new.

In section 10 we discuss Cayley's and Salmon's irrational tact-invariant of nets of quadrics which vanishes on the locus of quadrics with singular discriminant curve.

We obtain two syzygies for irrational invariants, see (8.1) and (10.1). The second of the syzygies is based on the expression of the Cayley-Salmon irrational tact-invariant in terms of rational invariants and irrational invariants of low degree.

In section 11 we make an attempt to give an algebraic (i.e. not analytic) foundation for Thomae's formula in genus 3.

Finally in section 12 we construct quadratic line complexes corresponding to a pair (C, θ) of a quartic curve together with an even theta characteristic.

Acknowledgement.

The first discussions on the topic of this paper were initiated by Igor Dolgachev about ten years ago. It is a pleasure to thank him. The author also wishes to thank the organizers of the Korea-Japan Conference on Algebraic Geometry (July 5-9,2004), especially JongHae Keum and Shigeyuki Kondo.

During the conference, some conversations with the hero of the event were helpful. We saw the picture of a Kummer surface on the conference poster. When I explained the construction of the quartic surface $\Theta(u_0, u_1, u_2, u_3) = 0$, Igor remarked that the surface is not general because of the difference between the moduli for surfaces and for curves. Later he found that the surface was discovered by S. Gundelfinger. Writing the final version of the paper, I found the following nice gift for him: a plane nonsingular quartic together with an even theta defines a quadratic line complex (see 12.4 below and Appendix 3), whence one can produce a Kummer surface.

Some parts of the text are pieces of Igor's letter. I mark all the places. Especially, it is necessary to say that the proof of Thomae's formulas of section 11 belongs to Igor.

After revising the initial text, I would like to thank the referee for his great work, numerous informal suggestions and linguistic corrections.

I would like to express my gratitude to the creators of Macaulay-2. This program was used in all symbolic calculations.

2. An irrational invariant of order $\frac{3}{2}$

As is well-known any theta characteristic of a plane quartic defines a representation of the quartic C in a symmetric determinantal form

$$(2.1) \qquad \det(y_0 \cdot M_0 + y_1 \cdot M_1 + y_2 \cdot M_2) = 0,$$

where M_0, M_1, M_2 are 4×4 symmetric matrices, $(y_0 : y_1 : y_2)$ are homogeneous coordinates on the projective plane. In other words, M_0, M_1, M_2 define three space quadrics or three quaternary quadratic forms

$$(2.2) \quad M_0 = M_0(x_0, x_1, x_2, x_3), \ M_1 = M_1(x_0, x_1, x_2, x_3), \ M_2 = M_2(x_0, x_1, x_2, x_3).$$

These three quadrics generate a net of quadrics, the plane coordinates $(y_0 : y_1 : y_2)$ are homogeneous parameters for the elements of the net, and the quartic C is the degeneration locus of the net (the *Hessian curve*).

In coordinate-free form, we assume that $\mathbb{P}^2 = \mathbb{P}(V), \mathbb{P}^3 = \mathbb{P}^3(W)$ for some linear spaces of dimension 3 and 4, respectively. Then the variety $\mathcal{N}$ of nets of quadrics in $\mathbb{P}^3$ is the projective space $\mathbb{P}(V^* \otimes S^2W^*)$ and the discriminant is a rational map of degree 36

$$\delta : \mathbb{P}(V^* \otimes S^2W^*) \to \mathbb{P}(S^4V^*).$$

It is given by the linear system of polynomials of degree 4 in coordinates on V^* and S^2W^*. The map is equivariant with respect to the natural action of $\mathrm{SL}(V) \times \mathrm{SL}(W)$ on the domain of the map and $\mathrm{SL}(V)$ on the target of the map. An *invariant of a net of quadrics* of order n is a $\mathrm{SL}(V) \times \mathrm{SL}(W)$-invariant element of $S^n(V^* \otimes S^2W^*)^* \cong S^n(V \otimes S^2W)$. An *invariant of a quartic* of order n is a $\mathrm{SL}(V)$-invariant element of $S^n(S^4V^*)^* \cong S^n(S^4V)$. The rational quotient of $\mathbb{P}(V^* \otimes S^2W^*)/\mathrm{SL}(W)$ is birationally isomorphic to the space of pairs (C, θ), where C is a nonsingular plane quartic and θ is an even theta characteristic on it. So it is a finite cover of degree 36 of an open subspace of $\mathbb{P}(S^4V^*)$ of nonsingular quadrics. Thus any invariant on the space of nets of quadrics can be viewed as a $\mathrm{SL}(V)$-invariant of the space of pairs (C, θ) as above.

Under the pull-back with respect to δ, an invariant of a quartic of order n becomes an invariant of nets of quadrics of order $4n$. We will say such an invariant is *rational*. Of course not all invariants of nets are rational. We will say that an invariant of order n on the space of nets of quadrics which does not come in this way is an *irrational invariant* of quartics equipped with an even theta characteristic of order $n/4$.

George Salmon describes in [**Sa1**], n° 235 an invariant $\Lambda(M_0, M_1, M_2)$ of three symmetric 4×4 matrices, using the Pfaffian of the following skew-symmetric matrix of twelfth order,

$$\Lambda(M_0, M_1, M_2) - \mathrm{Pfaffian} \begin{pmatrix} 0 & -M_2 & M_1 \\ M_2 & 0 & -M_0 \\ -M_1 & M_0 & 0 \end{pmatrix}.$$

$\Lambda(M_0, M_1, M_2)$ is an invariant of the net. It is an irrational invariant of quartics of order $\frac{3}{2}$.

REMARK 2.1. One can generalize the symmetric determinant representation to a nonsingular curve of any degree d equipped with a non-vanishing even theta characteristic. All of above extends to this case immediately. In this way we obtain irrational invariants of pairs (C, θ) of order n/d.

REMARK 2.2. G. Salmon writes that invariant Λ was overlooked in some early editions of his textbook [**Sa1**]. In those editions, he assumed that a base of any net of quaternary quadratic forms is reducible to a triple of sums of five squares of linear forms, that is to

$$M_1 = A_1X^2 + B_1Y^2 + C_1Z^2 + D_1U^2 + E_1V^2,$$
$$M_2 = A_2X^2 + B_2Y^2 + C_2Z^2 + D_2U^2 + E_2V^2,$$
$$M_3 = A_3X^2 + B_3Y^2 + C_3Z^2 + D_3U^2 + E_3V^2,$$

where X, Y, Z, U, V are quaternary linear forms. According to Salmon's recital, W. Frahm noted this mistake [**Fra**], and then E. Toeplitz presented invariant Λ in [**Toe**]. Non-vanishing of Λ is an obstruction for the mentioned reduction. Therefore we will name Λ as the *Toeplitz invariant*.

Toeplitz gave another way of obtaining Λ using a cubic line complex. This way is also described in Salmon's book. A short description is as follows.

Three quadrics define a cubic line complex. A line l belongs to the complex if and only if for three pairs of the points of intersection of l with three quadrics, the simultaneous three-linear invariant of three binary quadratic forms vanishes. More explicitly, if (t_0, t_1) are homogeneous coordinates on l, and the points of intersection of l with $M_k = 0$ are the zeros of quadratic form

$$A_{0k}t_0^2 + 2A_{1k}t_0t_1 + A_{1k}t_1^2, \quad k = 0, 1, 2,$$

then the mentioned three-linear invariant is the determinant of the coefficients A_{ik}. The cubic complex is mentioned in n° 238b of Salmons book [**Sa1**], vol I. There is a tradition to refer to D. Montesano in this connection.

Thus we have a cubic complex

$$(2.3) \qquad T(p_{ij}) = 0, \quad T(p_{ij}) = \Sigma C(ij, kl, mn)p_{ij}p_{kl}p_{mn},$$

where p_{ij} are the Plücker line coordinates. For any line complex there exists a quadratic invariant, see our Supplement in section 13. If we take the value of the quadratic invariant for the Toeplitz cubic line complex, then we get (up to an integer non-zero multiplier) the Toeplitz invariant Λ.

REMARK 2.3. If we use three general quaternary quadratic forms with letter coefficients a_{ij}, b_{ij}, c_{ij}, 30 letters totally, then we get the Pfaffian as a homogeneous huge polynomial of degree 6 containing 1410 distinct monomials of form

$$C(ij, kl, mn, pq, rs, uv)a_{ij}a_{kl}b_{mn}b_{pq}c_{rs}c_{uv},$$

where the coefficients $C(ij, kl, mn, pq, rs, uv)$ are taken from $\{-2, -1, 1, 2\}$. For some special quadratic forms, the expressions of Λ are shorter.

REMARK 2.4. In a footnote, G. Salmon writes about a symbolic representation of the Toeplitz invariant. Sometimes we will use the symbolic method also. Therefore it would be helpful to show the method for a simple construction.

For the net of quadrics

$$y_0 M_0(x_0, x_1, x_2, x_3) + y_1 M_1(x_0, x_1, x_2, x_3) + y_1 M_1(x_0, x_1, x_2, x_3)$$

we will use parallel symbolic notations

$$(ay)(px)^2, \quad (by)(qx)^2, \quad (cy)(rx)^2,$$
$$(Ay)(Px)^2, \quad (By)(Qx)^2, \quad (Cy)(Rx)^2,$$

where, for example,

$$(ay) = a_0y_0 + a_1y_1 + a_2y_2, \quad (px) = p_0x_0 + p_1x_1 + p_2x_2 + p_3x_3,$$

and the coefficient $m_{k,ij}$ of M_k is expressible as $a_k p_i p_j$. Then we have the following expression for the Toeplitz invariant.

$$2^7 \cdot 3^2 \cdot \Lambda = (abc)(ABC)(pqPQ)(prPR)(qrQR),$$

where (abc) is the determinant of the coefficients of three ternary linear forms a, b, c, $(pqPQ)$ is the determinant of the coefficients of four quaternary linear forms p, q, P, Q.

Another symbolic descriptions of the Toeplitz invariant are following. If three ternary quadratic forms are written as

$$M_0 = (px)^2 = (Px)^2, \quad M_1 = (qx)^2 = (Qx)^2, \quad M_2 = (rx)^2 = (Rx)^2,$$

then

$$(pqPQ)(prPR)(qrQR) = 8\Lambda,$$
$$(pqrR)(pqPQ)(rPQR) = 4\Lambda,$$
$$(pqrP)(pqQR)(rPQR) = -2\Lambda.$$

REMARK 2.5. We would like to consider the Toeplitz invariant as a quadratic plane complex in the nine-dimensional projective space. We will consider quaternary quadrics as the points of the nine dimensional projective space $\mathbb{P}^9$. According to the indexing of the coefficients of a quaternary quadric, we will write ten homogeneous coordinates in $\mathbb{P}^9$ as t_{ij}, $ij \in \{$ 00, 11, 22, 33, 01, 02, 03, 12, 13, 23 $\}$. Three quadrics M_0, M_1, M_2 are three points of $\mathbb{P}^9$. They define a plane. Every plane in $\mathbb{P}^9$ has 120 Plücker coordinates $p(ij, kl, mn)$. The expression of the Toeplitz invariant in terms of the Plücker plane coordinates is the following.

$$\Lambda = p(00, 33, 03) \cdot p(11, 22, 12) -$$
$$p(00, 03, 12) \cdot p(12, 13, 23) + p(00, 03, 13) \cdot p(11, 22, 23) -$$
$$p(00, 22, 02) \cdot p(11, 33, 13) - p(00, 02, 12) \cdot p(11, 33, 23) -$$
$$p(00, 03, 23) \cdot p(11, 12, 23) + p(00, 02, 23) \cdot p(11, 13, 23) -$$
$$p(11, 01, 03) \cdot p(22, 33, 02) + p(00, 01, 12) \cdot p(22, 33, 13) +$$
$$p(00, 11, 01) \cdot p(22, 33, 23) + p(11, 33, 01) \cdot p(22, 02, 03) +$$
$$p(11, 03, 13) \cdot p(22, 02, 03) - p(11, 01, 03) \cdot p(22, 03, 23) +$$
$$p(00, 03, 13) \cdot p(22, 12, 13) - p(00, 11, 03) \cdot p(22, 13, 23) +$$
$$p(00, 01, 13) \cdot p(22, 13, 23) - p(22, 01, 12) \cdot p(33, 01, 03) +$$
$$p(22, 01, 02) \cdot p(33, 01, 13) + p(11, 02, 12) \cdot p(33, 02, 03) +$$
$$p(11, 01, 02) \cdot p(33, 02, 23) - p(00, 22, 01) \cdot p(33, 12, 13) +$$
$$p(00, 02, 12) \cdot p(33, 12, 13) + p(00, 11, 02) \cdot p(33, 12, 23) -$$
$$p(00, 01, 12) \cdot p(33, 12, 23) - p(33, 01, 23) \cdot p(01, 02, 12) -$$
$$p(33, 02, 13) \cdot p(01, 02, 12) + p(33, 03, 12) \cdot p(01, 02, 12) +$$
$$p(22, 01, 23) \cdot p(01, 03, 13) - p(22, 02, 13) \cdot p(01, 03, 13) +$$
$$p(22, 03, 12) \cdot p(01, 03, 13) + p(01, 03, 23) \cdot p(01, 12, 23) -$$
$$p(01, 02, 23) \cdot p(01, 13, 23) + p(01, 13, 23) \cdot p(02, 03, 12) +$$
$$p(11, 01, 23) \cdot p(02, 03, 23) - p(11, 02, 13) \cdot p(02, 03, 23) -$$
$$p(11, 03, 12) \cdot (02, 03, 23) - p(02, 03, 13) \cdot p(02, 12, 13) +$$
$$p(01, 02, 13) \cdot p(02, 13, 23) - p(01, 02, 23) \cdot p(03, 12, 13) +$$
$$p(02, 03, 12) \cdot p(03, 12, 13) - p(01, 03, 12) \cdot p(03, 12, 23) -$$
$$p(00, 01, 23) \cdot p(12, 13, 23) - p(00, 02, 13) \cdot p(12, 13, 23).$$

Because of presence of some quadratic relations connecting the Plücker plane coordinates, the expression is not unique. Moreover, the author is not sure that the expression is shortest. In a sense, this identity is a kind of Laplace's development for our special Pfaffian. Thus if a net does not belong to the plane complex, then the net does not contain a base whose elements are representable as linear combinations of squares of five linear forms.

REMARK 2.6. A few words about the evectant of the Toeplitz invariant. In the classical theory of invariants, *evectant* means a new form (or a set of forms) produced with the help of partial derivatives of an invariant with respect to the coefficients of the initial form. For our case, we use the invariant Λ. The normalized partial derivatives of the invariant with respect to the letter coefficients a_{ij} of M_0 are

$$\frac{\partial\Lambda}{\partial a_{00}}, \quad \frac{\partial\Lambda}{\partial a_{11}}, \quad \frac{\partial\Lambda}{\partial a_{22}}, \quad \frac{\partial\Lambda}{\partial a_{33}},$$

$$\frac{1}{2}\cdot\frac{\partial\Lambda}{\partial a_{01}}, \quad \frac{1}{2}\cdot\frac{\partial\Lambda}{\partial a_{02}}, \quad \frac{1}{2}\cdot\frac{\partial\Lambda}{\partial a_{03}},$$

$$\frac{1}{2}\cdot\frac{\partial\Lambda}{\partial a_{12}}, \quad \frac{1}{2}\cdot\frac{\partial\Lambda}{\partial a_{13}}, \quad \frac{1}{2}\cdot\frac{\partial\Lambda}{\partial a_{23}}.$$

We consider the parial derivatives as coefficients of a quaternary quadratic form $N_0(u_0, u_1, u_2, u_3)$. It is not hard to calculate all the partial derivatives using the identity of previous remark 2.5. In such a way, we get three quadratic forms N_0, N_1, N_2, they form a base of a contragredient net of quadrics $z_0 N_0 + z_1 N_1 + z_2 N_2$.

The symbolic process for producing this contravariant net is given by the following formula

$$z_0 N_0 + z_1 N_1 + z_2 N_2 = (abc)(ABz)(pqPQ)(prPu)(qrQu).$$

EXAMPLE 2.7. It is possible to write down the equation of a general quartic as

$$F(x_0, x_1, x_2) = 0,$$

where the left hand side is the following determinant

$$(2.4) \quad \begin{vmatrix} 0 & ax_0 + bx_1 + cx_2 & fx_0 + gx_1 + hx_2 & px_0 + qx_1 + rx_2 \\ ax_0 + bx_1 + cx_2 & 0 & x_2 & x_1 \\ fx_0 + gx_1 + hx_2 & x_2 & 0 & x_0 \\ px_0 + qx_1 + rx_2 & x_1 & x_0 & 0 \end{vmatrix}.$$

We will denote a 4×4 symmetric matrix M and corresponding quaternary quadratic form of variables u_0, u_1, u_2, u_3 by the same symbol M. To see this let us consider three quaternary quadratic forms M_0, M_1, M_2 defined by the matrices corresponding to the latter determinantal equation, then one can write

$$M_0 = 2(ax_0x_1 + ex_0x_2 + px_0x_3 + x_2x_3),$$

$$M_1 = 2(bx_0x_1 + fx_0x_2 + qx_0x_3 + x_1x_3),$$

$$M_2 = 2(cx_0x_1 + gx_0x_2 + rx_0x_3 + x_1x_2).$$

The geometric meaning of such a canonical form is the following. Three general quaternary quadratic forms define eight points in the projective space. If we assume that four of these points are the vertices of the coordinate tetrahedron, then we get that pure quadratic monomials $A_{ii}x_i^2$, $0 \le i \le 3$, are absent in each form, that is the matrices M_0, M_1, M_2 have zero diagonals. Further, let us consider the parts of quadratic forms containing three mixed monomials without x_0, that is x_1x_2, x_1x_3, x_2x_3. After an obvious linear recombinations of the basic quadratic forms M_0, M_1, M_2, we can get that two first mixed monomials are absent in M_0, and the first one has coefficient 2, the first and the third monomials are absent in M_1, the second and the third monomials are absent in M_2.

Sometimes we will use further possible normalization, whose meaning is that point $(1 : 1 : 1 : 1)$ belongs to the intersection of three quadrics. Such an additional normalization was considered by Riemann for the first time, see [**Ri**], pp. 496–497, formulas (17). For the latter case, we will use words *Riemann's normalization*.

For our example we have the following expression for the invariant Λ

$$\Lambda(M_0, M_1, M_2) = a(g^2 - q^2) + f(p^2 - c^2) + r(b^2 - e^2)+$$

(2.5) $$(bcg + bgp + egp + bpq) - (ceg + bcq + ceq + epq).$$

EXAMPLE 2.8. It is convenient to have a general point of the variety of nets with vanishing Toeplitz invariant. If we put

$$e = b, \quad q = g, \quad p = b - c, \quad f = -2g$$

in example 2.7, then we obtain such a point.

REMARK 2.9. Λ^2 is an irrational invariant of order 3. Indeed, if Λ^2 is rational, then it belongs to the linear space of invariants of order 3, but the latter space is generated by a unique invariant $I_3(C)$ (see [Sa2]). It is not hard to see that $I_3(C)$ and and Λ^2 are not proportional, because their sizes are not equal. For the above example 2.7, formula for $I_3(C)$ is given in Appendix.

Moreover, in example 2.8 we have $\Lambda = 0$, but the values of all non-zero rational invariants of degrees 3, 6, 9, 12 for the plane quartic from example 2.8 are calculable, they do not vanish, therefore $\Lambda, ..., \Lambda^8$ are irrational.

3. Quartic surfaces associated to a plane quartic curve equipped with an even theta characteristic

Igor Dolgachev communicated that the quartic surface $\Theta(u_0, u_1, u_2, u_3)$ described below was first found by S. Gundelfinger in [**Gu**].

The simplest definition of the Gundelfinger quartic is symbolic. Using parallel symbolic notations $(ay)(px)^2$, $(by)(qx)^2$, $(cy)(rx)^2$, $(Ay)(Px)^2$, $(By)(Qx)^2$, $(Cy)(Rx)^2$ from 2.4 for our net of quadrics we put

(3.1) $$\Theta(u_0, u_1, u_2, u_3) =$$
$$k(abc)(ABC)(pqRu)(prQu)(qrPu),$$

where k is normalizing rational coefficient. Another construction of the $\Theta(u)$ is based on some ideas from Salmon's *Conic Sections* [**Sa3**], n° 389. G. Salmon (with a reference to J. Sylvester) constructs a combined invariant of three plane conics. This invaraint is an invariant of the net generated by these three conics. If the conics are defined by quadratic forms

$$A_{00}t_0^2 + A_{11}t_1^2 + A_{22}t_2^2 + 2A_{01}t_0t_1 + 2A_{02}t_0t_2 + 2A_{12}t_1t_2,$$
$$B_{00}t_0^2 + B_{11}t_1^2 + B_{22}t_2^2 + 2B_{01}t_0t_1 + 2B_{02}t_0t_2 + 2B_{12}t_1t_2,$$
$$C_{00}t_0^2 + C_{11}t_1^2 + C_{22}t_2^2 + 2C_{01}t_0t_1 + 2C_{02}t_0t_2 + 2C_{12}t_1t_2,$$

where (t_0, t_1, t_2) are homogeneous coordinates on the projective plane, and

$$m(ij, kl, pq) = \begin{vmatrix} A_{ij} & A_{kl} & A_{pq} \\ B_{ij} & B_{kl} & B_{pq} \\ C_{ij} & C_{kl} & C_{pq} \end{vmatrix},$$

then the Sylvester-Salmon invariant θ is defined by the following Salmon's formula.

$$\theta = m(00, 11, 22)^2 - 8m(12, 02, 01)^2+$$

$$4m(00,11,22)m(12,02,01) + 4m(00,11,12)m(00,22,12)+$$
$$4m(11,22,02)m(11,00,02) + 4m(22,00,01)m(22,11,01)+$$
$$8m(00,12,02)m(11,12,02) + 8(m00,12,01)m(22,12,01)+$$
$$8m(22,02,01)m(11,02,01) - 8m(00,02,01)m(11,22,12)-$$
$$8m(11,01,12)m(22,00,02) - 8m(22,12,02)m(00,11,01).$$

One can consider $m(ij,kl,pq)$ as the Plücker coordinates of planes in the five-dimensional projective space of plane conics, and the the condition of vanishing of the invariant θ defines a quadratic plane complex in $\mathbb{P}^5$ (cf. Remark 2.5).

We have three *quaternary* quadrics M_0, M_1, M_2. Let us take the plane

$$u_0 x_0 + u_1 x_1 + u_2 x_2 + u_3 x_3 = 0,$$

where (u_0, u_1, u_2, u_3) are the dual coordinates. The intersections of the plane with three quadrics are three plane conics. The condition of vanishing of the invariant θ has order 4 with respect to the dual coordinates. The evaluation of θ for the three conics is proportional to the Gundelfinger quaternary quartic $\Theta(u_0, u_1, u_2, u_3)$.

REMARK 3.1. This recital belongs to Igor Dolgachev.

We will rewrite the determinantal equation in the following form

$$\begin{vmatrix} L_{00} & L_{01} & L_{02} & L_{03} \\ L_{10} & L_{11} & L_{12} & L_{13} \\ L_{20} & L_{21} & L_{22} & L_{23} \\ L_{30} & L_{31} & L_{32} & L_{33} \end{vmatrix} = 0,$$

where $L_{ij} = L_{ij}(y_0, y_1, y_2)$ is a linear form, $L_{ij} = L_{ji}$, $i, j \in \{0, 1, 2, 3\}$. Such a form produces a 3-parameter family of contact cubics (i.e. plane cubic curves everywhere tangent to C). Recall that a determinantal representation defines a linear isomorphism between the space of cubics in $\mathbb{P}^2$ and the space of quadrics in $\mathbb{P}^3$. The curve C is embedded onto a curve X of degree 6 in $\mathbb{P}^3$ by the linear system $|\theta(1)|$. The curve X is the locus of singular points of quadrics from the net (the Steinerian curve, also called the Jacobian curve of the net). The isomorphism is obtained by restriction of quadrics to X. The image of the variety of quadrics of rank 1 (i.e. double planes) is the family of contact cubics. Another way to see the contact cubics is by restricting the net of quadrics to a plane H to get a net of conics. Its Hessian curve is the contact cubic corresponding to this plane.

If $(u_0 : u_1 : u_2 : u_3)$ are homogeneous coordinates in the dual projective space $\check{\mathbb{P}}^3 = \mathbb{P}(W^*)$, then contact cubic curves are given by the following equation

(3.2)
$$\begin{vmatrix} L_{00} & L_{01} & L_{02} & L_{03} & u_0 \\ L_{10} & L_{11} & L_{12} & L_{13} & u_1 \\ L_{20} & L_{21} & L_{22} & L_{23} & u_2 \\ L_{30} & L_{31} & L_{32} & L_{33} & u_3 \\ u_0 & u_1 & u_2 & u_3 & 0 \end{vmatrix} = \sum A_{ij}(y) u_i u_j = 0,$$

where $A_{ij}(y)$ are cofactors of the matrix $A(y) = (L_{ij}(y))$. For any point $y \in C$, the matrix $A(y)$ is of corank 1, hence the cofactor matrix is of rank 1 and the equation (3.2) represents the double plane $(\sum a_i u_i)^2$, where $(a_0 : a_1 : a_2 : a_3) \in \mathbb{P}^3$ is the image of y on the sextic model X of C.

The universal family of contact cubics is a hypersurface $\mathcal{W}$ in $\mathbb{P}^2 \times \check{\mathbb{P}}^3$ of bidegree $(3, 2)$ with equation (3.2).

The first projection

$$(3.3) \qquad\qquad\qquad q : \mathcal{W} \to \mathbb{P}^2$$

is a quadric bundle. Its fibre over a point $y \in \mathbb{P}^2$ is the quadric in $\check{\mathbb{P}}^3$ with equation $\sum A_{ij}(y)u_iu_j$ from (3.2). The discriminant curve of this quadric bundle is of course the curve C. The fibres over points of C are quadrics of rank 1. It follows from the theory of determinantal representations (see [**Be1**]) that there exists a rational map from $\mathcal{W}$ to the universal family of the net defined by (C, θ) which on nonsingular fibres is given by taking the dual quadric.

The second projection

$$(3.4) \qquad\qquad\qquad p : \mathcal{W} \to \check{\mathbb{P}}^3$$

is a family of plane cubics. Its fibre over a point $u = (u_0 : u_1 : u_2 : u_3) \in \check{\mathbb{P}}^3$ is the contact cubic which cuts out the divisor $2D$ on C such that the image of D on X is the plane section corresponding to the linear function $\sum u_ix_i$.

Varying the pairs (C, θ), or, equivalently, nets of quadrics, we have a family of varieties $\mathcal{W}_\lambda$ parametrized by the space of nets $V^* \otimes S^2W^*$. We can view it as a subvariety in $\mathbb{P}(V) \times \mathbb{P}(W) \times \mathbb{P}(V^* \otimes S^2W^*)$ of multi-degree $(3, 2, 3)$.

Recall that the algebra of projective invariants on the space of ternary cubics is generated by the Clebsch invariants $S = S_4$ and $T = T_6$ of orders 4 and 6. The locus of zeros of S_4 (resp. T_6) is the closure of the orbit of harmonic cubics (resp. anharmonic cubics) (i.e. cubics with Weierstrass equation $y^2 + x^3 + x = 0$ (resp. $y^2 + x^3 + 1 = 0$). If we evaluate S_4 (resp. T_6) on the cubic curve defined by (3.2) we get a polynomial $S_4(u_0, \ldots, u_3)$ of degree 8 (resp. a polynomial $T_6(u_0, \ldots, u_3)$ of degree 12) in (u_0, u_1, u_2, u_3) whose coefficients are homogeneous polynomials of degree 12 (resp. 18) on the space of $V^* \otimes S^2W^*$ of nets of quadrics.

LEMMA 3.2. *The polynomial* $T(u) = T_6(u_0, u_1, u_2, u_3)$ *is reducible,*

$$T(u) = R(u_0, u_1, u_2, u_3) \cdot \Theta(u_0, u_1, u_2, u_3),$$

where $\Theta(u_0, u_1, u_2, u_3)$ *is of degree 4 in variables* u_0, u_1, u_2, u_3, *and of order 6 in parameters of nets of quadrics. More precisely, by a convenient normalization of the invariants* S, T, *we have identity*

$$T = (4\Theta^2 - 3S)\Theta.$$

PROOF. Recall that there is universal family of pairs $(B, \alpha))$ of plane cubics together with a nonzero 2-torsion divisor class defined as a hypersurface $\mathcal{V}$ in the weighted projective space $\mathbb{P}(1^{10}, 2)$ defined by the equation

$$(3.5) \qquad\qquad \Theta^3 - S(A_0, \ldots, A_9)\Theta - T(A_0, \ldots, A_9) = 0,$$

where S, T are the Clebsch invariants written as polynomials in coefficients A_i of a general cubic form (see [**Gi1**], pp. 140–141). This family is just the pull-back of the universal family of elliptic curves over the modular curve $X_1(2) = H/\Gamma_1(2)$ parametrizing elliptic curves together with a non-zero 2-torsion point under the map $\mathbb{P}^9 \to \mathbb{P}(2, 3) = \overline{H/\Gamma}$ defined by the invariants S, T. Let us show that our family (3.4) is a family of cubic curves together with a non-zero 2-torsion divisor class. This will define a rational map

$$\check{\mathbb{P}}^3 \to \mathcal{V}, \ u = (u_0 : u_1 : u_2 : u_3) \mapsto (\Theta(u) : A_0(u) : \ldots : A_9(u)),$$

where $A_i(u)$ are quadratic polynomials defining the map $\mathbb{P}^3 \to \mathbb{P}^9$ and $\Theta(u)$ is a quartic polynomial. From equation (3.5) we infer that

$$T(u) = \Theta(u)(\Theta(u)^2 - S(u))$$

which proves the assertion. To show that any fibre of the family (3.4) is equipped with a canonical 2-torsion divisor class we consider the restriction of the quartic C to any contact cubic B from the family. Let d be the divisor of degree 3 cut out on B by a line ℓ. We have

$$C \cap B = 2(p_1 + \ldots + p_6) \sim 4\ell \cap B = 4d,$$

hence $2(p_1 + \ldots + p_6 - 2d) \sim 0$. To show that the divisor class $p_1 + \ldots + p_6 - 2d$ is not trivial assume the contrary. Then there exists a conic K which cuts out the divisor $p_1 + \ldots + p_6$ on B. This implies that the quartic curve C belongs to the pencil spanned by the conic taken with multiplicity 2 and the union of B and a line. The line intersects C at two points $a + b$ with multiplicity 2, i.e.. a bitangent to C. Thus

$$C \cap K \sim p_1 + \ldots + p_6 + a + b \sim 2K_C.$$

Since $\mathcal{N} \sim p_1 + \ldots + p_6 - K_C$ is an even theta charcaterritic, we get $\mathcal{N} \sim a + b$, a contradiction. $\square$

REMARK 3.3. One can also give a computational proof using Macaulay-2. For a simplification of calculation, it is enough to prove Lemma for the case of example 2.7. In this case it is not hard to calculate S and T and to check that the polynomial Θ written down in Appendix is a divisor of T, and that the factorization of T is the same as in the formulation of the lemma.

REMARK 3.4. This new remark and the final question are of Igor Dolgachev again.

Recall that a general net L of conics in $\mathbb{P}^2$ defines two nonsingular cubics curves. The first curve (the Hessian cubic) F lies in L and parametrizes singular conics in the net. It comes equipped with a non-trivial 2-torsion divisor class α (an even theta characteristic in the case of curves of genus 1). The second one (the Steinerian cubic) F' lies in $\mathbb{P}^2$ and parametrizes singular points of conics from the net. There is a natural map from $F \to F'$ which assigns to a singular conic its singular point. The net L defines a non-trivial 2-torsion divisor class α on F and the map $F \to F'$ is the unramified double cover corresponding to the quotient map $F \to F' = F/(\alpha)$. Now if $T_6(F) = 0$, the curve F is isomorphic to a curve with Weierstrass equation $y^2 + x^3 + x = 0$, hence acquires an automorphism g of order 4 which leaves a pair of 2-torsion divisor classes invariant, and permutes the remaining two. If α is invariant, then the automorphism g descends to F' and hence $T_6(F') = 0$. This shows that the locus of zeroes of $T_6(u_0, u_1, u_2, u_3)$ in the family of contact cubics consists of two components. One parametrizes the pairs such that the corresponding plane defines a net of conics whose Hessian and Steinerian curves are harmonic cubics. This is our quartic surface Θ. The other component is an octic surface. It parametrizes planes which define a nets of conics with harmonic Hessian curve but non-harmonic Steinerian curve. As we saw in the proof of Lemma 3.2, it belongs to the pencil spanned by Θ^2 and $S_4(u_0, u_1, u_2, u_3)$. There are many octic surfaces which one can associate to the net (e.g. the Hessian of Θ, the dual of the octic scroll of trisecants of the sextic model of C, the union of

8 planes corresponding to the Cayley octad of base points of the net) (see [**Ed1**]). Which one is our octic?

REMARK 3.5. The answer is as follows. Indeed, some integer linear combination of $S(u_0, u_1, u_2, u_3)$ and $\Theta^2(u_0, u_1, u_2, u_3)$ defines the union of 8 planes. The coefficients of the combination depend on the ways of normalization of the forms S, Θ. For the case of example 2.7, four of the planes are the coordinates planes. The factorization with the help of Macaulay-2 shows that $u_0 u_1 u_2 u_3$ is a divisor of the linear combination. If we use additional Riemann's normalization mentioned in example 2.7, then we get new divisor $(u_0 + u_1 + u_2 + u_3)$ of the linear combination. Each fixation of an additional base point of the net of quadrics produces a new factor of the combination.

We would like to construct another quartic surface $\Omega(x_0, x_1, x_2, x_3) = 0$ in the initial projective space. The construction is similar to the construction of Steinerian curve associated to three conics. Recall an explicit formula for Steinerian curve. If three conics mentioned in the beginning of the section are given, if (s_0, s_1, s_2) are the dual coordinates, then the curve is defined by the following cubic form

$$\begin{vmatrix} A_{00} & A_{11} & A_{22} & A_{12} & A_{02} & A_{01} \\ B_{00} & B_{11} & B_{22} & B_{12} & B_{02} & B_{01} \\ C_{00} & C_{11} & C_{22} & C_{12} & C_{02} & C_{01} \\ 2s_0 & 0 & 0 & 0 & s_2 & s_1 \\ 0 & 2s_1 & 0 & s_2 & 0 & s_0 \\ 0 & 0 & 2s_1 & s_1 & s_0 & 0 \end{vmatrix}.$$

If six quaternary quadratic forms of variables (u_0, u_1, u_2, u_3) with coefficients

$$A_{ij}, B_{ij}, C_{ij}, D_{ij}, E_{ij}, F_{ij}$$

respectively are given, then they define the following quaternary quartic of the dual coordinates (x_0, x_1, x_2, x_3)

$$(3.6) \quad \begin{vmatrix} A_{00} & A_{11} & A_{22} & A_{33} & A_{01} & A_{02} & A_{03} & A_{12} & A_{13} & A_{23} \\ B_{00} & B_{11} & B_{22} & B_{33} & B_{01} & B_{02} & B_{03} & B_{12} & B_{13} & B_{23} \\ C_{00} & C_{11} & C_{22} & C_{33} & C_{01} & C_{02} & C_{03} & C_{12} & C_{13} & C_{23} \\ D_{00} & D_{11} & D_{22} & D_{33} & D_{01} & D_{02} & D_{03} & D_{12} & D_{13} & D_{23} \\ E_{00} & E_{11} & E_{22} & E_{33} & E_{01} & E_{02} & E_{03} & E_{12} & E_{13} & E_{23} \\ F_{00} & F_{11} & F_{22} & F_{33} & F_{01} & F_{02} & F_{03} & F_{12} & F_{13} & F_{23} \\ 2x_0 & 0 & 0 & 0 & x_1 & x_2 & x_3 & 0 & 0 & 0 \\ 0 & 2x_1 & 0 & 0 & x_0 & 0 & 0 & x_2 & x_3 & 0 \\ 0 & 0 & 2x_2 & 0 & 0 & x_0 & 0 & x_1 & 0 & x_3 \\ 0 & 0 & 0 & 2x_3 & 0 & 0 & x_0 & 0 & x_1 & x_2 \end{vmatrix}.$$

We take the following six quadratic forms of variables (u_0, u_1, u_2, u_3). Three of them are mentioned in 2.6, they are N_0, N_1, N_2, other three are the convolutions of M_0, M_1, M_2 with the Gundelfinger quartic.

$$A = N_0, \quad B = N_1, \quad C = N_2,$$
$$D = < M_0, \Theta >, \quad E = < M_1, \Theta >,$$
$$F = < M_2, \Theta > .$$

Recall that the result of the convolution of a quadratic form of variables

$$(x_0, x_1, x_2, x_3)$$

and a quartic of the contragredient variables

$$(u_0, u_1, u_2, u_3)$$

is a quadratic form of the latter variables. Using these six quadratic forms and the above 6×6 determinant, we obtain a quaternary quartic

$$\Omega(x_0, x_1, x_2, x_3).$$

We will name it as the Steinerian quartic.

4. A Cremona transformation and projective transformations associated to a plane quartic curve equipped with an even theta characteristic

Three quadratic quaternary forms define four cubic quaternary forms. The construction of these four is the following. The Jacobian matrix of the quadratic forms is of size 4×3. Four 3×3 minors of the Jacobian matrix taken with alternating signs are the mentioned four cubic forms. These four forms define a space cubo-cubic involutive transformation. Let us consider the web

$$u_0 X_0 + u_1 X_1 + u_2 X_2 + u_3 X_3$$

generated by the four cubic forms X_0, X_1, X_2, X_3. The symbolic description of the web is the following. If the net of quadrics $y_0 M_0 + y_1 M_1 + y_2 M_2$ is presented in symbolically notations by $(ay)(px)^2$, $(by)(qx)^2$, $(cy)(rx)^2$, then the web of quaternary cubics is given by

$$(pqru)(abc)(px)(qx)(rx).$$

REMARK 4.1. Using the simplifications presented in the case of 2.7, it is not hard to get the explicit formulas for four quaternary cubics. The calculation gives us the opportunity to find an octic surface $J(x_0, x_1, x_2, x_3)$. Formally, the octic is the Jacobian determinant of four cubic forms. The geometric meaning of the surface is the following: $J(x_0, x_1, x_2, x_3) = 0$ is a ruled surface, the rectilinear generatrices of the surface have point transforms by the cubo-cubic transformation, they are exceptional lines. These generatrices are the trisecants of a curve X of degree 6 and genus 3. The curve was mentioned in 3.1. The fact that the transformation is involutive is expressed explicitly by the following identities

$$(4.1) \qquad X_k(X_0, X_1, X_2, X_3) = -x_k \cdot J(x_0, x_1, x_2, x_3), \quad k = 0, 1, 2, 3.$$

It is not hard to see that each base point of the net of quadrics belongs to the domain of definition of the transformation, and such a point is fixed with respect to the transformation. Indeed, it is sufficient to verify this assertion for the case of example 2.7, it is sufficient to take the coordinate point (1:0:0:0) and it is easy to see that X_1, X_2, X_3 vanish at the point, X_0 does not vanish.

There are some quartic surfaces invariant with respect to the transformation, that is if a point belongs to the surface, then the image of the point belongs to the same surface. We will indicate six such surfaces. They are defined by the following quaternary quartics.

$$(4.2) \qquad
\begin{aligned}
E_{01} &= X_1 x_0 - X_0 x_1, & E_{02} &= X_2 x_0 - X_0 x_2, \\
E_{03} &= X_3 x_0 - X_0 x_3, & E_{12} &= X_2 x_1 - X_1 x_2, \\
E_{23} &= X_3 x_2 - X_2 x_3, & E_{31} &= X_1 x_3 - X_3 x_1.
\end{aligned}$$

Identities (4.1) show that every of the quartics is preserved by the cubo-cubic transformation. Moreover, every quartic of the linear system spanned by the six quartics is preserved also. Geometric meaning of the forms is simple, E_{ij} are the Plücker coordinates for the line connecting point (x_0, x_1, x_2, x_3) and the transform of the point.

For (C, θ), one can construct a covariant quartic surface which is invariant with respect to the transformation. The quaternary quartic is a linear combination of the six quartics E_{ij}. If we take the convolutions of the six quartics with the Gundelfinger quartic (3.1), then we obtain six quantities depending on the coefficients of the initial net. That is the quantities are

$$(4.3) \qquad s_{01} =< E_{01}, \Theta >, \quad s_{02} =< E_{02}, \Theta >,$$

$$s_{03} =< E_{03}, \Theta >, \quad s_{12} =< E_{12}, \Theta >,$$

$$s_{23} =< E_{23}, \Theta >, \quad s_{31} =< E_{31}, \Theta > .$$

Sometimes we will consider the quantities as the coefficients of the equation of a linear line complex

$$s_{01}p_{01} + s_{02}p_{02} + s_{03}p_{03} + s_{12}p_{12} + s_{23}p_{23} + s_{31}p_{31} = 0.$$

This complex is a covariant of the initial net of quadrics. If we form the combined invariant of two linear complexes, the we get a covariant quartic

$$(4.4) \qquad \Phi = s_{01}E_{23} + s_{02}E_{31} + s_{03}E_{12} + s_{31}E_{02} + s_{12}E_{03} + s_{23}E_{01}$$

REMARK 4.2. Another point of view on the objects associated with the Cremona transformation is as follows. The blow-up of the base curve X for the homaloidal web produces a Fano threefold. The cubo-cubic transformation induces a biregular involution of the threefold. The Picard group of the threefold is of rank two, the group is freely generated by the image H of plane and by the inverse image E of the curve. An element $xH + yE$ of the Picard group is transformed to $(3x + 8y)H - (x + 3y)E$ by the involution . This linear transformation of the Picard group has two primitive eigen vectors. They are $(2H - E)$ and $(4H - E)$. The first vector corresponds to proper value (-1), and it does not correspond to an effective divisor, because curve X is not disposed in a quadric. But the second vector defines the anticanonical linear system of the threefold, therefore the linear system is fixed, it contains the images of quartics mentioned in 4.1.

REMARK 4.3. The cubo-cubic transformation induces an involutive automorphism of every quartic of the linear system of quartics generated by E_{ij}'s. In this linear system, it is not hard to find quartics having wider group of automorphisms. One can find a quartic with an additional involution. If a quartic has a double point, then the reflection with respect to the point is an involution of the quartic. The fixed points of the cubo-cubic transformations belong to the base points set of the linear system. We have enough parameters for formation of a double point in the set of the fixed points. Moreover, we have enough parameters for producing the quartic of the system having a double point in a general point of the space. In the latter case two involutions generate an infinite group of automorphisms of the quartic. G. Fano had some interest to the quartics having infinite group of automorphisms. He published an article (see [**Fa**]) on the subject in the scientific pontifical proceedings of hard forties. In the text, Fano also considered automorphisms of the indefinite integer quadratic forms of the Picard lattices of the surfaces.

It was long before the K3 industrial boom. All the above mentioned involutions of quartics are restrictions to the quartics of some space Cremona transformations, these transformations belong to the so called decomposition group of the quartics, see the definition of such a group in [**Gi3**].

We would like to construct a projective space transformation corresponding in an equivariant way to our curve with a theta characteristic. such a transformation is defined by a 4×4 matrix p. The matrix will be described with the help of a bilinear form $\Sigma p_{ij} x_i u_j$. We have the web of cubic forms

$$u_0 X_0(x) + u_1 X_1(x) + u_2 X_2(x) + u_3 X_3(x).$$

Let us consider the second homaloidal web of cubics

$$x_0 U_0(u) + x_1 U_1(u) + x_2 U_2(u) + x_3 U_3(u)$$

produced with the help of the contravariant net of quadrics $z_0 N_0 + z_1 N_1 + z_2 N_2$ from Remark 2.6. The matrix p is the Gram convolution of the webs, that is

(4.5) $$p_{ij} = < X_j(x), U_i(u) >, \quad i, j \in \{0, 1, 2, 3\}.$$

Shortly, the projective transformation is the Gram convolution of two Cremona cubo-cubic transformations of the mutually dual projective spaces.

We would like to construct a plane collineation corresponding in an equivariant way to the quartic curve with a theta characteristic. Such a transformation is defined by a 3×3 matrix t. The matrix will be described with the help of a bilinear form $\Sigma t_{ij} y_i z_j$.

First step. We need an auxiliary plane conic $D(y_0, y_1, y_2) = 0$. This conic is defined as the convolution of the square of the initial net of quadrics with the Gundelfinger quartic,

(4.6) $$D(y_0, y_1, y_2) = < (y_0 M_0 + y_1 M_1 + y_2 M_2)^2, \Theta > .$$

Second step. Using the symbolic method, we construct the following tri-quadratic form

$$G(z; x; u) = (abz)(cdz)(qx)(sx)(pqru)(pqsu),$$

where the initial net of quadrics is presented as

$$(ay)(px)^2, (by)(qx)^2, (cy)(rx)^2, (dy)(sx)^2,$$

see 2.4.

Third step. The convolution of the tri-quadratic form with the ternary quadratic form 4.6 cancels three variables and produces a biquadratic biquaternary biform,

$$K(x; u) = < D(y_0, y_1, y_2), G(z_0, z_1, z_2; x_0, x_1, x_2, x_3; u_0, u_1, u_2, u_3) > .$$

The last step. Two convolutions with two reciprocally dual nets of quadrics give us the matrix defining a projective transformation of the projective plane

(4.7) $$t_{ij} = << M_i, K >, N_j >,$$

$$t(y; z) = << y_0 M_0 + y_1 M_1 + y_2 M_2, K(x; u) >, z_0 N_0 + z_1 N_1 + z_2 N_2 >,$$

where N_i are defined in 2.6.

5. An irrational invariant of order $\frac{9}{2}$

So far we have two invariants. The first one is the Toepltz invariant Λ of degree $3/2$, The second one is the rational invariant I_3. Λ^2 is an irrational invariant of degree 3. We will use the following linear combination

$$(5.1) \qquad J_3 = \frac{1}{4}(I_3 - 5\Lambda^2).$$

For the case of example 2.7, the coefficients of J_3 are integers without a nontrivial common divisor. Such a polynomial is primitive, and the use of a multiplier such as $1/4$ in (5.1) will be named as primitivization. Sometimes we will denote a primitivization multiplier by an indefinite rational constant c. The subtraction in (5.1) is used because J_3 is shorter than I_3. We will use some similar normalizations for our future invariants. The motivations of such normalizations are the same and they will be omitted. Every time we will get something relatively short and primitive.

For the construction of an invariant Q of order $\frac{9}{2}$, one can use the matrices of two projective transformations from the previous section. The traces of the matrices give us new invariant

$$(5.2) \qquad Q = \frac{1}{3}(c \cdot (p_{00} + p_{11} + p_{22} + p_{33}) - 120\Lambda^3),$$

where p_{ij} is taken from 4.5, c is a rational number, or

$$Q = \frac{1}{240}(c \cdot (t_{00} + t_{11} + t_{22}) - 60 \cdot \Lambda^3),$$

where t_{ij} is taken from (4.7), c is a rational number.

The third way of obtaining Q is the calculation of the quadratic invariant of the linear line complex whose coefficients are s_{ij} from (4.3).

$$(5.3) \qquad Q = \frac{1}{80}(c \cdot (s_{01}s_{23} + s_{02}s_{31} + s_{03}s_{12}) - 9\Lambda^3 - 16J_3\Lambda),$$

where J_3 is taken from 5.1.

EXAMPLE 5.1. For some illustrations and calculations it is convenient to have an example where both the irrational invariants Λ and Q vanish. Such an example is helpful for obtaining syzygies modulo the ideal generated by these two invariants. The example is produced from 2.7 with the help of the following additional relations eliminating letters $a, f, e, p, c,$.

$$a = r, \quad f = r, \quad q = g, \quad p = c, \quad e = b.$$

For this case $\Lambda = 0$, $Q = 0$, but I_3 is not zero. I_3 is proportional to

$$(r^3 - bcg) \cdot (4r^3 - bcg - r \cdot (b^2 + c^2 + g^2)).$$

For general b, c, g, r corresponding quartic is non-singular. But the quartic defined by the contravariant net is a double conic.

6. An irrational invariant of order 6

We have rational invariant C_6 of the quartic, C_6 is the so called *catalecticant*. Using with the above mentioned invariants, we have the following linearly independent invariants of degree 6.

$$\Lambda^4, \quad J_3\Lambda^2, \quad J_3^2, \quad Q\Lambda, \quad C_6.$$

Furthermore, out of the linear space spanned by the five invariants, there exists an additional invariant X of degree 6.

We fix three following elements in the space.

$$K_6 = \frac{1}{72}(5C_6 - I_3^2),$$

$$L_6 = \frac{1}{5}(9K_6 + 2J_3^2),$$

$$M_6 = \frac{1}{2}(L_6 + 3Q\Lambda).$$

New invariant X is produced with the help of quadratic form D from 4.6,

$$X = \frac{1}{2^8 \cdot 3^2}(c \cdot \text{Discr}(D) - 2\Lambda^4 + 4J_3\Lambda^2 + 40L_6 - 144M_6).$$

REMARK 6.1. There exists more complicated way of producing X. We will briefly describe it. For a general quaternary quartic $f(u_0, u_1, u_2, u_3)$, the simplest invariant has order 4 in the coefficients of f. The symbolic expression of the invariant is $(\alpha\beta\gamma\delta)^4$, where

$$f = \alpha^4 = \beta^4 = \gamma^4 = \delta^4,$$

$(\alpha\beta\gamma\delta)$ is the determinant of quaternary linear forms $\alpha, \beta, \gamma, \delta$. This invariant for the Gundelfinger quartic (3.1) is a linear combinations of the above mentioned invariants of degree 6, X has a non-zero coefficient in the combination, therefore X is expressible in terms of the invariant for the Gundelfinger quartic and other invariants.

Subsection 12.4 below contains a construction for X based on a special quadratic line complex.

7. Irrational invariants of order $\frac{15}{2}$

So far we have five linearly independent invariants of degree $\frac{15}{2}$. They are obtained from the invariants of degree 6 with the multiplication by Λ. We construct two invariants T and I. Invariant I is due to Salmon, therefore we will name it as the Salmon invariant. The construction of T is also based on some Salmon's ideas.

We begin with a general remark concerning non-linear line complexes. Such a complex can not be written down by a unique way as a polynomial of the Plücker line coordinates because there exists the Plücker quadratic relation for the coordinates. Despite the fact, it is possible to define the convolution of two line complexes of the same degree. It is necessary to use a preliminary special procedure of normalization for the equations of both the complexes. The procedure is described in the second volume of Salmon's book [**Sa1**] on solid geometry, Chap. VII, n°350,n°351. See also our section 13. After such a normalization, the convolution of two line complexes of the same degree is definable correctly, the convolution gives us a combined bilinear invariant for these two complexes.

Let us take the Toeplitz cubic line complex mentioned in 2.2 and the cube of the linear line complex whose coefficients are presented in (4.3). The convolution (more precisely, the combined bilinear invariant) of the two cubic line complexes is invariant T. For example 5.1 T does not vanish, therefore it does not belong to the ideal generated by the invariants of degree 6. We would like to normalize the invariant. If C is the convolution, then we put

$$T = \frac{1}{384}(3 \cdot C + 81 \cdot \Lambda^5 + 216 \cdot J_3 \cdot \Lambda^3 + 128 \cdot J_3^2 \cdot \Lambda - 128 \cdot L_6 \cdot \Lambda + 384 \cdot M_6 \cdot \Lambda),$$

where L_6 and M_6 are defined in section 6.

The idea of the next invariant I is rather general. In fact, we can define it for nets of quadrics in any $\mathbb{P}^n$. Let $A = (L_{ij})$ be a matrix of linear forms defining the net. As in (3.2) we define the determinant

$$(7.1) \qquad G(y_0, y_1, y_2, U_0, \ldots, U_n) = \begin{vmatrix} L_{00} & L_{01} & \ldots & L_{0n} & U_0 \\ L_{10} & L_{11} & \ldots & L_{1n} & U_1 \\ . & . & \ldots & . & . \\ L_{n0} & L_{n1} & \ldots & L_{nn} & U_n \\ U_0 & U_1 & \ldots & U_n & 0 \end{vmatrix}.$$

For any point with coordinates $(u_0 : \ldots : u_n)$ in the dual projective space $\check{\mathbb{P}}^n$ the equation $G(y_0, y_1, y_2, u_0, \ldots, u_n) = 0$ defines a contact curve of degree n of the Hessian curve C of degree $n+1$ of the net of quadrics. If $\mathbb{P}^2 = \mathbb{P}(V)$ and $\mathbb{P}^n = \mathbb{P}(W)$, the polynomial G can be considered an element of $S^2 W^* \otimes S^n V^*$, hence it defines a linear map $\Phi : S^2 W \to S^n V^*$ of linear spaces of the same dimension. So it is natural to view its determinant as an invariant of a net. The definition of the determinant depends on a choice of bases in the spaces. That is we must fix a sign. We will make such a fixation below in 7.4, using another definition of I. The determinant is a polynomial of degree $n(n+1)(n+2)/2$ in coordinates on the space of quadrics $V^* \otimes S^2 W^*$. Thus it is an irrational (we will see it later) invariant on the space of curves of degree $n+1$ of order $n(n+2)/2$. In the case of quartics we get the promised invariant I of degree $15/2$.

Theorem 7.1 below and the preliminary comments belong to Igor Dolgachev.

Recall that a net of quadrics with nonsingular Hessian curve does not contain quadrics of corank > 1. It is known that the locus of quadrics of corank > 2 is equal to the singular locus of the discriminant hypersurface of all quadrics in $\mathbb{P}^n$ and it is of codimension 2. A net of quadrics defines a plane section of the discriminant hypersurface. If the plane section intersects the discriminant hypersurface at its singular point, the Hessian curve acquires a singular point corresponding to a quadric from the net of corank > 1. On the other hand, if the plane section is touching the discriminant hypersurface at its nonsingular point, the Hessian curve is singular at this point, but the corresponding quadric is still of corank 1.

THEOREM 7.1. *Let N be a net of quadrics in $\mathbb{P}^n$ with irreducible reduced Hessian curve. Consider the following properties:*

(i) *the net N contains a quadric of rank > 1;*
(ii) $I = 0$;
(iii) *the curve C is singular.*

Then

$$(i) \Longrightarrow (ii) \Longrightarrow (iii).$$

PROOF. (i)$\Rightarrow$ (ii) In this case the matrix of linear forms $A = (L_{ij})$ is of corank ≥ 2 at some point $x \in C$. Since

$$G(x_0, x_1, x_2, U_0, \ldots, U_n) = \sum A_{ij}(x_0, x_1, x_2) U_i U_j,$$

where A_{ij} are $n \times n$-minors of the matrix A, we obtain that all A_{ij} vanish at the point x. This shows that the image of the linear map $S^2 V \to S^n E^*$ defined by

G is contained in the space of of polynomials vanishing at x. So the map is not surjective, and hence the invariant I^2 vanishes.

(ii) $\Rightarrow$ (iii) Suppose C is nonsingular. Let X be its Steinerian model in $\mathbb{P}^n$ (the Jacobian curve of the net). It is known (see [**Be1**]) that the restriction map

$$\mathbb{P}(S^2 V^*) \to \mathbb{P}(S^n E^*),$$

where

$$S^2 V^* = H^0(\mathbb{P}^n, \mathcal{O}_{\mathbb{P}^n}(2)), \quad S^n E^* = H^0(X, \mathcal{O}_X(2)) = H^0(C, \mathcal{O}_C(n)),$$

is a bijection (see [**Be1**]). The composition of this map with the Veronese map $\mathbb{P}(V^*) \to \mathbb{P}(S^2 V^*)$ is our map $H \mapsto B_H$ which assigns to a hyperplane H the corresponding contact curve B_H. Thus $I^2 \neq 0$ contradicting the assumption. $\square$

Let Δ be the discriminant invariant of a plane quartic. Pulling it back to the space of nets of quadrics we obtain an invariant on the space of quartics which vanishes on nets of quartics with singular Hessian curve. The previous theorem shows that the irrational invariant I^2 vanishes on a closed subset of this locus. However this does not prove that I^2 divides the discriminant. The following result proves this assertion.

THEOREM 7.2. *The irrational invariant I^2 divides the discriminant invariant.*

PROOF. It is enough to check it for a general family of net of quadrics represented by a example 2.7. To compute I^2 we order the ten partial derivatives as follows.

$$\partial^3/\partial x_0^3, \partial^3/\partial x_1^3, \partial^3/\partial x_2^3, \partial^3/\partial x_0^2 \partial x_1, \partial^3/\partial x_0^2 \partial x_2,$$
$$\partial^3/\partial x_1^2 \partial x_0, \partial^3/\partial x_1^2 \partial x_2, \partial^3/\partial x_2^2 \partial x_0, \partial^3/\partial x_2^2 \partial x_1, \partial^3/\partial x_0 \partial x_1 \partial x_2,$$

and

$$\partial^2/\partial u_0^2, \partial^2/\partial u_1^2, \partial^2/\partial u_2^2, \partial^2/\partial u_3^2, \partial^2/\partial u_0 \partial u_1,$$
$$\partial^2/\partial u_0 \partial u_2, \partial^2/\partial u_0 \partial u_3, \partial^2/\partial u_1 \partial u_2, \partial^2/\partial u_1 \partial u_3, \partial^2/\partial u_2 \partial u_3.$$

After the calculation and the factorization of the 10×10 determinant, we get the following expression

$$I = (be - af) \cdot (ar - cp) \cdot (gq - fr) \cdot$$
$$(rb^2 - fc^2 + (g - q)bc) \cdot$$
$$(fp^2 - aq^2 + (b - e)pq) \cdot$$
$$(ag^2 - re^2 + (p - c)eg).$$

First, we prove that I is a divisor of the discriminant. To do this we need to show that the vanishing of every factor of I ensures singularity of the curve. In the case of example 2.7, the equation of the curve can be rewritten in one of the following forms

$$x_0^2 \xi_0^2 + x_1^2 \xi_1^2 + x_2^2 \xi_2^2 -$$
$$2(x_1 x_2 \xi_1 \xi_2 + x_0 x_2 \xi_0 \xi_2 + x_0 x_1 \xi_0 \xi_1) = 0,$$

where

$$\xi_0 = ax_0 + bx_1 + cx_2,$$
$$\xi_1 = fx_0 + gx_1 + hx_2,$$
$$\xi_2 = px_0 + qx_1 + rx_2,$$

or in form
$$(-x_0\xi_0 + x_1\xi_1 + x_2\xi_2)^2 - 4x_1x_2\xi_1\xi_2 = 0,$$
or in form
$$(x_0\xi_0 - x_1\xi_1 + x_2\xi_2)^2 - 4x_0x_2\xi_0\xi_2 = 0,$$
or in form
$$(x_0\xi_0 + x_1\xi_1 - x_2\xi_2)^2 - 4x_0x_1\xi_0\xi_1 = 0.$$
Here $(af - be) = 0$ means that three linear form ξ_0, ξ_1, x_2 have a common zero, therefore in this case the first equation indicates that the curve is singular, $(ar - cp) = 0$ means that three linear form ξ_0, ξ_2, x_1 have a common zero, $(fr - gq) = 0$ means that three linear form ξ_1, ξ_2, x_0 have a common zero.

Furthermore
$$(rb^2 - fc^2 + (g - q)bc) = 0$$
means that two lines $x_0 = 0$, $\xi_0 = 0$, and conic
$$(x_0\xi_0 + x_1\xi_1 - x_2\xi_2) = 0$$
have a common point, this point must be singular on the quartic according to the fourth form of the curve equation. In this place one can use these two lines, the conic $x_0\xi_0 - x_1\xi_1 + x_2\xi_2 = 0$ and the third form of equation. The consideration of the last two factors of I is similar.

Thus it is proved that I is a divisor. Every of the factors of I changes his sign by an admissible transposition of three basic quadratic forms. Such a transposition does not change the discriminant, therefore every factor enters twice in the discriminant. $\qquad\square$

REMARK 7.3. The irrationality of I^2 is a consequence of the irreducibility of the discriminant (as a polynomial in the coefficients of the quartic). Indeed, if I belongs to the space of invariants of the curve, then it is a proper divisor of the discriminant.

Because of the importance of the Salmon invariant, we will describe some other definitions of I.

REMARK 7.4. The following identity
$$-1280 \cdot I(y_0M_0 + y_1M_1 + y_0M_0) = \Lambda(z_0N_0 + z_1N_1 + z_0N_0)$$
gives us an opportunity of an alternative definition of the Salmon invariant I for the net of ternary quadrics. That is the value of the invariant Λ for the contravariant net coincides (up to an integer multiplier) with the invariant I of the initial net. Certainly, here, we are enforced to fix some sign of an expression for I. We see that in the comparison with determinants, Pfaffian is more convenient tool for the sign fixation.

REMARK 7.5. Every quadric M in $\mathbb{P}^3$ defines a quadratic complex consisting of the lines tangent to M, cf. 12.3 below. We will denote the complex by $C(M, M)$, the double M is used because of the existence of the following natural generalization of the construction. Two distinct quadrics M and M' define a quadratic complex $C(M, M')$ consisting of the lines such that two pairs of their intersection points with the quadrics are mutually harmonic. Three given quadrics M_0, M_1, M_2 define six quadratic line complexes $C(M_0, M_0)$, $C(M_1, M_1)$, $C(M_2, M_2)$, $C(M_0, M_1)$, $C(M_0, M_2)$, $C(M_1, M_2)$. Taking the convolutions of these complexes with the Toeplitz cubic line complex (2.3), we get six linear line complexes K_{00}, K_{11}, K_{22}, K_{01}, K_{02}, K_{12}. A line complex is defined by its six coefficients. Therefore six linear line complexes have a simple combined multi-linear invariant, it is the determinant of their coefficients. The combined invariant of the six linear line complexes $C(M_i, M_j)$ coincides (up to a multiplier) with the invariant I. Thus we have a representation of I in the form of a determinant of sixth order.

REMARK 7.6. There exists a way of producing of invariant T without using the theory of line complexes. If we take the connvolution of quartic (4.4) with the square of the contravariant net from 2.6, then we get a contravariant ternary conic. The convolution of the conic with conic 4.6 is an invariant V. It is possible to use this invariant for a definition of T (if we have I defined). Indeed, we have the following identity

$$9 \cdot V = 2^{12} \cdot 3^2 \cdot Q \cdot J_3 + 9 \cdot 2^{10} \cdot 3^2 \cdot 7 \cdot T + 2^{15} \cdot 3^2 \cdot 5^2 \cdot I +$$
$$2^7 \cdot (5 \cdot C_6 - 2^4 J_3^2 - 2^5 \cdot 3^2 \cdot 5 \cdot X)\Lambda +$$
$$2^7 \cdot 3^2 \cdot 29 \cdot Q \cdot \Lambda^2 - 2^7 \cdot 31 \cdot J_3 \cdot \Lambda^3 - 2^5 \cdot 19 \cdot \Lambda^5.$$

We will not use the identity further. We write it down for the indication of the way how to avoid the use of the Toeplitz cubic complex. The proof of the identity is obtained with the help of Macaulay-2.

REMARK 7.7. For the special case 5.1, I is completely decomposable,

$$I = (b+c)(b-c)(b-g)(b+g)(b+r)(b-r)(c-g)(c+g)(c-r)(c+r)(g-r)(g+r)r^3.$$

Moreover, for this case invariants I and T are proportional, but they are not proportional for example 2.8.

REMARK 7.8. It is interesting to follow the decompositions of I by some changes of the writings of the net of quadrics. Let us begin with a slightly general net in the comparison with example 2.7. Assume that

$$M_0 = 2(ax_0x_1 + ex_0x_2 + px_0x_3 + kx_2x_3),$$
$$M_1 = 2(bx_0x_1 + fx_0x_2 + qx_0x_3 + lx_1x_3),$$
$$M_2 = 2(cx_0x_1 + gx_0x_2 + rx_0x_3 + mx_1x_2).$$

For this case

$$I = k^4l^4m^4(be - af)(gq - fr)(cp - ar)\times$$
$$(flp^2 - akq^2 + bpqk - epql)\times$$
$$(rme^2 - akg^2 + cegk - egpm)\times$$
$$(rmb^2 - flc^2 + bcgl - bcqm).$$

Riemann's normalization mentioned in example 2.7 is expressed here by the following substitution

$$a = -(e + p + k), \quad b = -(f + q + l), \quad c = -(g + r + m).$$

After the substitution we get the net of quadrics having basic point (1:1:1:1) besides the coordinate points. Then the decomposition of I will be transformed in the following form.

$$I = k^4l^4m^4(kg - em)(gq - fr)(kq - pl)\times$$
$$(eq + kq - pf)(pg + kg - er)(fr - qg - lg - qm - lm)\times$$
$$(pf + kf - eq - el)(er + kr - pm - pg)(fm + qm - lg - lr).$$

Let us fix sixth base point $(U : V : W : T)$. General net of quadrics having the six base points is determined by the above three base forms M_0, M_1, M_2 with the following coefficients

$$k = DU(W - T), \quad l = DU(T - V), \quad m = DU(V - W),$$
$$a = AU(W - T), \quad f = BU(T - V), \quad r = CU(V - W),$$
$$e = AU(T - V) + DT(U - W), \quad p = AU(V - W) + DW(T - U),$$
$$b = BU(W - T) + DT(V - U), \quad q = BU(V - W) + DV(U - T),$$
$$c = CU(W - T) + DW(U - V), \quad g = CU(T - V) + DV(W - U),$$

where A, B, C, D are four free homogeneous parameters. For the latter case the decomposition of I is as follows.

$$I = D^{21}TVW(U - V)(T - U)(U - W)(T - W)^5(T - V)^5(V - W)^5 U^{15} \times$$
$$(U(AV + BW) - (A + B + D)UT + DT^2) \times$$
$$(U(AV + CT) - (A + C + D)UW + DW^2) \times$$
$$(U(BW + CT) - (B + C + D)UV + DV^2) \times$$
$$(U(AV^2 + BW^2) - (AV + BW)UT + D(T - U)VW) \times$$
$$(U(AV^2 + CT^2) - (AV + CT)UW + D(W - U)VT) \times$$
$$(U(BW^2 + CT^2) - (BW + CT)UV + D(V - U)WT) \times$$
$$(AUV(U - V) + BUW(U - W) - (A + B + D)U^2T +$$
$$(A + D)UVT + (B + D)UWT - DVWT) \times$$
$$(BUW(U - W) + CUT(U - T) - (B + C + D)U^2V +$$
$$(B + D)UVW + (C + D)UVT - DVWT) \times$$
$$(AUV(U - V) + CUT(U - T) - (A + C + D)U^2W +$$
$$(A + D)UVW + (C + D)UWT - DVWT).$$

Let us consider the case where seven base points of the net are fixed. Assume that seventh base point is $(x : y : z : s)$. It is well known that if we fix seven base points, then eighth base point of the net is determined uniquely.

Another point of view is associated with the theory of space Cremona transformations. Six points $(1 : 0 : 0 : 0)$, $(0 : 1 : 0 : 0)$, $(0 : 0 : 1 : 0)$, $(0 : 0 : 0 : 1)$, $(1 : 1 : 1 : 1)$, $(U : V : W : T)$ define an involutive Cremona transformation such that if point $(x : y : z : s)$ is the seventh base point of a net of quadrics, then the eighth base point $(X_0 : X_1 : X_2 : X_3)$ is the image of $(x : y : z : s)$ by the transformation. The explicit formulas for the transformation are following.

$$X_0 = D_0 W_1 W_2 W_3,$$
$$X_1 = D_1 W_0 W_2 W_3,$$
$$X_2 = D_2 W_1 W_0 W_3,$$
$$X_3 = D_3 W_0 W_1 W_2,$$

where D_0, D_1, D_2, D_3 are four linear forms of the coordinates x, y, z, s of the initial point,

$$D_0(x, y, z, s) = (W - T)y + (T - V)z + (V - W)s,$$
$$D_1(x, y, z, s) = (T - W)x + (U - T)z + (W - U)s,$$
$$D_2(x, y, z, s) = (V - T)x + (T - U)y + (U - V)s,$$

$$D_3(x, y, z, s) = (V - W)x + (W - U)y + (U - V)z,$$

W_0, W_1, W_2, W_3 are four quadratic forms of the coordinates,

$$W_0(x, y, z, s) = T(W - V)yz + W(V - T)ys + V(T - W)zs,$$

$$W_1(x, y, z, s) = T(U - W)xz + W(T - U)xs + U(W - T)zs,$$

$$W_2(x, y, z, s) = T(V - U)xy + V(U - T)xs + U(T - V)ys,$$

$$W_3(x, y, z, s) = W(V - U)xy + V(U - W)xz + U(W - V)yz.$$

The degree of the Cremona transformation is equal to 7. The transformation is not a covariant of our net of quadrics, but we can use some parts of the formulas for the transformation.

Let us take three following basic quadratic forms for our net.

$$M_0 = x_0 W_1(x, y, z, s) D_0(x_0, x_1, x_2, x_3) - x D_0(x, y, z, s) W_1(x_0, x_1, x_2, x_3),$$

$$M_1 = x_0 W_2(x, y, z, s) D_0(x_0, x_1, x_2, x_3) - x D_0(x, y, z, s) W_2(x_0, x_1, x_2, x_3),$$

$$M_2 = x_0 W_3(x, y, z, s) D_0(x_0, x_1, x_2, x_3) - x D_0(x, y, z, s) W_3(x_0, x_1, x_2, x_3).$$

It is not hard to verify that these three forms vanish at the eight basic points. The invariant I for the net generated by these three quadrics has the following decomposition

$$I = x^{21} U^{21} D_0^{21} (W - T)^{11} (V - T)^{11} (V - W)^{11} \times$$

$$(Wy - Vz)(Us - Tx)(Ws - Tz)(Uy - Vx)(Vs - Ty)(Uz - Wx) \times$$

$$yzs(x - s)(z - y)(x - y)(s - z)(x - z)(s - y)(U - T)(U - V)(U - W).$$

8. An irrational invariant of order 9

Recall the description of rational invariants of degree 9. One can take the following base for the linear space of rational invariants of degree 9.

$$I_3^3, \quad I_3 C_6, \quad Z_9, \quad E_9,$$

where Z_9 is the convolution of two ternary conics, one of the conics depends on (y_0, y_1, y_2), the second conic depends on (z_0, z_1, z_2), see [**Sa2**], E_9 is the convolution of two ternary quartics, one of them is the Clebsch covariant for C, the second is the result of a bimultiplication of C and its Clebsch covariant. The bimultiplication of quartics is described in [**Gi2**].

THEOREM 8.1. *The rational invariant Z_9 is expressible in the terms of rational and irrational invariants of lower degree.*

PROOF. The following identity is valid.

(8.1) $\qquad Z_9 = 2^9 \cdot J_3^3 + 2^5 \cdot 3^5 \cdot Q^2 + 2^6 \cdot 3^5 \cdot (T + Q \cdot J_3 + 2^7 \cdot I)\Lambda -$

$2^7 \cdot 3^6 \cdot X \cdot \Lambda + 2^4 \cdot 3^3 \cdot 7 L_6 \cdot \Lambda^2 + 2^8 \cdot 3 \cdot 7 \cdot J_3^2 \cdot \Lambda^2 + 2^3 \cdot 3 \cdot 7^3 \cdot J_3 \cdot \Lambda^4 + 2 \cdot 5 \cdot 7^3 \cdot \Lambda^6$

Its verification for the case of 2.7 is simple but far of our old mathematical traditions. Macaulay-2 shows that the size of the difference between the left hand side and the right hand side is zero. $\qquad\qquad\qquad\qquad\qquad\qquad\qquad\qquad\qquad\qquad\square$

REMARK 8.2. One can consider the identity as a *syzygy for irrational invariants*. That is if we study the field of fractions of the graded algebra of invariants as an algebraic extension of the field of fractions of the graded algebra of rational invariants, then the identity is the first non-trivial equation from the series of equations defining the extension. The extension has degree 36, the collection of our irrational invariants is relatively large, therefore obtaining the first identity is far from the final explicit description of the extension.

Let us consider the linear space of all (rational an irrational) invariants of degree 9. So far, in the space we have the subspace spanned by E_9 and by the degree 9 products of the invariants of lower degree. We will construct a new invariant of degree 9 out of the subspace. There are three ways.

First, one can take the trace of the second exterior power of 3×3 matrix t from (4.7).

Second, one can take the trace of the second exterior power of 4×4 matrix p from (4.5).

Third, we take the following convolution of two ternary quartics

$$Y =< \text{The Clebsch covariant of } (C), \det(z_0N_0 + z_1N_1 + z_2N_2) > .$$

The third way is most convenient for calculations.

REMARK 8.3. For the special case 5.1, in the space of invariants of degree 9 we see five linearly independent elements

$$J_3^3, \quad J_3C_6, \quad J_3X, \quad E_9, \quad Y.$$

It shows that Y does not belong to the subspace spanned by E_9 and by the degree 9 products of the invariants of lower degree.

9. An irrational invariant of order $\frac{21}{2}$

Our goal in this short section is to write down the formula for an invariant W of degree 21/2 not belonging to the subspace spanned by the degree 21/2 products of the invariants of lower degree. Let us take the Steinerian quaternary quartic Ω(see 3.6 and the comments to the determinant defing the quartic). The new invariant W is the convolution of the quartic with the Gundelfinger quartic

$$W = \frac{1}{32} < \Omega(x_0, x_1, x_2, x_3), \Theta(u_0, u_1, u_2, u_3) > .$$

10. The tact-invariant

The following preliminary remark to new subject belongs to Igor Dolgachev.

REMARK 10.1. The general theory of symmetric determinant representation of irreducible plane curves (see [**Be2**]) makes the difference between determinantal representations of a singular curve. A first construction is based on a choice of a "honest" theta characteristic of the canonical sheaf ω_C, i.e. an invertible sheaf θ such that $\theta^2 \cong \omega_C$. This gives a determinantal representation by a net of quadrics which does not contain quadrics of corank > 1. The second one chooses a torsion-free non-invertible sheaf $\mathcal{E}$ such that $\mathcal{E}xt^1(\mathcal{E}, \omega_C) \cong \mathcal{E}$, and $h^0(\mathcal{E}) = 0$. It is defined by taking a normalization $p : \bar{C} \to C$, and defining $\mathcal{E}$ to be equal to $p_*\theta$, where θ is an even non-vanishing theta characteristic on $\bar{C}$ (see [**Be1**]). The corresponding net of quadrics contains a quadric of corank > 1. In fact, the ideal generated by the minors of the matrix defining the determinantal representation of C is equal to the conductor ideal of the singular point of C. So we expect that the locus of singular plane curves of degree $n + 1$ defines two irreducible component in the locus of nets of quadrics. One is given by the invariant I, the other one is the *tact-invariant* of order 12.

About the etymology: according to A. Cayley, the tact-invariant means the taction (=tangency) invariant. This tact-invariant is described in § 237 of the book [Sa1] and denoted by letter J. Geometric meaning of J is the following: J vanishes if and only if two of the eight basic points of the quadric pencil coincide. In other words, $J = 0$ if and only if the base scheme locus of the net of quadrics in $\mathbb{P}^3$ given by three quadrics

$$M_0(x_0, x_1, x_2, x_3) = 0, \quad M_1(x_0, x_1, x_2, x_3) = 0, \quad M_1(x_0, x_1, x_2, x_3) = 0$$

is singular. A way to compute J in practice is as follows. Using the elimination process of a pair of variables x_i, x_j leads to a resultant $R_{ij}(x_k, x_l)$, the latter is a binary form of other variables x_k, x_l. The discriminant $\mathrm{Discr}(R_{ij})$ of R_{ij} contains J. A comparison of the discriminants $\mathrm{Discr}(R_{ij})$ for different pairs (i, j) gives us J as the greatest common divisor. For special equations of the quartic, it is necessary to be cautious, because some factors of the auxiliary discriminants may be overlooked after some natural cancellations. For a control over such small factors, we may use the Jacobian criterion of singularity. We will realize the process of control in the situation of example 2.7.

Let us form the Jacobian matrix $D(M_0, M_1, M_2)/D(x_0, x_1, x_2, x_3)$ for three quadratic forms M_0, M_1, M_2 mentioned in the example. The matrix has size 3×4. A point in the net basic set is singular if and only if the Jacobian matrix has rank 2 in the point. In the coordinate basic point $(1 : 0 : 0 : 0)$, $(0 : 1 : 0 : 0)$, $(0 : 0 : 1 : 0)$, $(0 : 0 : 0 : 1)$ the Jacobian matrix has only one nonzero minor of order 3, they are

$$d, a, f, r$$

respectively, where d is the determinant of the coefficients of three linear forms ξ_0, ξ_1, ξ_2, that is

$$d = \begin{vmatrix} a & b & c \\ e & f & g \\ p & q & r \end{vmatrix}.$$

Thus, in the case of example 2.7, the product $afrd$ is a divisor of J. Because of the alternating property of determinants, it is clear that $(afrd)^2$ is a divisor. Hence $J = (afrd)^2 J'$. As to the factor J', then we indicate that it is irreducible (as a form of degree 12 of variables $a, b, c, f, g, h, p, q, r$). The size of J' is equal to 1506. The process of elimination gives us a resultant-like representation of J'.

$$J' = \frac{1}{16F^2} \begin{vmatrix} 4A & 3B & 2C & G & 0 & 0 \\ 0 & 4A & 3B & 2C & G & 0 \\ 0 & 0 & 4A & 3B & 2C & G \\ B & 2C & 3G & 4E & 0 & 0 \\ 0 & B & 2C & 3G & 4E & 0 \\ 0 & 0 & B & 2C & 3G & 4E \end{vmatrix},$$

where F^2 is a parasitic factor of the resultant,

$$F = b^2 r - c^2 f + bcg - bcq,$$

$$A = ar^2 - cpr, \quad E = af^2 - bef,$$
$$B = cpq + bcr - cgp - cer + bpr + 2agr - 2aqr - c^2 q,$$
$$G = bcf + cef + beg - bfp - beq + 2afq - 2afg - b^2 g,$$
$$C = bcg - ceg + cfp + bgp + bcq + ceq -$$
$$bpq + ber - 2agq - 2afr + ag^2 + aq^2 - b^2 r - c^2 f.$$

REMARK 10.2. If we use Riemann's normalization of 2.7, then J' is reducible, the tact-invariant acquires an additional factor, the factor is L^2, where

$$L = bc + ce + bg + eg + bp + gp + cq + eq + pq + 2b + 2c + 2e + 2g + 2p + 2q + 4.$$

Then

$$J = ((e + p + 1)(b + q + 1)(c + g + 1)dL)^2 J'',$$

where J'' is an irreducible polynomial of letters b, c, e, g, p, q. A description of the decompositions of J similar to remark 7.8 contains more complexities, we omit it here.

THEOREM 10.3. *The irrational tact-invariant J divides the discriminant invariant Δ of plane quartics.*

PROOF. As we have already explained in this section that there are two reasons which cause the Hessian curve to acquire a singular point. The first reason is the existence of a quadric of corank > 1 in the net. The invariant I^2 is responsible for this. The second reason is the occurence of a base point on the Steinerian curve. The invariant J is responsible for this. This shows that the locus of zeroes of J is contained in the locus of zeroes of the pre-image of the discriminant invariant on the space of nets of quadrics. Since we have shown by computation that J is irreducible, we obtain the assertion of the theorem.   □

COROLLARY 10.4. *(Salmon's theorem)*

$$I^2 \cdot J = k\Delta,$$

for some constant k.

PROOF. We have seen already that I^2 and J divide Δ. It remains to compare the orders. The order of the discriminant is equal to $27 = 15 + 12$.   □

The irrational tact-invariant has degree 12. Recall the description of rational invariants of degree 12. One can take the following base for the linear space of these rational invariants.

$$I_3^4, \quad I_3^2 C_6, \quad I_3 Z_9, \quad I_3 E_9, \quad C_6^2, \quad A_{12}, \quad K_{12},$$

where

$$K_{12} = \frac{1}{12} I_3 (\text{the Clebsch covariant of } C),$$

A_{12} is produced with the help of the contavariant ternary quadratic form E of order 4 with respect to the coefficients of C, the form is mentioned in [**Sa2**].

$$A_{12} = -\frac{1}{12} \text{Discriminant } (E) + \frac{4}{3} K_{12}.$$

Below we will show that J is a linear combination of the rational invariants and the monomials composed from some rational or irrational invariants of degree less than 12, that is the tact-invariant is eliminable off the eventual list of the basic invariants for C, θ.

THEOREM 10.5. *The tact-invaraint is expressible in the terms of rational invariants and irrational invariants of lower degree.*

PROOF. The proof is similar to the proof of 8.1. The following identity is valid.

$$(10.1) \qquad 2^{12} \cdot 3^5 \cdot J = 2^2 \cdot K_{12} - 3^2 \cdot A_{12} + 2^6 \cdot 3^4 \cdot 11 \cdot Q^2 \cdot J_3 -$$
$$3 \cdot 11 \cdot 17 \cdot J_3^2 \cdot C_6 + 2^7 \cdot 3^3 \cdot J_3^2 \cdot X - 2^4 \cdot 5 \cdot C_6^2 +$$
$$2^{10} \cdot 3^5 \cdot Q \cdot T + 2^{11} \cdot 3^5 \cdot 67 \cdot Q \cdot I + 2^9 \cdot 3^3 \cdot C_6 \cdot X -$$
$$2^{12} \cdot 3^4 \cdot X^2 - 2^2 \cdot 3 \cdot 7 \cdot J_3 \cdot E_9 + 3^3 \cdot J_3 \cdot Y -$$
$$3^3 \cdot (3^2 \cdot W + 2^4 \cdot 89 \cdot J_3^2 \cdot Q + 2^{10} \cdot 3 \cdot 83 \cdot I \cdot J_3 +$$
$$2^6 \cdot 3 \cdot 13 \cdot T \cdot J_3 + 2^6 \cdot 5 \cdot Q \cdot C_6 + 2^{13} \cdot 3^2 \cdot Q \cdot X) \cdot \Lambda +$$
$$2^2 \cdot (2^6 \cdot 3^3 \cdot 5 \cdot J_3^3 + 2^2 \cdot 3^4 \cdot 29 \cdot C_6 \cdot J_3 + 2^7 \cdot 3^4 \cdot 59 \cdot X \cdot J_3 +$$
$$2 \cdot 3^4 \cdot Y + 2^5 \cdot 3^6 \cdot 5 \cdot Q^2 + 3^2 \cdot 23 \cdot E_9) \cdot \Lambda^2 +$$
$$2^4 \cdot 3^4 \cdot (3491 \cdot Q \cdot J_3 + 2 \cdot 3 \cdot 5 \cdot 7^2 \cdot T + 2^3 \cdot 3 \cdot 67 \cdot 107 \cdot I) \cdot \Lambda^2 -$$
$$2^3 \cdot 3^4 \cdot (71 \cdot J_3^2 + 2^7 \cdot 5 \cdot 17 \cdot X + 5 \cdot 11 \cdot C_6) \cdot \Lambda^4 +$$
$$2^4 \cdot 3^4 \cdot 601 \cdot Q \cdot \Lambda^5 - 2^4 \cdot 3^4 \cdot 23 \cdot J_3 \cdot \Lambda^6 + 2^2 \cdot 3^3 \cdot 5 \cdot 23 \cdot \Lambda^8.$$

Macaulay-2 shows that the size of the difference between the left hand side and the right hand side is zero. $\qquad \square$

REMARK 10.6. The identity 10.1 is not a syzygy for irrational invariants, but it is not hard to get a new syzygy with the help of (10.1). Indeed, if F is the right hand side of 10.1, then using Salmon's identity 10.4 we get

$$(10.2) \qquad 2^{12} \cdot 3^5 \cdot I^2 \cdot F = k \cdot \Delta,$$

where Δ is the discriminant of our quartic.

11. An algebraic foundation of Thomae's formulas for genus 3

Thomae's formulas for the products of of even theta constants for hyperelliptic curves are rather well-known. But as it seems, the simplest non-hyperelliptic case (that is the case of plane quartics) is almost abandoned now. The first (and maybe the last) geometer who paid attention to Thomae's formulas for genus 3 was Felix Klein, see [**Kl**], §23, *Das Product der Nullwerte der 36 geraden Thetafunktionen.* Klein expressed a power of the discriminant of plane quartic as a product of theta constants. We would like to find an analogue of Thomae's formula for nonsingular plane quartics.

Further preliminary comments and the proof of two Thomae's formulas for genus 3 belong to Igor Dolgachev.

Recall first Thomae's formulas in the case of hyperelliptic curves C of genus g or, equivalently, binary forms of degree $2g + 2$.

Let P_1^{2g+2} be the GIT-quotient of the product $(\mathbb{P}^1)^{2g+2}$ by the group SL_2 acting diagonally (with respect to the symmetric linearization). Let $\Delta_g \subset P_1^{2g+2}$ be the closed subset whose complement is the orbit space of distinct point sets. It is well-known that the complement $U_1^{2g+2} = P_1^{2g+2} \setminus \Delta_g$ can be identified with an irreducible component $\mathcal{H}yp_g^o$ of the moduli space of hyperelliptic curves with level 2 structure. The stabilizer subgroup of $\mathrm{Sp}_{2g+2}(2)$ of this component is isomorphic to the symmetric group Σ_{2g+2}. We have an isomorphism

$$(11.1) \qquad \phi_g : P_1^{2g+2}/\Sigma_{2g+2} \cong \mathrm{Proj}(S^{2g+2}((\mathbb{C}^2)^*)^{\mathrm{SL}_2(2)}).$$

The space on the right can be viewed as a compactification $\overline{\mathcal{Hyp}}_g$ of $\mathcal{Hyp}_g$. The image of the zero locus of Δ_g under ϕ_g is the zero locus of the discriminant invariant D_{2g+2} of binary forms of degree $2g+2$

$$(11.2) \qquad D_{2g+2} = \prod_{1 \le i < j \le 2g+2} (i,j)^2,$$

written in terms of bracket functions (ij) (maximal minors of the $2 \times 2g+2$-matrix whose columns are the projective coordinates of the zeroes of a binary form).

It is known that a hyperelliptic curve of genus g has $\frac{1}{2}\binom{2g+2}{g+1}$ non-vanishing even theta characteristics parametrized by subsets S of cardinality $g+1$ of the set $B_g = \{1, \ldots, 2g+2\}$ up to a complementary set. Let $\theta_S(0)$ be the value at zero of the corresponding theta function $\theta_S(z)$ with theta characteristic S. Thomae formulas are

$$(11.3) \qquad \theta_S^4(0) = d \prod_{i<j, i,j \in S} (i,j) \prod_{i<j, i,j \notin S} (i,j),$$

where d is a constant independent of S (see [**Mu**]). For genus $g = 2$ we get by multiplying

$$(11.4) \qquad \prod_S \theta_S^2(0) = c \prod_{1 \le i < j \le 6} (i,j)^2 = cD_6,$$

for some constant c independent of S (equal to d^{-5} [**Gr**]).

Let $A = S^6((\mathbb{C}^2)^*)^{\mathrm{SL}_2}$ be the algebra of invariants of binary sextics. It is known from 19th century that its subalgebra $A^{(2)}$ of invariants of even degree is the polynomial algebra generated by invariants I_2, I_4, I_6, I_{10} of degrees indicated by the subscript. Under the isomorphism ϕ_2 from (11.1) the ring of invariants $A^{(2)}$ becomes isomorphic to the ring $M_2(\mathrm{Sp}_4(\mathbb{Z}))^{\mathrm{even}}$ of modular forms of even degree on the Siegel space $\mathcal{Z}_2$ with respect to the Siegel group $\mathrm{Sp}_4(\mathbb{Z})$. The invariant I_{10} is the discriminant (11.2) and the corresponding modular form is the Igusa's χ_{10} equal to the square of a cusp form Δ_5 of weight 5 with a nontrivial character (see [**Fr**]). The latter is equal to the product of all even theta constants $\theta_S(0)$ (considered as functions on $\mathcal{Z}_2$). Thus

$$(11.5) \qquad \chi_{10} = \prod_S \theta_S^2(0) = cD_6$$

under the isomorphism $S^6(\mathbb{C}^2)^*)^{\mathrm{SL}_2} \cong M_2(\mathrm{Sp}_4(\mathbb{Z}))^{\mathrm{even}}$.

In genus 3 we obtain

$$\prod_S \theta_S^8(0) = c \prod_{1 \le i < j \le 8} (i,j)^{30} = cD_8^{15},$$

This suggests that there exists a modular form Θ on the Siegel space $\mathcal{Z}_3$ with a nontrivial character such that the restriction of Θ to the locus of periods of hyperelliptic curves corresponds to the invariant D_8 of degree 14 of binary forms of degree 8 under the isomorphism ϕ_3.

Under the Veronese map $\mathbb{P}^1 \to \mathbb{P}^{g+1}$ the variety P_1^{2g+2} becomes isomorphic to the subvariety R_g^{2g+2} of the GIT-quotient $P_g^{2g+2} = (\mathbb{P}^{g+1})^{2g+2}//\mathrm{SL}_2$. The image $(R_g^{2g+2})^o$ of U_1^{2g+2} in R_g^{2g+2} parametrizes orbits of sets of distinct $2g+2$ points lying on the Veronese curve of degree $g+1$. The functions $\theta_S(0)^4$ map $\mathcal{Hyp}_g^o$ to the projective space $\mathbb{P}^N$, where $N = \binom{2g+2}{g+1}/2 - 1$. The image lies in the subspace of dimension $N_1 = \frac{1}{2}\binom{2g+2}{g+1}/(g+2) - 1$ (the number of linear independent fourth powers of theta constants). On the other hand the bracket functions $(i_1, \ldots, i_{g+1})$ on the space P_g^{2g+2} (maximal minors of the $(g+1) \times (2g+2)$- matrix of projective coordinates of points from a point set) define a map to the space $\mathbb{P}^{N_1-1}$ (the number N_1 is equal to the number of standard tableaux defining a basis in the space of bracket functions). Using the isomorphism

$$(R_g^{2g+2}) \cong \overline{Hyp}_g(2)$$

Thomae's formulas imply that the linear systems generated by the functions $\theta_S^4(0)$ coincide with the linear system generated by bracket functions

$$(i_1, \ldots, i_{g+1})(j_1, \ldots, j_{g+1})$$

formed by two complementary ordered subsets of B_g. Observe that this function vanishes on the set of points in P_{g+1}^{2g+2} representing the point sets $(p_1, \ldots, p_{2g+2})$ such that the subset $S = p_{i_1}, \ldots, p_{i_{g+1}}$ or $\bar{S} = p_{j_1}, \ldots, p_{j_{g+1}}$ is linearly dependent in $\mathbb{P}^{g+1}$. If the point set lies on a Veronese curve, then this could happen only if $p_i = p_j$ for $i, j \in S$ or $i, j \in \bar{S}$. This agrees with Thomae's formula (11.3). We would like to find an analogue of Thomae's formula for non-hyperelliptic curves of genus 3, i.e. nonsingular plane quartics. In the space P_3^8 consider the closed codimension 3 subvariety S_3^8 equal to the closure of orbits of an ordered base locus of a general net of quadrics (the variety of self-associated point sets, [**DO**]). Let $(S_3^8)^o$ be the open subset of $P_3^8 \setminus R_3^8$ whose complement consists of points represented by point sets $(p_1, \ldots, p_8)$ such that no four points lie in a plane.

Let $\mathcal{N}_3$ be the moduli space of nets of quadrics in $\mathbb{P}^3 = \mathbb{P}(W)$, viewed as as the GIT-quotient of the Grassmannian $G(3, S^2W^*)$ by the group SL_4. According to [**Wa**], a net of quadrics is unstable if and only if all quadrics in the net are singular. So these nets must be excluded. Strictly semi-stable points correspond to nets of quadrics whose Hessian curve contains a multiple component or tacnodes. There is only one minimal closed orbit of dimension 0. It is represented by nets of quadrics through a twisted cubic. Its Hessian curve is a double conic.

A net of quadrics is called *regular* if its Hessian curve is nonsingular. A point set from S_3^8 defines a regular net of quadrics if and only if it belongs to $(S_3^8)^o$ (see [**DO**]). Obvioulsy a regular net is a stable net. Let $(\mathcal{N}_3)^{\mathrm{reg}}$ be the open subset of $\mathcal{N}_3$ parametrizing orbits of regular nets of quadrics. Assigning to a point set the linear system of quadrics passing through this set we obtain a regular map

(11.6) $$\Phi : (S_3^8)^o \to (\mathcal{N}_3)^{\mathrm{reg}}$$

It is a Galois cover with the Galois group isomorphic to the symmetric group Σ_8. Taking the Hessian curve we get a regular map

$$h : \mathcal{N}_3^{\mathrm{reg}} \to |\mathcal{O}_{\mathbb{P}^2}(4)|^{\mathrm{ns}} /\!/ \mathrm{SL}_3.$$

It is a finite cover of degree 36 equal to the number of even theta characteritics. The composition map

$$h \circ \Phi : (S_3^8)^o \to |\mathcal{O}_{\mathbb{P}^2}(4)|^{\mathrm{ns}} / \mathrm{SL}_3$$

is a finite map of degree 8!36. This number is the order of the group $\mathrm{Sp}_6(2)$ and it is not a coincidence. This group acts on $(S_3^8)^o$ by Cremona transformations and the map $h \circ \Phi$ is a Galois map with Galois group isomorphic to $\mathrm{Sp}_6(2)$ (see [**DO**]). The cover h corresponds to a subgroup of $\mathrm{Sp}_6(2)$ isomorphic to Σ_8.

Let us enlarge $(S_3^8)^o$ by considering the subset $(S_3^8)'$ of $S_3^8 \setminus R_3^8$ represented by point sets $(p_1, \ldots, p_8)$ such that no three points coincide. The locus of points in $(S_3^8)'$ where $p_i = p_j$ for some $i \neq j$, is a codimension 3 subvariety Δ_{ij}. Obviously, the 28 subvarieties Δ_{ij} are disjoint in $(S_3^8)'$. Let $\widetilde{S_3}^8$ be the blow-up of $(S_3^8)'$ along the union of the Δ_{ij}'s. The map Φ extends to a regular map

$$\tilde{\Phi} : \widetilde{S_3}^8 \to \mathcal{N}_3$$

and taking the Hessian curve we get a regular map

$$(11.7) \qquad\qquad h \circ \tilde{\Phi} : \widetilde{S_3}^8 \to |\mathcal{O}_{\mathbb{P}^2}(4)|^{nodal}/\mathrm{SL}_3$$

with target map equal to the moduli space of nodal quartics. A point in an exceptional divisor Δ_{ij} corresponds to a line containing the point $p_i = p_j$ which is a common tangent to all quadrics from the net.

Now we are in business. The locus of singular curves in $|\mathcal{O}_{\mathbb{P}^2}(4)|^{nodal}/\mathrm{SL}_3$ is given by the discriminant invariant polynomial D of degree 27. Its pre-image in $\mathcal{N}_3$ consists of 2 irreducible components Δ_1 and Δ_2. The first component is the locus of zeroes of the invariant I^2 and consists of nets of quadrics containing a quadric of corank 2. The second component is the locus of zeroes of the tacinvariant J and consists of nets of quadrics containing a base point on its Steinerian curve. The pre-image of Δ_1 in $\widetilde{S_3}^8$ is the union of 35 divisors parametrizing points sets with two complementary subsets of 4 linearly dependent points. The preimage of Δ_2 is the union of 28 divisors, the exceptional divisors of the blow-up $\widetilde{S_3}^8 \to (S_3^8)^o$. The number $28 + 35 = 63$ is the number of nonzero vectors in $\mathbb{F}_2^6$. It is not a coincidence (see the explanation in [**DO**]). The pre-image of the zero of the discriminant invariant of quartics is the union of 35 divisors D_{ijkl} and 28 divisors D_{ij}. This is an analogue of (11.5) (taking into account (11.4)).

The map (11.7) is a ramified cover with the branch divisor equal to the variety $V(D)$ of singular quartics. Its ramification divisor is the union of the divisors D_{ijkl} and D_{ij} with ramification of index 2. The stabilizer subgroup of D_{ijkl} (resp. D_{ij}) in Σ_8 is the subgroup $\Sigma_{4,4}$ (resp. $\Sigma_{2,6}$), where $\Sigma_{a,b}$ is the subgroup of permutations preserving the partition of $\{1, \ldots, 8\}$ into the sum of two subsets of cardinalities a and b. The inertia subgroup of D_{ijkl} (resp. D_{ij}) is the group of order 2 generated by the prermutation $(5, 6, 7, 8, 1, 2, 3, 4)$ (resp. (12)). Let $\mathcal{N}_3^{\mathrm{nod}}$ be the subvariety of $\mathcal{N}_3$ parametrizing nets with nodal discriminant curve. The cover (11.7) factors through the cover

$$(11.8) \qquad\qquad h' : \mathcal{N}_3^{\mathrm{nodal}} \to |\mathcal{O}_{\mathbb{P}^2}(4)|^{nodal}/\mathrm{SL}_3$$

It is a ramified cover of degree 36. Its ramification divisor is the divisor Δ_2, the image of the divisors D_{ij}. The map is unramified at the divisor Δ_1. The restriction map $\Delta_1 \to V(D)$ (resp. $\Delta_2 \to V(D)$ is of degree 16 (resp. 10). Note that $36 = 16 + 2 \times 10$. The fibres of the first map are parametrized by nonvanishing even theta characteristics on the normalization of the discriminant curve. The fibres of the second map are parametrized by "honest" theta characteristics on the discriminant curve (see [**Ha**]).

In terms of invariants I^2, J and D this can be interpreted as follows.

THEOREM 11.1.

$$\prod_{even\ \theta} I(C, \theta) = k \cdot (\text{Discriminant}(C))^{10},$$

$$\prod_{even\ \theta} J(C, \theta) = k' \cdot (\text{Discriminant}(C))^{16},$$

where k, k' are constants depending on the way of the normalizations of the discriminant and the quantities $I(C, \theta)$.

PROOF. This is just the usual formula for the norm of an ideal in a finite extension of fields (in our case corresponding to the maps $\Delta_i \to V(D)$). □

It is easy to see that these equalities agree with Salmon's Theorem 10.4.

12. Quadratic line complexes attached to a quartic with an even theta

In Introduction we promised to construct quadratic complexes. We will begin with some general constructions. In the final of the section 12.4, we will construct a special quadratic line complex for a quartic with an even theta. In the general situation, for a plane curve C of degree $n + 1$, for its even θ and for a contravariant $E = E(z_0, z_1, z_2)$ of order $(n - 1)$, where (z_0, z_1, z_2) are contragredient variables for (y_0, y_1, y_2), we will construct a quadratic complex $K(C, \theta, E)$.

Below, the description of the general situation is a product of our collaboration with Igor Dolgachev, this text is his recital.

Let N be a regular net of quadrics in $\mathbb{P}^n$ defined by an even non-vanishing theta characteristic θ. We have the multiplication map

(12.1) $|\theta(1)| \times |\theta(1)| \to |\mathcal{O}_{\mathbb{P}^2}(n)|,$

where C is the Hessian curve. It can be viewed as follows. A point in $|\theta(1)|$ is a hyperplane H in $\mathbb{P}^n$ which defines a contact curve B_H of degree n. If a pair of contact curves B_H, B'_H cut out the divisors $2D_H, 2D_{H'}$ on C, then the divisor $D_H + D_{H'}$ is cut out by a curve $B_{H,H'}$ equal to the image of the pair (H, H') under the map (12.1). Thus the curves $B^2_{H,H'}$ and $B_H B_{H'}$ cut out the same divisor on C, and hence

(12.2) $B^2_{H,H'} = B_H B_{H'} + UC$

for some curve U of degree $2n - (n + 1) = n - 1$.

Recall that the equation for a contact curve B_H is given in (3.2), where we use the coordinates $(u_0, \ldots, u_n)$ of H. The equation (12.2) corresponds to the following beautiful identity of determinants due to Hesse [**He**]

$$\begin{vmatrix} L_{00} & L_{01} & \ldots & L_{0n} & u_0 \\ L_{10} & L_{11} & \ldots & L_{1n} & u_1 \\ . & . & \ldots & . & . \\ L_{n0} & L_{n1} & \ldots & L_{nn} & u_n \\ u_0 & u_1 & \ldots & u_n & 0 \end{vmatrix} \times \begin{vmatrix} L_{00} & L_{01} & \ldots & L_{0n} & v_0 \\ L_{10} & L_{11} & \ldots & L_{1n} & v_1 \\ . & . & \ldots & . & . \\ L_{n0} & L_{n1} & \ldots & L_{nn} & v_n \\ v_0 & v_1 & \ldots & v_n & 0 \end{vmatrix} -$$

$$\begin{vmatrix} L_{00} & L_{01} & \ldots & L_{0n} & u_0 \\ L_{10} & L_{11} & \ldots & L_{1n} & u_1 \\ . & . & \ldots & . & . \\ L_{n0} & L_{n1} & \ldots & L_{nn} & u_n \\ v_0 & v_1 & \ldots & v_n & 0 \end{vmatrix}^2 = \begin{vmatrix} L_{00} & L_{01} & \ldots & L_{0n} \\ L_{10} & L_{11} & \ldots & L_{1n} \\ . & . & \ldots & . \\ L_{n0} & L_{n1} & \ldots & L_{nn} \end{vmatrix} \times U,$$

where U is a polynomial of entries of the matrices, participating in the identity,

$$U = U(M_0, M_1, M_2, y_0, y_1, y_2, u_0, u_1, ..., u_n, v_0, v_1, ..., v_n),$$

U depends quadratically on $u_0, u_1, ..., u_n$ and on $v_0, v_1, ..., v_n$. Moreover, we can express the latter dependence through Plücker's determinants $p_{ij} = u_i v_j - u_j v_i$, $i \neq j$. Thus

(12.3) $$U = U(M_0, M_1, M_2, y_0, y_1, y_2, p_{ij}).$$

The dependence of V on y_0, y_1, y_2 is of degree $(n - 1)$, therefore we have now a family of quadratic line complexes parameterized by the projective plane.

THEOREM 12.1. *Let $(Q_y)_{y \in \mathbb{P}^2}$ be a regular net of quadrics in $\mathbb{P}^n$ with Hessian curve C . For any $y \in \mathbb{P}^2 \setminus C$ let G_y be the subset of the Grassmannian of codimension 2 subspaces in $\mathbb{P}^n$ which are tangent to the quadric Q_y. Then, considered as the Grassmannian of lines in the dual projective space, it is a quadratic line complex given by the equation (12.3).*

PROOF. Let $\mathbb{P}^n = \mathbb{P}(V)$, then the coordinates $(u_0, \dots, u_n)$ of planes in $\mathbb{P}^n$ is a basis in V, so the Plücker coordinates $p_{ij} = u_i v_j - u_j v_i$ are coordinates in $\Lambda^2 V^*$. Thus the equation (12.3) is a family of quadric hypersurfaces in the Grassmannian of lines in the dual projective space $\mathbb{P}(V^*)$. By the duality it is the same as a quadric hypersurface in the Grassmannian of codimension 2 subspaces in $\mathbb{P}^n(V)$. Let $L = H \cap H'$, where H, H' are hyperplanes with coordinates $\bar{u} = (u_0, \dots, u_n)$ and $\bar{v} = (v_0, \dots, v_n)$. A hyperplane H is tangent to the quadric Q_y if and only if equation (3.2) holds. A codimension 2 subspace $H \cap H'$ is tangent to Q_y if and only if the corresponding line is tangent to the dual quadric $\check{Q}_y : \sum A_{ij}(y) \xi_i \xi_j = 0$, where $A_{ij}(y)$ are cofactors of the matrix $(L_{ij}(y))$. The condition for this is

$$\det \begin{pmatrix} \sum A_{ij}(y) u_i u_j & \sum A_{ij}(y) u_i v_j \\ \sum A_{ij}(y) u_i v_j & \sum A_{ij}(y) v_i v_j \end{pmatrix} = 0.$$

The left-hand side is the equation $U(y) = 0$. $\square$

EXAMPLE 12.2. If $n = 2$, we get from [**He**]

$$U = L_{00} p_{12}^2 + L_{11} p_{02}^2 + L_{22} p_{01}^2 + 2 L_{12} p_{02} p_{01} + 2 L_{02} p_{12} p_{01} + 2 L_{01} p_{12} p_{02}.$$

Identifying the Plücker coordinates with the coordinates in the plane where the conics of the net, we see that this equation is the equation of the original net of conics.

EXAMPLE 12.3. The equation of the quadratic complex of tangent lines to a quadric is called in classic literature a *complex equation of a quadric* (see [**Je**], n° 46, 88). If a quadric Q is given by an equation $a_0 T_0^2 + \dots + a_3 T_3^2 = 0$, then the equation is

(12.4) $$\sum_{0 \leq i < j \leq 3} a_i a_j p_{ij}^2 = 0.$$

Its Hessian divisor is given by $(4 a_0 a_1 a_2 a_3 a_4 t_0^2 - t_1^2)^3$ in coordinates t_0, t_1 on the pencil generated by the Klein quadric $p_{12} p_{34} - p_{13} p_{24} + p_{14} p_{23} = 0$ and the quadric (12.4). In the GIT-moduli space of pencils of quadrics in $\mathbb{P}^5$ (or, equivalently, moduli space of binary sextics), this quadratic complex represents the unique closed semi-stable orbit.

How can we get one complex produced with the help of an invariant process? If we take a contravariant E, then the convolution of U with E removes the dependence on y, and we get a quadratic complex

$$K(M_0, M_1, M_2, E) = <U, E>.$$

Let us consider the case of plane quartics with more details. According to [**Sa2**], n° 300, a quartic admits some contravariant conics, the simplest of them has coefficents as polynomials of the fourth order in the coefficients of the original curve C. It is obtained by contracting the equation of the quartic C with the Clebsch contravariant sextic (see [**DK**]). Let $E = \sum e_{ij} z_i z_j$ be this simplest contravariant conic. The convolution of E and U is

$$K = (\sum e_{ij} \partial^2 / \partial y_j \partial y_j) V.$$

Thus we get a quadratic complex

$$K = K(M_0, M_1, M_2, p_{01}, p_{02}, p_{03}, p_{12}, p_{13}, p_{23})$$

which is determined by (C, θ, E). We used the simplest contravariant conic here, because for this case, using the data of Example 2.7, it is not hard to check that the polynomial K is not identically zero.

Instead of E, for the quartic case, one can take the irrational contravariant conic mentioned in 7.6.

REMARK 12.4. We have the Toeplitz cubic line complex $T(p_{kl})$ from (2.3) . Moreover, six quantities s_{ij} from (4.3) define a linear complex. The convolution of the linear complex with the Toeplitz cubic line complex is a quadratic line complex L. More explicitly,

$$L = (\sum s_{ij} \frac{\partial}{\partial p_{ij}}) T(p),$$

$$L = s_{01} \frac{\partial T}{\partial p_{01}} + s_{02} \frac{\partial T}{\partial p_{02}} + s_{03} \frac{\partial T}{\partial p_{03}} + s_{12} \frac{\partial T}{\partial p_{12}} + s_{31} \frac{\partial T}{\partial p_{31}} + s_{23} \frac{\partial T}{\partial p_{23}}.$$

Thus we get a quadratic complex

$$L = L(M_0, M_1, M_2, p_{01}, p_{02}, p_{03}, p_{12}, p_{13}, p_{23})$$

which is determined by (C, θ).

A quadratic complex has a quadratic invariant, see section 13 below. Taking the value of the quadratic invariant for L, we get an irrational invariant of degree 6, one can use this invariant for a definition of X from section 6.

The coefficients of L are calculable for the case of example 2.7. For the case of example 5.1, the equation of the quadratic complex L is not too long, it is given in Appendix 3.

REMARK 12.5. The final questions are of Igor Dolgachev. Recall that a quadratic complex is defined by a pencil of quadrics in $\mathbb{P}^5$ and hence defines a set of 6 points in $\mathbb{P}^1$ (the set of singular quadrics in the pencil). This in its turn defines a genus 2 curve. Thus we obtain non-trivial rational maps from the moduli space of nets of quadrics in $\mathbb{P}^3$ (or pairs (C, θ)) to the moduli space of quadratic line complexes (or, equivalently, the moduli space of binary sextics). Are the maps dominant? What are their fibers?

13. Supplement. The quadratic invariant of a line complex

We would like to define the notion of quadratic invariant for a line complex and the notion of bilinear invariant for two line complexes.

Let us begin with some general considerations. Let $x_1, x_2, ..., x_n$ be a set of variables, $u_1, u_2, ..., u_n$ be the set of the dual variables for $x_1, x_2, ..., x_n$, that is we have the following convolution identities $< x_i, u_j >= \delta_j^i$. Let

$$Q = Q(x_1, x_2, ..., x_n) = \sum_{i,j} a_{ij} x_i x_j$$

be a non-degenerate quadratic form,

$$D(Q) = \mathrm{Discriminant}(Q),$$

$$\widehat{Q} = \widehat{Q}(u_1, u_2, ..., u_n) = \sum_{i,j} A_{ij} u_i u_j$$

be the reciprocal form for Q. Quadratic form Q determines a Q-Laplacian, that is the following differential operator

$$\Delta_Q = \widehat{Q}(\partial/\partial x_1, ..., \partial/\partial x_n) = \sum_{i,j} A_{ij} \frac{\partial}{\partial x_i} \frac{\partial}{\partial x_j}.$$

It is clear that

$$\Delta_Q(Q) = 2nD(Q).$$

For any homogeneous polynomial $M = M(x_1, x_2, ..., x_n)$, it is not hard to verify that

$$\Delta_Q(MQ) = 2(n + 2\deg(M))D(Q)M + Q\Delta_Q(M).$$

Hence

$$\Delta_Q(Q^p) = 2p(n + 2p - 2)D(Q)Q^{p-1},$$

and

(13.1) $\Delta_Q(MQ^p) = Q^p \Delta_Q(M) + 2p(n + 2p - 2 + 2\deg(M))D(Q)MQ^{p-1}.$

We say that two homogeneous polynomials

$$G_m(x_1, x_2, ..., x_n) \quad \text{and} \quad H_m(x_1, x_2, ..., x_n)$$

of the same order m are Q-equivalent, if there exists a form J_{m-2} of order $m - 2$ such that

$$G_m - H_m = J_{m-2}Q.$$

A homogeneous polynomial $H(x_1, x_2, ..., x_n)$ is called Q-harmonic if

$$\Delta_Q(H) = 0.$$

LEMMA 13.1. *If a homogeneous polynomial is Q-harmonic and Q-equivalent to zero, then the polynomial is equal to zero.*

PROOF. Let $M = M(x_1, x_2, ..., x_n)$ be a form, $d = \deg(M)$ be its order. We have to show that if $\Delta_Q(MQ^p) = 0$, $p > 0$ and Q is not a divisor of M, then $M = 0$.
Identity (13.1) indicates that if MQ^p is harmonic, then Q is a divisor of M. $\square$

LEMMA 13.2. *For every homogeneous polynomial M, there exists a Q-harmonic homogeneous polynomial $H = H_Q(M)$ such that H and M are Q-equivalent. For a given G, such a Q-harmonic H is unique.*

PROOF. The last assertion follows from 13.1.

We denote $d = \deg(M)$. If a form H of order d is written in the following form

$$H = M + M_1 Q + M_2 Q^2 + M_3 Q^3 + ... + M_p Q^p + ...,$$

where $\deg(M_p) = d - 2p, 1 \leq p \leq [d/2]$, then $\Delta_Q H$ admits the following writing

$$\Delta_Q H = \Delta_Q(M) + 2D(Q)(n + 2d - 4)M_1 +$$

$$Q(\Delta_Q(M_1) + 4D(Q)(n + 2d - 6)M_2) +$$

$$Q^2(\Delta_Q(M_2) + 6D(Q)(n + 2d - 8)M_3) + ... +$$

$$Q^{p-1}(\Delta_Q(M_{p-1}) + 2pD(Q)(n + 2d - 2p - 2)M_p) + ...$$

Therefore if we take

$$M_1 = -\frac{\Delta_Q(M)}{2D(Q)(n + 2d - 4)},$$

$$M_p = -\frac{\Delta_Q(M_{p-1})}{2D(Q)(n + 2d - 2p - 2)}, \quad 2 \leq p \leq [d/2].$$

then we obtain a harmonic form H which is Q-equivalent to M.   □

For any form $M(x_1, x_2, ..., x_n)$ we will denote by $N_Q(M)(u_1, u_2, ..., u_n)$ the following form

$$N_Q(M)(u_1, ..., u_n) = M(\partial \widehat{Q}/\partial u_1, ..., \partial \widehat{Q}/\partial u_n).$$

A quadratic Q-invariant of the form M is the following convolution

(13.2) $I_Q(M) = < H_Q(M), N_Q(H_Q(M)) >,$

where $H(M)$ is the harmonic representative of M mentioned in 13.2.

Let us consider a line complex defined by equation $M = 0$, where

$$M = M(p_{01}, p_{02}, p_{03}, p_{12}, p_{31}, p_{23})$$

is a form of the Plücker coordinates. Let $q_{01}, q_{02}, q_{03}, q_{12}, q_{31}, q_{23}$ be the dual Plücker coordinates. The Plücker quadric is

$$Q = p_{01}p_{23} + p_{02}p_{31} + p_{03}p_{12},$$

its reciprocal is defined by

$$-16\widehat{Q} = q_{01}q_{23} + q_{02}q_{31} + q_{03}q_{12}.$$

Formula (13.2) defines the quadratic invariant of the line complex.

The bilinear combined invariant of two line complexes of the same degree is the convolution $< H_Q(M), H_{\widehat{Q}}(N) >$, where $M(p) = 0$ is the equation of one of the complexes in the Plücker coordinates, $N(q) = 0$ is the equation of other complex in the dual Plücker coordinates.

REMARK 13.3. The construction of the harmonic representative for an equation of a line complex was written down in vol. II, Chapter VII, n°350, n°351 of Salmon's book [**Sa1**]. Salmon uses words *normal form of the complex equation*.

With my point of view, this place in the book is a germ of the Hodge-de Rham theory of harmonic forms. Indeed, 13.2 is an analogue of the Hodge-de Rham theorem. Moreover, the proof of 13.2 gives us identity

$$H_Q(M) - M = QP,$$

where
$$P = M_1 + M_2 Q + M_3 Q^2 + \ldots + M_p Q^{p-1} + \ldots.$$
One can rewrite the identity as
$$H_Q(M) - M = Q\Delta_Q(N),$$
where
$$N = N_1 + N_2 Q + N_3 Q^2 + \ldots + N_p Q^{p-1} + \ldots,$$
$$N_1 = -\frac{M}{2D(Q)(n+2d-4)},$$
$$N_p = \frac{M_p - \Delta_Q(N_{p-1})}{2D(Q)(n+2d-2p-2)}, \quad 2 \le p \le [d/2].$$

We see that N is produced from M with the help of the action of an operator G, $N = G(M)$, this G is an analogue of the Green operator.

14. Appendix

1. Formula of $I_3(C)$ for the case of Example 2.7.

$$I_3 =$$

$$-8bcegpq+$$
$$64a^2 f^2 r^2+$$
$$32afr(bgp + ceq)+$$
$$9(b^2 g^2 p^2 + c^2 e^2 q^2)-$$
$$64afr(ber + cfp + agq)-$$
$$10(cpb^2 q^2 + bec^2 g^2 + gqe^2 p^2)+$$
$$14(c^2 f^2 p^2 + a^2 g^2 q^2 + b^2 e^2 r^2)-$$
$$10(afc^2 g^2 + fre^2 p^2 + arb^2 q^2)+$$
$$32(abeggr + acfgpq + bcefpr)-$$
$$24afr(bcg + ceg + bpq + epq + egp + bcq)+$$
$$5(c^4 f^2 + a^2 g^4 + f^2 p^4 + e^4 r^2 + b^4 r^2 + a^2 q^4)+$$
$$24afr(a(q^2 + g^2) + f(c^2 + p^2) + r(b^2 r + e^2))+$$
$$6(bfgp^3 + abpg^3 + efqc^3 + aceq^3 + gprb^3 + cqre^3)+$$
$$2(aepg^3 + bfqc^3 + bfqp^3 + aepq^3 + cgrb^3 + cgre^3)-$$
$$2((cp + ar)e^2 g^2 + (gq + fr)b^2 c^2 + (be + af)p^2 q^2)-$$
$$12(pf^2 c^3 + cf^2 p^3 + qa^2 g^3 + br^2 e^3 + ga^2 q^3 + r^2 b^3)-$$
$$6(cp(b^2 g^2 + e^2 q^2) + be(g^2 p^2 + c^2 q^2) + gq(b^2 p^2 + c^2 e^2))+$$
$$5(b^2 c^2 g^2 + c^2 e^2 g^2 + e^2 g^2 p^2 + b^2 c^2 q^2 + b^2 p^2 q^2 + e^2 p^2 q^2)+$$
$$8(bcepg^2 + begqc^2 + cgpqb^2 + cgpqe^2 + begqp^2 + bcepq^2)-$$
$$14(bcfgp^2 + efpqc^2 + abpqg^2 + acegq^2 + egprb^2 + bcqre^2)+$$
$$18(abgpq^2 + bfgpc^2 + aceqg^2 + cefqp^2 + bgpre^2 + ceqrb^2)+$$
$$14(af(g^2 p^2 + c^2 q^2) + ar(b^2 g^2 + e^2 q^2) + rf(b^2 p^2 + c^2 e^2))-$$
$$2(fcp(eg + bq)(c + p) + agq(bc + ep)(g + q) + ebr(cg + pq)(b + e))+$$
$$10((rb^2 - aq^2)bq(p - c) + (fc^2 - ag^2)cg(e - b) + (fp^2 - re^2)pe(g - q))-$$
$$12(a(cfp + ber)(g^2 + q^2) + f(ber + agq)(c^2 + p^2) + r(cfp + agq)(e^2 + b^2)).$$

2. Equation of the Gundelfinger quartic surface.

$$\Theta(u_0, u_1, u_2, u_3) =$$

$$u_0^4 + (fr - gq)^2 u_1^4 + (ar - cp)^2 u_2^4 + (af - be)^2 u_3^4 -$$

$$2((g+q)u_1 + (c+p)u_2 + (b+e)u_3)u_0^3 +$$

$$2((fr-gq)(g+q)u_1^3 + (ar-cp)(c+p)u_2^3 + (af-be)(b+e)u_3^3)u_0 +$$

$$2(cq - gp - br + er)((fr - gq)u_1^2 - (af - cp)u_2^2)u_1 u_2 +$$

$$2(ce - ag - bp + aq)((ar - cp)u_2^2 - (af - be)u_3^2)u_2 u_3 +$$

$$2(bg - cf + fp - eq)((fr - gq)u_1^2 - (af - be)u_3^2)u_1 u_3 +$$

$$((g+q)^2 - 2(fr - gq))u_0^2 u_1^2 +$$

$$((c+p)^2 - 2(ar - cp))u_0^2 u_2^2 +$$

$$((b+e)^2 - 2(af - be))u_0^2 u_3^2 +$$

$$((cq - gp - br + er)^2 - 2(fr - gq)(ar - cp))u_1^2 u_2^2 +$$

$$((bg - cf + fp - eq)^2 - 2(af - be)(fr - gq))u_1^2 u_3^2 +$$

$$((ce - ag - bp + aq)^2 - 2(ar - cp)(af - be))u_2^2 u_3^2 +$$

$$2(pq - cg + 2(gp + cq) - (b+e)r)u_0^2 u_1 u_2 +$$

$$2(eg - bq + 2(bg + eq) - (c+p)f)u_0^2 u_1 u_3 +$$

$$2(bc - ep + 2(ce + bp) - (g+q)a)u_0^2 u_2 u_3 +$$

$$2((er - gp)(g+q) + (cq - br)(g - q) - (fr - qg)(c - p))u_0 u_1^2 u_2 +$$

$$2((fp - eq)(g+q) + (fc - bg)(g - q) - (fr - gq)(b - e))u_0 u_1^2 u_3 +$$

$$2((br - cq)(c+p) + (gp - er)(c - p) - (ar - cp)(g - q))u_0 u_1 u_2^2 +$$

$$2((aq - pb)(c+p) + (ag - ce)(c - p) + (ar - cp)(b - e))u_0 u_2^2 u_3 +$$

$$2((cf - bg)(b+e) + (eq - fp)(b - e) + (af - be)(g - q))u_0 u_1 u_3^2 +$$

$$2((ag - ec)(b+e) + (aq - bp)(b - e) + (af - be)(c - p))u_0 u_2 u_3^2 +$$

$$2((fp - eq)(gp - er) + (cf - bg)(br - cq) - 2fr(ce + bp) +$$

$$(ceq - bgp)(g - q) + (cfp - agq + ber + afr)(g+q))u_1^2 u_2 u_3 +$$

$$2((ce - ag)(gp - er) + (aq - bp)(cq - br) - 2ar(bg + eq) +$$

$$(bgp - ceq)(c - p) + (agq - cfp + ber + afr)(c+p))u_1 u_2^2 u_3 +$$

$$2((fp - eq)(aq - bp) + (bg - cf)(ag - ce) - 2af(gp + cq) +$$

$$(ceq - bgp)(b - e) + (agq - ber + cfp + afr)(b+e)))u_1 u_2 u_3^2 +$$

$$2(8(ber + cfp + agq) - a(q+g)^2 - f(c+p)^2 - r(b+e)^2 +$$

$$(cg(b+e) + bq(c+p) + ep(g+q)) - 3(bgp + ceq) - 12afr)u_0 u_1 u_2 u_3.$$

3. Quadratic form defining the quadratic line complex of remark 12.4 for the case of Example 5.1.

$$L = r(bg - cr)^2(b^2 - g^2)p_{02}^2 -$$

$$r(bc - gr)^2(b^2 - c^2)p_{01}^2 -$$

$$r(br - cg)^2(c^2 - g^2)p_{03}^2 +$$

$$r(3bc^2g - b^2cr - 2c^3r - cg^2r - bgr^2 + 2cr^3)(b^2 - g^2)p_{01}p_{03} +$$

$$r(2b^3r - 3b^2cg + bc^2r + bg^2r + cgr^2 - 2br^3)(c^2 - g^2)p_{01}p_{02} -$$

$$(3bcg^2 - b^2gr - c^2gr - 2g^3r - bcr^2 + 2gr^3)(b^2 - c^2)p_{02}p_{03} -$$

$$2r^2(2b^2c^2 - b^2g^2 - c^2g^2 - b^2r^2 - c^2r^2 + 2g^2r^2)p_{01}p_{23} +$$

$$2r^2(b^2c^2 - 2b^2g^2 + c^2g^2 + b^2r^2 - 2c^2r^2 + g^2r^2)p_{02}p_{31}+$$
$$2r^2(b^2c^2 + b^2g^2 - 2c^2g^2 - 2b^2r^2 + c^2r^2 + g^2r^2)p_{03}p_{12}-$$
$$(b^3c^2g + b^2c^3r + 2b^2cg^2r - 2bc^2gr^2 + bg^3r^2 - 4b^2cr^3 - 3cg^2r^3 + 4cr^5)p_{01}p_{12}-$$
$$(b^3cg^2 + 2b^2c^2gr + b^2g^3r + bc^3r^2 - 2bcg^2r^2 - 4b^2gr^3 - 3c^2gr^3 + 4gr^5)p_{02}p_{12}-$$
$$(b^2c^3g + b^3c^2r + 2bc^2g^2r - 2b^2cgr^2 + cg^3r^2 - 4bc^2r^3 - 3bg^2r^3 + 4br^5)p_{01}p_{31}-$$
$$(b^2cg^3 + b^3g^2r + 2bc^2g^2r - 2b^2cgr^2 + c^3gr^2 - 3bc^2r^3 - 4bg^2r^3 + 4br^5)p_{02}p_{23}-$$
$$(bc^3g^2 + 2b^2c^2gr + c^2g^3r + b^3cr^2 - 2bcg^2r^2 - 3b^2gr^3 - 4c^2gr^3 + 4gr^5)p_{03}p_{31}-$$
$$(bc^2g^3 + 2b^2cg^2r + c^3g^2r + b^3gr^2 - 2bc^2gr^2 - 3b^2cr^3 - 4cg^2r^3 + 4cr^5)p_{03}p_{23}.$$

References

[Be1] A. Beauville, *Variétés de Prym et Jacobiennes intermédiaires*, Ann. Scient. Éc. Norm. Sup., (4), **10** (1977), 309–391.

[Be2] A. Beauville, Determinantal hypersurfaces, Ann. Scient. Michigan Math. J., **48** (2000), 39–64.

[DO] I. Dolgachev, D. Ortland, *Point sets in projective spaces and theta functions*, Astérisque, v. **165**, 1988.

[DK] I. Dolgachev, V. Kanev, *Polar covariants of plane cubics and quartics*, J. Algebra, **98** (1993), 216–301.

[Ed1] W. Edge, *Notes on a net of quadric surfaces. IV. Combinantal invariants of low order*, Proc. London Math. Soc., (2), **47** (1941), 123–141.

[Ed2] W. Edge, *The Klein group in three dimensions*, Math. Ann., **15** (1979).

[Fa] D. Fano, *Superficie di 4° ordine contenenti una rete di curve di genere 2*, Pont. Acad. Sci. Comment. **7** (1943), 185–205.

[Fra] W. Frahm, *Bemerkung über das Flächennetz zweiter Ordnung*, Mathematische Annalen, **7** (1874), 635–638.

[Fr] E. Freitag, *Siegelsche Modulfunktionen*, Springer-Verlag, 1983.

[Gi1] M. Gizatullin, *On some tensor representations of the Cremona group of the projective plane*, In London Mathematical Society, Lecture Note Series, v. **264**, 1998, 111–150.

[Gi2] M. Gizatullin, *Bialgebra and geometry of plane quartics*, Asian Journal of Mathematics, **5**, No. 3, (2001), 3–50.

[Gi3] M. Gizatullin, *The groups of decomposition, inertia and ramification in birational geometry*, Algebraic geometry and its Applications, Aspects of mathematics, vol. **E25** , (1994), 39–45.

[Gr] D. Grant, *A generalization of Jacobi's derivative formula to dimension 2*, J. für Reine und Angew. Math., **392** (1988), 125–136.

[Gu] S. Gundelfinger, *Ueber das simultane System von drei ternären quadratischen Formen*, J. für Reine und Angew. Math., **80** (1875), 73–85.

[Ha] J. Harris, *Theta characteristics on algebraic curves*, Trans. A.M.S., **271** (1982), 611–638.

[He] L. Hesse, *Ueber die Determinanten und ihre anwendungen in der Geometrie, insbesondere auf Curven vierten Ordnung*, J. für Reine und Angew. Math., **49** (1855), 243–264.

[Je] C. Jessop, *The line complex*, Cambridge Univ. Press, 1903 (reprinted by Chelsea, 1969).

[Kl] F. Klein, *Zur Theorie der Abel'schen Functionen*, Mathematische Annalen, **36** (1889/90), , or *Gesammelte mathematische Abhandlungen*, Bd. 3, Berlin, Springer, 1921-23, 388–474.

[Mu] D. Mumford, *Lectures on Theta, II*, Progress in Math. v. **43**, Birkhüser, 1984.

[Ri] B. Riemann, *Zur Theorie der Abel'schen Functionen*, in *Gesammelte mathematische Werke*, 2. Auflage, 1892, 487–504.

[Sa1] G. Salmon, *Analytic geometry of three dimensions*, Vol. I, 1870,vol. II, Chelsea Publ. Co., 1958, 1963.

[Sa2] G. Salmon, *A treatise on the higher plane curves*, Hodges, Foster and Figgis, Dublin 1879 (reprinted by Chelsea Publ. Co., 1960).

[Sa3] G. Salmon, *A treatise on conic sections*, London, 1878.

[Toe] E. Toeplitz, *Über ein Flächennetz zweiter Ordnung*, Mathematische Annalen, **11** (1877), 365–397.

[Wa] C. T.C. Wall, *Nets of quadrics,and theta characteritics of singular curves*, **289**, Trans. Royal Soc. London, Ser. A, **289** (1978), 229–269.

Chair of Mathematical Analysis, The Institute of Mathematics, Physics and Informatics, Samara State Pedagogical University, Maxim Gorki Street, building 65/67, 443099 Samara, RUSSIA

E-mail address: gizmarat@yandex.ru

Contemporary Mathematics
Volume **422**, 2007

A Rationality Criterion for Projective Surfaces - partial solution to Kollár's Conjecture

JongHae Keum

Dedicated to Igor Dolgachev on his sixtieth birthday

ABSTRACT. Kollár's conjecture states that a complex projective surface S with quotient singularities and with $H^2(S, \mathbb{Q}) \cong \mathbb{Q}$ should be rational if its smooth part S^0 is simply connected.

We confirm the conjecture under the additional condition that either S has at least 5 singular points or the exceptional divisor in a minimal resolution of S has at most 3 components over each singular point of S.

1. Introduction

In his study of Seifert structures on simply connected rational homology spheres, János Kollár suggested the following conjecture ([**Ko**] Conjecture 4.17, or Conjecture 9.2 its differential geometric equivalent.):

CONJECTURE 1.1. *Let S be a projective surface with quotient singularities such that*

(1) $H^2(S, \mathbb{Q}) \cong \mathbb{Q}$,
(2) $\pi_1(S^0) = \{1\}$, *where S^0 is its smooth part.*

Then S is rational.

In this paper we confirm the conjecture under the additional condition that either S has at least 5 singular points or the exceptional divisor in a minimal resolution of S has at most 3 components over each singular point of S. More precisely, we prove the following:

THEOREM 1.2. *Let S be a projective surface with quotient singularities such that*

(1) $H^2(S, \mathbb{Q}) \cong \mathbb{Q}$,
(2) $H_1(S^0, \mathbb{Z}) = 0$,
(3) *either S has at least 5 singular points or the inverse image $f^{-1}(p)$ has at most 3 components for each singular point p in S, where $f : S' \to S$ is a minimal resolution.*

2000 *Mathematics Subject Classification.* Primary: 14J26, 14J27, 14J29.
Key words and phrases. projective surface, quotient singularity, rationality.
Research supported by KOSEF grant R01-2003-000-11634-0.

Then S is rational.

Note that the condition (2) $H_1(S^0, \mathbb{Z}) = 0$ is weaker than $\pi_1(S^0) = \{1\}$.

We also remark that if S is non-singular and satisfies the conditions (1) and (2) of Theorem 1.2, then S is either the complex projective plane $\mathbb{CP}^2$ or a surface of general type with $q = p_g = 0$, $3c_2 = c_1^2 = 9$, so called a fake projective plane. Recently G. Prasad and S.-K. Yeung have shown that no fake projective plane with $H_1(S, \mathbb{Z}) = 0$ exists [**PY**].

Throughout this paper, we work over the field $\mathbb{C}$ of complex numbers.

Acknowledgements. I like to thank János Kollár for useful conversations through e-mails. I am also grateful to the referee for many helpful comments.

2. Preliminaries

LEMMA 2.1. *Let V be an irreducible reduced complex analytic space, and V^0 its smooth part. Let $f : V' \to V$ be a resolution of singularities. Then*

(1) *The inclusion $V^0 \subset V'$ gives surjective homomorphisms*

$$\pi_1(V^0) \to \pi_1(V'), \quad H_1(V^0, \mathbb{Z}) \to H_1(V', \mathbb{Z}).$$

(2) *If f has connected fibres, it induces surjective homomorphisms*

$$\pi_1(V') \to \pi_1(V), \quad H_1(V', \mathbb{Z}) \to H_1(V, \mathbb{Z}).$$

PROOF. It suffices to prove the assertions for fundamental groups.

The first assertion follows from the fact that the complement $V' \setminus V^0$ has real codimension ≥ 2.

If f has connected fibres, a loop in V can be lifted to a loop in V'. □

When the resolution is projective, the condition in (2) is always satisfied by Zariski's Main Theorem.

The following also can be proved by a standard argument. For a proof, we refer the reader, e.g. to [**Ko**] Proposition 4.14.

PROPOSITION 2.2. *Let S be a projective surface with quotient singularities such that $H^2(S, \mathbb{Q}) \cong \mathbb{Q}$ and $H^1(S, \mathbb{Q}) = 0$. Let $f : S' \to S$ be a resolution of singularities. Then*

(1) $H^1(S, \mathcal{O}_S) = H^1(S', \mathcal{O}_{S'}) = 0.$
(2) $H^2(S, \mathcal{O}_S) = H^2(S', \mathcal{O}_{S'}) = 0.$

Recall the definition of *Kodaira (logarithmic) dimension.* Let V^0 be a non-singular variety and let V be a *smooth completion* of V^0, i.e., V is nonsingular projective and $D := V \setminus V^0$ is an integral reduced divisor with simple normal crossings. If $H^0(V, m(K_V + D)) = 0$ for all $m \geq 1$, the *Kodaira (logarithmic) dimension* $\kappa(V^0) = -\infty$. Otherwise, $|m(K_V + D)|$ gives rise to a rational map φ_m for some m and the *Kodaira dimension* $\kappa(V^0)$ is the maximum of $\dim(\varphi_m(V^0))$.

The Kodaira dimension of V^0 does not depend on the choice of the completion V [**I**]. Also $\kappa(V^0)$ takes value in $\{-\infty, 0, 1, \ldots, \dim V^0\}$.

Obviously, $\kappa(V) \leq \kappa(V^0)$.

PROPOSITION 2.3. *Let S be a projective surface with quotient singularities such that $H^2(S, \mathbb{Q}) \cong \mathbb{Q}$ and $H_1(S^0, \mathbb{Z}) = 0$. Let $f : S' \to S$ be a minimal resolution. Then one of the following cases occurs.*

(1) S is rational.
(2) S' is a surface, not necessarily minimal, with $q = p_g = 0$, $\kappa(S') = 1$, $H_1(S', \mathbb{Z}) = 0$ and $\kappa(S^0) = 2$.
(3) S' is a surface of general type, not necessarily minimal, with $q = p_g = 0$, $H_1(S', \mathbb{Z}) = 0$.

PROOF. Since $H_1(S^0, \mathbb{Z}) = 0$, $H_1(S', \mathbb{Z}) = H_1(S, \mathbb{Z}) = 0$ by Lemma 2.1. By Proposition 2.2, S' is a surface with $q = p_g = 0$, $H_1(S', \mathbb{Z}) = 0$. By classification theory (see [**BHPV**]), either S' is rational or $\kappa(S') \geq 1$. Since $H^2(S, \mathbb{Q}) \cong \mathbb{Q}$, S has Picard number 1 and $H^2(S, \mathbb{Q})$ is positive definite. Thus for every curve C on S, $C^2 \geq 0$. In particular, S is relatively minimal, i.e. there is no curve C with $K_S \cdot C < 0$, $C^2 < 0$.

Assume $\kappa(S') = 1$. If $\kappa(S^0) = 1$, then there is an elliptic fibration on S ([**Ka1**] Theorem 2.3 or [**M1**] Ch.II Theorem 6.1.4 or [**KZ**] Theorem 4.1.), thus $\mathrm{Pic}(S)$ has rank at least 2, a contradiction. Thus $\kappa(S^0) = 2$ and the assertion follows. $\square$

Replacing the condition $H_1(S^0, \mathbb{Z}) = 0$ by $\pi_1(S^0) = \{1\}$, one gets the following:

COROLLARY 2.4. Let S be a projective surface with quotient singularities such that $H^2(S, \mathbb{Q}) \cong \mathbb{Q}$ and $\pi_1(S^0) = \{1\}$. Let $f : S' \to S$ be a minimal resolution. Then one of the following cases occurs.

(1) S is rational.
(2) S' is a simply connected surface, not necessarily minimal, with $q = p_g = 0$, $\kappa(S') = 1$, and $\kappa(S^0) = 2$.
(3) S' is a simply connected surface of general type, not necessarily minimal, with $q = p_g = 0$.

So far, no example S satisfying the condition of Corollary 2.4 and belonging to the case (2) or (3) has been found, and it is not likely such an example exists. On this basis J. Kollár suggests his conjecture.

LEMMA 2.5. Let S be a normal compact surface with rational singularities, and $f : S' \to S$ a resolution of singularities. Let $\mathcal{R} \subset H^2(S', \mathbb{Z})$ be the subgroup generated by the cohomology classes of the exceptional curves of f. Set

$$\overline{\mathcal{R}} := \{\alpha \in H^2(S', \mathbb{Z}) : m\alpha \in \mathcal{R} \quad \text{for} \quad \text{some} \quad m \in \mathbb{Z} \setminus \{0\}\}.$$

Then the following are equivalent

(1) $H_1(S^0, \mathbb{Z}) = 0$.
(2) $q(S') = 0$ and $\mathcal{R} = \overline{\mathcal{R}}$.

PROOF. Assume (1). By Lemma 2.1, $H_1(S', \mathbb{Z}) = 0$, and hence $q(S') = 0$. By the universal coefficient theorem, $H^2(S', \mathbb{Z})$ is torsion free, so is $\overline{\mathcal{R}}$. Since $q(S') = 0$, $\mathrm{Pic}(S')$ can be regarded as a primitive sublattice of $H^2(S', \mathbb{Z})$. Since $\mathcal{R} \subset \mathrm{Pic}(S')$, $\overline{\mathcal{R}} \subset \mathrm{Pic}(S')$. If $\mathcal{R} \neq \overline{\mathcal{R}}$, then there would be a finite étale cover of S^0, thus $H_1(S^0, \mathbb{Z}) \neq 0$, a contradiction.

Assume (2). Since $q(S') = 0$, $\mathrm{Pic}(S')$ embeds in $H^2(S', \mathbb{Z})$. If $H_1(S^0, \mathbb{Z}) \neq 0$, then there is a finite étale cover of S^0, thus there exist an element $\alpha \in \mathrm{Pic}(S')$ and an integer $m > 1$ such that $m\alpha$ is either trivial or linearly equivalent to an effective divisor supported in the exceptional set of f, but α is not. Since $\mathrm{Pic}(S') \subset H^2(S', \mathbb{Z})$, this implies that $\overline{\mathcal{R}} \neq \mathcal{R}$. $\square$

DEFINITION 2.6. Let $p \in F$ be a normal surface singularity. Then F is a cone over a real 3-manifold M called the *link*.

If the singularity is rational, then $H^1(M, \mathbb{Z}) = 0$ and $H^2(M, \mathbb{Z})$ is torsion.

For surfaces with $H^2(S, \mathbb{Q}) = \mathbb{Q}$, J. Kollár gives more precise information in terms of links.

PROPOSITION 2.7. ([**Ko**] Corollary 4.18) *Let S be a normal compact surface with rational singularities p_i with links M_i. Assume that $H_1(S, \mathbb{Z}) = 0$ and $H^2(S, \mathbb{Q}) = \mathbb{Q}$. Then the following are equivalent*

(1) $H_1(S^0, \mathbb{Z}) = 0$.
(2) *The Weil divisor class group $Weil(S) \cong \mathbb{Z}$.*
(3) *Each $H^2(M_i, \mathbb{Z})$ is cyclic, their orders m_i are pairwise coprime and there is a Weil divisor B' which generates $H^2(M_i, \mathbb{Z})$ for every i.*
(4) *There is a Weil divisor B with $B^2 = 1/\Pi m_i$.*
(5) *There is a Cartier divisor H and a Weil divisor B with $H^2 = \Pi m_i$, $B \cdot H = 1$.*

The folowing result due to Y. Miyaoka plays a crucial role in the proof of our main theorem.

THEOREM 2.8. ([**M2**] *Theorem 1.1) Let S be a projective surface with quotient singularities. Denote by $Sing(S)$ the set of singular points of S. Let $f : S' \to S$ be a minimal resolution and E be the inverse image $f^{-1}(Sing(S))$, a reduced integral divisor. Assume $K_{S'} + E$ has Zariski decomposition with positive part P and negative part $N + N'$, where N is supported away from E and N' is supported in E. Then we have the inequality*

$$\sum_{p \in Sing(S)} (e(E_p) - \frac{1}{|G_p|}) \le c_2(S') - \frac{1}{3}P^2 - \frac{1}{4}N^2,$$

where $e(E_p)$ is the Euler number of $E_p := f^{-1}(p)$ and G_p is the local fundamental group of p.

COROLLARY 2.9. *Let S be a projective surface with quotient singularities, and $f : S' \to S$ be a minimal resolution. Assume S is relatively minimal, i.e. there is no curve C with $K_S \cdot C < 0$, $C^2 < 0$. Assume $\kappa(S^0) \ge 0$. Then we have the inequality*

$$(2.1) \qquad \sum_{p \in Sing(S)} (e(E_p) - \frac{1}{|G_p|}) \le c_2(S') - \frac{1}{3}K_S^2.$$

PROOF. The canonical divisor K_S is numerically effective by [**MT**], Theorem 2.11 or [**KZ**], Theorem 2.1. Since a quotient singularity is just a log terminal singularity, we have

$$K_{S'} + E = f^*K_S + (E - D),$$

where $E - D$ is an effective $\mathbb{Q}$-divisor whose support is equal to E. Thus the positive part P of Zariski decomposition of $K_{S'} + E$ is f^*K_S and the negative part is $E - D$. $\qquad \square$

Let S be a projective surface with quotient singularities. Then one can write

$$(2.2) \qquad K_{S'} = f^*K_S - \sum_{p \in \text{Sing}(S)} D_p$$

where D_p is an effective $\mathbb{Q}$-divisor supported in $E_p = f^{-1}(p)$.

COROLLARY 2.10. *Let S be a projective surface with quotient singularities such that $H^2(S, \mathbb{Q}) \cong \mathbb{Q}$ and $H_1(S^0, \mathbb{Z}) = 0$. Let $f : S' \to S$ be a minimal resolution. Assume that S is not rational. Then we have the inequality*

$$(2.3) \qquad \sum_{p \in Sing(S)} e(E_p) - c_2(S') + \frac{1}{3} K_{S'}^2 \leq \sum_{p \in Sing(S)} \left(\frac{1}{|G_p|} + \frac{1}{3} D_p^2 \right).$$

PROOF. By Proposition 2.3, $\kappa(S^0) = 2$. Since $H^2(S, \mathbb{Q}) \cong \mathbb{Q}$, S has Picard number 1 and $H^2(S, \mathbb{Q})$ is positive definite. Thus for every curve C on S, $C^2 \geq 0$. In particular, S is relatively minimal. (This also follows from Kawamata's Cone Theorem. Indeed, the existence of a curve C with $K_S \cdot C < 0$, $C^2 < 0$ would imply the existence of an extremal contraction, which is either divisorial or gives a fibration, both contradicting to the fact that S has Picard number 1.) By Corollary 2.9, we get the inequality (2.1). It remains to see that

$$K_S^2 = (f^* K_S)^2 = K_{S'}^2 - \sum_{p \in Sing(S)} D_p^2.$$

$\square$

Now we prove one case of Theorem 1.2.

COROLLARY 2.11. *Let S be a projective surface with quotient singularities such that $H^2(S, \mathbb{Q}) \cong \mathbb{Q}$ and $H_1(S^0, \mathbb{Z}) = 0$. Assume that S is not rational. Then S has at most 4 singular points.*

PROOF. Let $f : S' \to S$ be a minimal resolution. By Corollary 2.9, we have the inequality (2.1). Note that

$$\sum_{p \in Sing(S)} e(E_p) = b_2(S') - 1 + |Sing(S)| = c_2(S') - 3 + |Sing(S)|.$$

Thus the inequality (2.1) becomes

$$(2.4) \qquad |Sing(S)| - 3 \leq \sum_{p \in Sing(S)} \frac{1}{|G_p|} - \frac{1}{3} K_S^2.$$

Let $p_1, ..., p_r$ be the singular points of S, and let $M_1, ..., M_r$ be the corresponding links. Since the singularities are rational, $H^2(M_i, \mathbb{Z})$ is isomorphic to the abelianization of the local fundamental group G_{p_i}. By Proposition 2.7, $H^2(M_i, \mathbb{Z})$ is cyclic, and their orders m_i are pairwise coprime. Let us assume that $m_1 < m_2 < ... < m_r$.

Assume $r = |Sing(S)| \geq 5$. If $m_1 > 1$, then m_i is greater than equal to the i-th prime number, thus

$$(2.5) \qquad \sum_{i=1}^{r} \frac{1}{|G_{p_i}|} \leq \sum_{i=1}^{r} \frac{1}{m_i} \leq r - 3.$$

If $m_1 = 1$, then $\frac{1}{|G_{p_1}|} < \frac{1}{2}$, thus

$$(2.6) \qquad \sum_{i=1}^{r} \frac{1}{|G_{p_i}|} \leq \frac{1}{|G_{p_1}|} + \sum_{i=2}^{r} \frac{1}{m_i} \leq \frac{1}{2} + \left(r - \frac{7}{2} \right) = r - 3.$$

Since $K_S^2 > 0$, both (2.5) and (2.6) lead to a contradiction to the inequality (2.4). Thus $|Sing(S)| \leq 4$. $\square$

REMARK 2.12. In the situation of Corollary 2.11, if $|\mathrm{Sing}(S)| = 4$, then two of the four singularities have the local fundamental group of order 2 and 3, respectively.

3. Proof of Main Theorem

In this section, we prove the other case of Theorem 1.2.

Fix a singular point $p \in S$, and let $E_1, ..., E_k$ ($k \le 3$) be the irreducible components of $E_p = f^{-1}(p)$. They form a string of smooth rational curves

$$(-n_1) \quad \text{or} \quad (-n_1) - (-n_2) \quad \text{or} \quad (-n_1) - (-n_2) - (-n_3)$$

where E_j is a $(-n_j)$-curve. Write

$$D_p = \sum_{j=1}^{k} a_j E_j.$$

Note that $0 \le a_j < 1$.

To use the inequality (2.3), we need to estimate $\frac{1}{|G_p|} + \frac{1}{3} D_p^2$.

LEMMA 3.1. Fix $p \in Sing(S)$. Assume that $f^{-1}(p)$ has 3 components E_1, E_2, E_3 with $E_i^2 = -n_i$. Assume that $n_1 + n_2 + n_3 \ge 10$. Then

$$\frac{1}{|G_p|} + \frac{1}{3} D_p^2 < -\frac{1}{2}.$$

PROOF. Since E_j is a $(-n_j)$-curve, $E_j \cdot K_{S'} = n_j - 2$. Intersecting E_j with $f^* K_S$ from (2.2), we see that

$$n_1 - 2 = a_1 n_1 - a_2$$

$$n_2 - 2 = -a_1 + a_2 n_2 - a_3$$

$$n_3 - 2 = -a_2 + a_3 n_3$$

Adding the equations, we get

$$\sum n_j - 6 = \sum a_j n_j - \sum a_j - a_2,$$

hence

$$\sum a_j(n_j - 2) = \sum n_j - 6 - a_1 - a_3 > \sum n_j - 8 \ge 2.$$

Since $D_p^2 = -D_p \cdot K_{S'} = -\sum a_j(n_j - 2)$, we have

$$\frac{1}{|G_p|} + \frac{1}{3} D_p^2 = \frac{1}{|G_p|} - \frac{1}{3} \sum a_j(n_j - 2) < \frac{1}{|G_p|} - \frac{2}{3} \le -\frac{1}{2}.$$

$\square$

For the cases where $n_1 + n_2 + n_3 \le 9$, we give exact estimates in Table 1.

LEMMA 3.2. Fix $p \in Sing(S)$. Assume that $f^{-1}(p)$ has 2 components E_1, E_2 with $E_i^2 = -n_i$. Assume that $n_1 + n_2 \ge 8$. Then

$$\frac{1}{|G_p|} + \frac{1}{3} D_p^2 < -\frac{1}{2}.$$

(n_1, n_2, n_3)	$|G_p|$	(a_1, a_2, a_3)	D_p^2	$\frac{1}{|G_p|} + \frac{1}{3}D_p^2$
$(2,2,2)$	4	$(0,0,0)/4$	0	$1/4$
$(2,2,3)$	7	$(1,2,3)/7$	$-3/7$	0
$(2,3,2)$	8	$(2,4,2)/8$	$-4/8$	$-1/24$
$(2,2,4)$	10	$(2,4,6)/10$	$-12/10$	$-3/10$
$(2,4,2)$	12	$(4,8,4)/12$	$-16/12$	$-13/36$
$(3,2,3)$	12	$(6,6,6)/12$	$-12/12$	$-1/4$
$(2,3,3)$	13	$(4,8,7)/13$	$-15/13$	$-4/13$
$(2,2,5)$	13	$(3,6,9)/13$	$-27/13$	$-8/13$
$(2,5,2)$	16	$(6,12,6)/16$	$-36/16$	$-11/16$
$(3,2,4)$	17	$(9,10,11)/17$	$-31/17$	$-28/51$
$(2,3,4)$	18	$(6,12,12)/18$	$-36/18$	$-11/18$
$(2,4,3)$	19	$(7,14,11)/19$	$-39/19$	$-12/19$
$(3,3,3)$	21	$(12,15,12)/21$	$-39/21$	$-4/7$

TABLE 1

PROOF. Intersecting E_j with f^*K_S from (2.2), we see that

$$n_1 - 2 = a_1 n_1 - a_2$$

$$n_2 - 2 = -a_1 + a_2 n_2$$

Adding the equations, we get

$$\sum n_j - 4 = \sum a_j n_j - \sum a_j,$$

hence

$$\sum a_j(n_j - 2) = \sum n_j - 4 - a_1 - a_2 > \sum n_j - 6 \geq 2.$$

Thus, we have

$$\frac{1}{|G_p|} + \frac{1}{3}D_p^2 = \frac{1}{|G_p|} - \frac{1}{3}\sum a_j(n_j - 2) < \frac{1}{|G_p|} - \frac{2}{3} \leq -\frac{1}{2}.$$

$\square$

For the cases where $n_1 + n_2 \leq 7$, we give exact estimates in Table 2.

(n_1, n_2)	$|G_p|$	(a_1, a_2)	D_p^2	$\frac{1}{|G_p|} + \frac{1}{3}D_p^2$
$(2,2)$	3	$(0,0)/3$	0	$1/3$
$(2,3)$	5	$(1,2)/5$	$-2/5$	$1/15$
$(2,4)$	7	$(2,4)/7$	$-8/7$	$-5/21$
$(3,3)$	8	$(4,4)/8$	$-8/8$	$-5/24$
$(2,5)$	9	$(3,6)/9$	$-18/9$	$-5/9$
$(3,4)$	11	$(6,7)/11$	$-20/11$	$-17/33$

TABLE 2

LEMMA 3.3. *Fix $p \in \text{Sing}(S)$. Assume that $f^{-1}(p)$ has 1 component E_1 with $E_1^2 = -n$. Let $d = |G_p|$. Then $d = n$ and*

$$\frac{1}{|G_p|} + \frac{1}{3}D_p^2 = \frac{1}{d} - \frac{(d-2)^2}{3d}$$

which equals to $\frac{1}{2}$ if $d = 2$, to $\frac{2}{9}$ if $d = 3$, and $\leq -\frac{1}{12}$ if $d \geq 4$.

PROOF. $D_p = \frac{d-2}{d}E_1$. □

If $H_1(S^0, \mathbb{Z}) = 0$, then by Lemma 2.1, $H_1(S', \mathbb{Z}) = 0$, hence $H^2(S', \mathbb{Z})$ is torsion free and becomes a lattice with intersection pairing.

From Lemma 2.5 and Proposition 2.7, we also have the following:

LEMMA 3.4. *Let S be a projective surface with quotient singularities satisfying the conditions (1) and (2) of Theorem 1.2. Write $\mathcal{R} = \oplus_p \mathcal{R}_p$ where $\mathcal{R}_p$ is the sublattice of $H^2(S', \mathbb{Z})$ generated by the components of $E_p = f^{-1}(p)$. Then*

    (1) *The numbers $|G_p/[G_p, G_p]| = |\det \mathcal{R}_p|$ are pairwise coprime.*
    (2) *There is an integer m such that $|\det \mathcal{R}|(f^* K_S)^2 = m^2$.*

PROOF. Here we give a short proof.

By Lemma 2.5, $\mathcal{R}$ is a primitive sublattice of $H^2(S', \mathbb{Z})$. Since $H^2(S', \mathbb{Z})$ is unimodular, we have an isomorphism between the discriminant groups

$$\text{disc}\mathcal{R} = \oplus_p \text{disc}\mathcal{R}_p \cong -\text{disc}\mathcal{R}^{\perp}.$$

Since $\mathcal{R}^{\perp}$ is of rank 1, $\text{disc}\mathcal{R}$ must be cyclic. This proves (1).

The divisor $(\det \mathcal{R})f^* K_S$ is an integral divisor belonging to $\mathcal{R}^{\perp}$, hence

$$(\det \mathcal{R})f^* K_S = mv$$

for some integer m, where v is a generator of $\mathcal{R}^{\perp}$. Since $v^2 = |\det \mathcal{R}|$, (2) follows. □

From now on, S denotes a projective surface satisfying the condition of Theorem 1.2, i.e. S is a projective surface with quotient singularities such that

    (1) $H^2(S, \mathbb{Q}) \cong \mathbb{Q}$,
    (2) $H_1(S^0, \mathbb{Z}) = 0$,
    (3) the inverse image $f^{-1}(p)$ has at most 3 components for each singular point p in S, where $f : S' \to S$ is a minimal resolution.

To get a contradiction, we also assume

    (4) S is not rational.

In this situation, by Corollary 2.10, we have the inequality (2.3). By the assumption (3), all singularities of S are cyclic.

We denote the left hand side and the right hand side of the inequality (2.3) by

$$\text{LHS} := \sum_{p \in \text{Sing}(S)} e(E_p) - c_2(S') + \frac{1}{3}K_{S'}^2$$

$$\text{RHS} := \sum_{p \in \text{Sing}(S)} \left(\frac{1}{|G_p|} + \frac{1}{3}D_p^2\right).$$

LEMMA 3.5. *Let S be a projective surface with quotient singularities satisfying the above conditions $(1) - (4)$. Assume that the number of singular points $|Sing(S)| \geq 3$. Then RHS $\leq \frac{9}{10}$.*

PROOF. From Lemma 3.1-3.3, we see that $\frac{1}{|G_p|} + \frac{1}{3}D_p^2 > 0$ for only one of the five types of singularities $(2,2,2)$, $(2,2)$, $(2,3)$, (2), (3). Also, by Lemma 3.4, the pair $(2,2,2)$ and (2) do not occur simultaneously. Neither the pair $(2,2)$ and (3).

If $|Sing(S)| \geq 3$, RHS takes its maximum value when $Sing(S) = (2) + (2,2) + (2,3)$, hence

$$\text{RHS} \leq \frac{1}{2} + \frac{1}{3} + \frac{1}{15} = \frac{9}{10}.$$

$\square$

Proof of Theorem 1.2. To get a contradiction, assume that S is not rational. By Proposition 2.3 it suffices to rule out the two cases

(2) S' is a surface, not necessarily minimal, with $q = p_g = 0$, $\kappa(S') = 1$, $H_1(S', \mathbb{Z}) = 0$ and $\kappa(S^0) = 2$.

(3) S' is a surface of general type, not necessarily minimal, with $q = p_g = 0$, $H_1(S', \mathbb{Z}) = 0$.

Since $q = p_g = 0$, by Bogomolov-Miyaoka-Yau inequality (or by Theorem 2.8) we have $c_2(S') \geq 3$.

If $c_2(S') = 3$, then $S' = S$ and S is non-singular, hence S is either the complex projective plane or a surface of general type with $q = p_g = 0$, $3c_2 = c_1^2 = 9$, so called a fake projective plane. The latter surface has the unit ball in $\mathbb{C}^2$ as its universal covering (this follows from the solution of S.-T. Yau [**Y**] to Calabi conjecture) hence has an infinite fundamental group. G. Prasad and S.-K. Yeung [**PY**] have shown that no fake projective plane with $H_1(S, \mathbb{Z}) = 0$ exists.

Thus we may assume that $c_2(S') \geq 4$ and S is singular.

By Corollary 2.10, we have the inequality (2.3). By Corollary 2.11, we also have $|Sing(S)| \leq 4$.

Case 1. $c_2(S') = 4$ and $K_{S'}^2 = 8$.

In this case $|Sing(S)| = 1$ and LHS $= 2 - 4 + \frac{8}{3} = \frac{2}{3}$, while RHS $\leq \frac{1}{2}$, a contradiction.

Case 2. $c_2(S') = 5$ and $K_{S'}^2 = 7$.

If $|Sing(S)| = 2$, then LHS $= 4 - 5 + \frac{7}{3} = \frac{4}{3}$, while RHS $\leq \frac{1}{2} + \frac{2}{9}$, a contradiction. If $|Sing(S)| = 1$, then LHS $= 3 - 5 + \frac{7}{3} = \frac{1}{3}$, while RHS $\leq \frac{1}{3}$, with equality only when $(n_1, n_2) = (2, 2)$. In this case $\det \mathcal{R} = 3$ and $(f^*K_S)^2 = K_{S'}^2 = 7$, a contradiction to Lemma 3.4(2).

Case 3. $c_2(S') = 6$ and $K_{S'}^2 = 6$.

If $|Sing(S)| = 3$, then LHS $= 6 - 6 + \frac{6}{3} = 2$, contradicts to Lemma 3.5. If $|Sing(S)| = 2$, then LHS $= 5 - 6 + \frac{6}{3} = 1$, while RHS $\leq \frac{1}{2} + \frac{1}{3}$, a contradiction. If $|Sing(S)| = 1$, then LHS $= 4 - 6 + \frac{6}{3} = 0$, hence we must have $(n_1, n_2, n_3) = (2, 2, 2)$ or $(2, 2, 3)$. In the first case (resp. the second) $\det \mathcal{R} = 4$ (resp. 7) and $(f^*K_S)^2 = 6$ (resp. $\frac{45}{7}$). Both contradict to Lemma 3.4(2).

Case 4. $c_2(S') = 7$ and $K_{S'}^2 = 5$.

If $|Sing(S)| \geq 3$, then LHS $\geq 7 - 7 + \frac{5}{3} = \frac{5}{3}$, contradicts to Lemma 3.5.

If $|\text{Sing}(S)| = 2$, then LHS $= 6 - 7 + \frac{5}{3} = \frac{2}{3}$, hence $(n_1, n_2, n_3) + (n_4) = (2, 2, 2) + (2)$ (no possible combination of type $(n_1, n_2) + (n_3, n_4)$). In this case $\det \mathcal{R}_{p_1} = 4$ and $\det \mathcal{R}_{p_2} = 2$, contradicting to Lemma 3.4(1).

Case 5. $c_2(S') = 8$ and $K_{S'}^2 = 4$.

If $|\text{Sing}(S)| \geq 3$, then LHS $\geq 8 - 8 + \frac{4}{3} = \frac{4}{3}$, contradicts to Lemma 3.5.
If $|\text{Sing}(S)| = 2$, then LHS $= 7 - 8 + \frac{4}{3} = \frac{1}{3}$, hence $(n_1, n_2, n_3) + (n_4, n_5) = (2, 2, 2) + (2, 2)$ or $(2, 2, 3) + (2, 2)$. In the first case (resp. the second) $\det \mathcal{R} = 12$ (resp. 21) and $(f^*K_S)^2 = 4$ (resp. $\frac{31}{7}$). Both contradict to Lemma 3.4(2).

Case 6. $c_2(S') = 9$ and $K_{S'}^2 = 3$.

If $|\text{Sing}(S)| \geq 3$, then LHS $\geq 9 - 9 + \frac{3}{3} = 1$, contradicts to Lemma 3.5.
If $|\text{Sing}(S)| = 2$, then LHS $= 8 - 9 + \frac{3}{3} = 0$, hence $(n_1, n_2, n_3) + (n_4, n_5, n_6) = (2, 2, 2) + (2, 2, 3)$ or $(2, 2, 2) + (2, 3, 2)$ or $(2, 2, 2) + (3, 2, 3)$. In the first case, $\det \mathcal{R} = 28$ and $(f^*K_S)^2 = \frac{24}{7}$, contradicts to Lemma 3.4(2). In the second case, $\det \mathcal{R} = 4 \cdot 8$, and in the third, $\det \mathcal{R} = 4 \cdot 12$, both contradict to Lemma 3.4(1).

Case 7. $c_2(S') = 10$ and $K_{S'}^2 = 2$.

If $|\text{Sing}(S)| = 4$, then LHS $= 11 - 10 + \frac{2}{3} = \frac{5}{3}$, contradicts to Lemma 3.5.
If $|\text{Sing}(S)| = 3$, then LHS $= \frac{2}{3}$, hence $(n_1, n_2, n_3) + (n_4, n_5, n_6) + (n_7) = (2, 2, 2) + (2, 2, 3) + (2)$ or $(2, 2, 2) + (2, 3, 2) + (2)$ (no possible combination of type $(n_1, n_2, n_3) + (n_4, n_5) + (n_6, n_7)$). In the first case, $\det \mathcal{R} = 56$ and $(f^*K_S)^2 = K_{S'}^2 - \sum D_p^2 = 2 + \frac{3}{7} = \frac{17}{7}$, contradicts to Lemma 3.4(2). In the second case, $\det \mathcal{R} = 4 \cdot 8 \cdot 2$, contradicts to Lemma 3.4(1).

Case 8. $c_2(S') = 11$ and $K_{S'}^2 = 1$.

If $|\text{Sing}(S)| = 4$, then LHS $= 12 - 11 + \frac{1}{3} = \frac{4}{3}$, contradicts to Lemma 3.5.
If $|\text{Sing}(S)| = 3$, then LHS $= \frac{1}{3}$, hence by Lemma 3.4(1), $(n_1, n_2, n_3) + (n_4, n_5, n_6) + (n_7, n_8) = (2, 2, 2) + (2, 2, 3) + (2, 2)$. Then, $\det \mathcal{R} = 4 \cdot 7 \cdot 3$ and $(f^*K_S)^2 = K_{S'}^2 - \sum D_p^2 = 1 + \frac{3}{7} = \frac{10}{7}$, contradicts to Lemma 3.4(2).

Case 9. $c_2(S') = 3s + 3$ and $K_{S'}^2 = 9 - 3s$ ($s \geq 3$).

In this case $|\text{Sing}(S)| \geq (b_2(S') - 1)/3 = 3s/3 = s$.
Assume $|\text{Sing}(S)| \geq s + 1$, then

$$
\begin{aligned}
\text{LHS} &= (b_2(S') - 1) + |\text{Sing}(S)| - c_2(S') + \frac{1}{3}K_{S'}^2 \\
&\geq 3s + (s + 1) - (3s + 3) + \frac{1}{3}(9 - 3s) = 1.
\end{aligned}
$$

Since $s \geq 3$, this contradicts to Lemma 3.5.
Assume $|\text{Sing}(S)| = s$, then

$$
\text{LHS} = (b_2(S') - 1) + |\text{Sing}(S)| - c_2(S') + \frac{1}{3}K_{S'}^2 = 0.
$$

Since $b_2(S') - 1 = 3s$, $\text{Sing}(S)$ consists of s singular points of length 3, thus by Lemma 3.4(1),

$$
\text{RHS} \leq \frac{1}{4} + 0 - \frac{4}{13} < 0,
$$

a contradiction.

Case 10. $c_2(S') = 3s + 4$ and $K_{S'}^2 = 8 - 3s$ ($s \geq 3$).

In this case $|\mathrm{Sing}(S)| \geq (b_2(S') - 1)/3 = (3s + 1)/3$, hence $|\mathrm{Sing}(S)| \geq s + 1$.

Since $|\mathrm{Sing}(S)| \leq 4$, $s = 3$ and $|\mathrm{Sing}(S)| = 4$. Thus

$$\mathrm{LHS} = (b_2(S') - 1) + |\mathrm{Sing}(S)| - c_2(S') + \frac{1}{3}K_{S'}^2 = \frac{2}{3}.$$

Since $b_2(S') - 1 = 1 + 3 + 3 + 3 = 2 + 2 + 3 + 3$, $\mathrm{Sing}(S)$ consists either of 1 singular point of length 1 and 3 singular points of length 3, or 2 singular points of length 2 and 2 singular points of length 3, thus by Lemma 3.4(1),

$$\mathrm{RHS} \leq \frac{1}{2} + 0 \quad \text{or} \quad \leq \frac{2}{9} + \frac{1}{4} + 0 \quad \text{or} \quad \leq \frac{1}{3} + \frac{1}{15} + \frac{1}{4} + 0,$$

all smaller than $\frac{2}{3}$, a contradiction.

Case 11. $c_2(S') = 3s + 2$ and $K_{S'}^2 = 10 - 3s$ ($s \geq 4$).

In this case $|\mathrm{Sing}(S)| \geq (b_2(S') - 1)/3 = (3s - 1)/3$, hence $|\mathrm{Sing}(S)| \geq s$.

Since $|\mathrm{Sing}(S)| \leq 4$, $s = 4$ and $|\mathrm{Sing}(S)| = 4$. Thus

$$\mathrm{LHS} = (b_2(S') - 1) + |\mathrm{Sing}(S)| - c_2(S') + \frac{1}{3}K_{S'}^2 = \frac{1}{3}.$$

Since $b_2(S') - 1 = 2 + 3 + 3 + 3$, $\mathrm{Sing}(S)$ consists of 1 singular point of length 2 and 3 singular points of length 3, thus by Lemma 3.4(1),

$$\mathrm{RHS} \leq \frac{1}{3} + \frac{1}{4} + 0 - \frac{4}{13} < \frac{1}{3},$$

a contradiction.

This completes the proof of Theorem 1.2.

COROLLARY 3.6. *Kollár's conjecture holds true if in addition either the surface S has at least 5 singular points or the exceptional divisor in a minimal resolution of S has at most 3 components over each singular point of S.*

4. Examples and Further Discussion

EXAMPLE 4.1. In [**Is**] M. Ishida discusses an elliptic surface Y with $p_g = q = 0$ with two multiple fibres, one of multiplicity 2 and one of muliplicity 3, and proves that the Mumford fake plane is its cover of degree 21, non-Galois. The surface Y is a Dolgachev surface [**BHPV**]. In particular, it is simply connected and of Kodaira dimension 1. Besides the two multiple fibres, its elliptic fibration $|F_Y|$ has 4 more singular fibres F_1, F_2, F_3, F_4, all of type I_3. It has also a sixtuple section E which is a (-3)-curve meeting one component of each of F_1, F_2, F_3 in 6 points, and two components of F_4 in 1 point and 5 points each. On Y one can find 9 smooth rational curves forming a configuration

$$(-2)\text{---}(-2) \quad (-2)\text{---}(-2) \quad (-2)\text{---}(-2) \quad (-2)\text{---}(-2)\text{---}(-3)$$

which can be contracted to 3 singular points of type $\frac{1}{3}(1, 2)$ and one singular point of type $\frac{1}{7}(1, 3)$. The resulting singular surface S satisfies the condition (1) and (3) of Theorem 1.2, but not (2). Indeed, $H_1(S^0, \mathbb{Z}) = \mathbb{Z}/3\mathbb{Z}$.

EXAMPLE 4.2. It is shown [**Ke**] that there is another Dolgachev surface X which is birational to a cyclic cover of degree 3 of Ishida surface Y. On X there are 9 smooth rational curves forming a configuration

$$(-2)\text{---}(-2)\text{---}(-3) \quad (-2)\text{---}(-2)\text{---}(-3) \quad (-2)\text{---}(-2)\text{---}(-3)$$

which can be contracted to 3 singular points of type $\frac{1}{7}(1,3)$. The resulting singular surface S satisfies the condition (1) and (3) of Theorem 1.2, but not (2). In this case, $H_1(S^0, \mathbb{Z}) = \mathbb{Z}/7\mathbb{Z}$.

Finally we consider surfaces S with rational double points only.

PROPOSITION 4.3. *Let S be a singular projective surface with rational double points such that $H^2(S, \mathbb{Q}) \cong \mathbb{Q}$ and $H_1(S^0, \mathbb{Z}) = 0$. Let $f : S' \to S$ be a minimal resolution. Then one of the following cases occurs.*

(1) *S is rational.*

(2) *S' is a minimal surface of general type, with $q = p_g = 0$, $H_1(S', \mathbb{Z}) = 0$, and*

   (2-1) *$K_{S'}^2 = 1$ and $\mathcal{R} \cong E_8$, or*
   (2-2) *$K_{S'}^2 = 2$ and $\mathcal{R} \cong E_7$, or*
   (2-3) *$K_{S'}^2 = 3$ and $\mathcal{R} \cong E_6$, or*
   (2-4) *$K_{S'}^2 = 4$ and $\mathcal{R} \cong D_5$, or*
   (2-5) *$K_{S'}^2 = 5$ and $\mathcal{R} \cong A_4$.*

PROOF. Since S has rational double points only, $f^* K_S = K_{S'}$. If K_S is anti-numerically effective, so is $K_{S'}$, hence S' is rational. If K_S is numerically effective, so is $K_{S'}$, hence S' is minimal. We use Proposition 2.3. We need to rule out the second possibility from Proposition 2.3. Suppose that the second case occur. Since $K_{S'}$ is a rational multiple of a fibre of the elliptic fibration, the exceptional divisor of f is supported in a union of fibres. This contradicts to $\kappa(S^0) = 2$.

Next Assume that S' is a minimal surface of general type. The divisor $K_{S'}$ is an integral divisor belonging to $\mathcal{R}^\perp$, hence $K_{S'} = mv$ for some integer m, where v is a generator of $\mathcal{R}^\perp$. Since $v^2 = |\det \mathcal{R}|$, $K_{S'}^2 = m^2 |\det \mathcal{R}|$. This leaves the five cases (2-1)-(2-5) and two more

   (2-6) $K_{S'}^2 = 6$ and $\mathcal{R} \cong A_1 \oplus A_2$, or
   (2-7) $K_{S'}^2 = 8$ and $\mathcal{R} \cong A_1$.

Both are ruled out by Theorem 1.2.   □

REMARK 4.4. If one loosens the bound to 4 on the number of components in Condition (3) of Theorem 1.2, one already encounters a non-trivial problem to rule out the possibility (2-5) from the above proposition.

Nevertheless, we propose a homological and stronger version of Kollár's conjecture.

CONJECTURE 4.5. *Let S be a projective surface with quotient singularities such that*

   (1) $H^2(S, \mathbb{Q}) \cong \mathbb{Q}$,
   (2) $H_1(S^0, \mathbb{Z}) = 0$,

Then S is rational.

References

[BHPV] W. Barth, K. Hulek, Ch. Peters, A. Van de Ven, *Compact Complex Surfaces*, second ed. Springer 2004.

[I]   S. Iitaka, *Algebraic geometry – An introduction to birational geometry of algebraic varieties, Graduate Texts in Mathematics*, **76**, Springer-Verlag, New York-Berlin, 1982.

[Is]   M. Ishida, *An elliptic surface covered by Mumford's fake projective plane*, Tohoku Math. J. **40** (1988), 367-398.

[Ka1]   Y. Kawamata, *On the classification of noncomplete algebraic surfaces*, Lecture Notes in Math. **732** (1979), 215-232, Springer, Berlin.

[Ka2]   Y. Kawamata, *The cone of curves of algebraic varieties*, Ann. of Math. **119** (1984), 603-633.

[Ke]   J. Keum, *A fake projective plane with an order 7 automorphism*, Topology **45** (2006), no. 5, 919-927.

[Ko]   J. Kollár, *Einstein metrics on five-dimensional Seifert bundles*, Jour. Geom. Anal. **15** (2005), no. 3, 445-476.

[KZ]   J. Keum, D.-Q. Zhang, *Algebraic surfaces with quotient singularities - including some discussion on automorphisms and fundamental groups*, Proceedings of "Algebraic Geometry in East Asia" Kyoto, 2001,

[M1]   M. Miyanishi, *Open Algebraic Surfaces*, CRM Monogragh Series, **12**, American Math. Soc. 2001.

[MT]   M. Miyanishi, S, Tsunoda, *Noncomplete algebraic surfaces with logarithmic dimension* $-\infty$ *and with nonconnected boundaries at infinity* , Japan. J. Math. (N.S.) **10** (1984), 195-242.

[M2]   Y. Miyaoka, *The maximal number of quotient singularities on surfaces with given numerical invariants*, Math. Ann. **268** (1984), 159-171.

[PY]   G. Prasad, S.-K. Yeung, *Fake projective planes*, math.AG/0512115.

[Y]   S.-T. Yau, *Calabi's conjecture and some new results in algebraic geometry*, Proc. Nat. Ac. Sc. USA **74** (1977), 1798-1799.

SCHOOL OF MATHEMATICS, KOREA INSTITUTE FOR ADVANCED STUDY, SEOUL 130-722, KOREA
E-mail address: `jhkeum@kias.re.kr`

Contemporary Mathematics
Volume **422**, 2007

The moduli space of 8 points on $\mathbb{P}^1$ and automorphic forms

Shigeyuki Kondō

Dedicated to Igor Dolgachev on his 60th birthday

ABSTRACT. First we give a complex ball uniformization of the moduli space of 8 ordered points on the projective line by using the theory of periods of $K3$ surfaces. Next we give a projective model of this moduli space by using automorphic forms on a bounded symmetric domain of type IV which coincides with the one given by cross ratios of 8 ordered points of the projective line.

.

1. Introduction

The main purpose of this paper is to give an application of the theory of automorphic forms on a bounded symmetric domain of type IV due to Gritsenko and Borcherds [**Bor**] for studying the moduli spaces. We consider the moduli space P_1^8 of semi-stable 8 ordered points on the projective line. It is known that P_1^8 is isomorphic to the Satake-Baily-Borel compactification of an arithmetic quotient of 5-dimensional complex ball by using the theory of periods of a family of curves which are the 4-fold cyclic covers of the projective line branched at eight points ([**DM**]). Recently Matsumoto and Terasoma [**MT**] gave an embedding of P_1^8 into $\mathbb{P}^{104}$ by using the theta constants related to the above curves. Their map coincides the one defined by the cross ratios of 8 points on the $\mathbb{P}^1$. Here they used the fact that the complex ball is canonically embedded in a Siegel upper half plane.

In this paper, instead of the periods of curves, we use the periods of $K3$ surfaces. In our case, the complex ball is embedded in a bounded symmetric domain of type IV. In fact, to each stable point from P_1^8, we associate a $K3$ surface with a non-symplectic automorphism of order 4 (§2). The period domain of these $K3$ surfaces is a 5-dimensional complex ball $\mathcal{B}$ (§3). This was essentially given in the paper [**Kon1**]. By using this, we shall see that P_1^8 is isomorphic to the Satake-Baily-Borel compactification $\bar{\mathcal{B}}/\Gamma(1-i)$ of $\mathcal{B}/\Gamma(1-i)$ where $\Gamma(1-i)$ is an arithmetic subgroup of a unitary group of a hermitian form of signature $(1,5)$ defined over the Gaussiann integers (Theorems 3.3, 4.6) . The symmetry group S_8 of degree 8 naturally acts on P_1^8. On the other hand, there exists an arithmetic subgroup Γ

1991 *Mathematics Subject Classification.* Primary 14H10; Secondary 14J28, 11F23.

Key words and phrases. Moduli, $K3$ surfaces, Automorphic forms.

Research of the author is partially supported by JSPS, A-14204001.

acting on $\mathcal{B}$ with $\Gamma/\Gamma(1-i) \cong S_8$. The above isomorphism $P_1^8 \cong \bar{\mathcal{B}}/\Gamma(1-i)$ is S_8-equivariant.

Next we apply the theory of automorphic forms [**Bor**] to this situation. The main idea comes from the paper of Allcock and Freitag [**AF**] in which they studied the same problem in the case of cubic surfaces. We shall show that there exists a 14-dimensional space of automorphic forms on $\mathcal{B}$ which gives an S_8-equivariant map from the arithmetic quotient $\bar{\mathcal{B}}/\Gamma(1-i)$ into $\mathbb{P}^{13}$ (Theorem 7.3). Under the identification $P_1^8 \cong \bar{\mathcal{B}}/\Gamma(1-i)$ we show that this map coincides with the one defined by the cross ratios of 8 points on the projective line (Theorem 7.2). Thus our map coincides the one given by Matsumoto and Terasoma [**MT**].

In this paper, a *lattice* means a $\mathbb{Z}$-valued non-degenerate symmetric bilinear form on a free $\mathbb{Z}$-module of finite rank. We denote by U the even lattice defined by the matrix $\begin{pmatrix} 0 & 1 \\ 1 & 0 \end{pmatrix}$, and by A_m, D_n or E_l the even negative definite lattice defined by the Dynkin matrix of type A_m, D_n or E_l respectively. Let L be a lattice. We denote by $O(L)$ the group of isomorphisms of L preserving the symmetric bilinear form. If m is an integer, we denote by $L(m)$ the lattice over the same $\mathbb{Z}$-module with the symmetric bilinear form multiplied by m. We also denote by $L^{\oplus m}$ the orthogonal direct sum of m copies of L and by L^* the dual of L.

2. $K3$ surfaces associated with 8 points on the projective line

2.1. In this section, we shall construct a $K3$ surface associated to distinct 8 points from P_1^8. In section 4, we shall generalize this to the cases of any stable points and semi-stable points from P_1^8. Let $\{(\lambda_i : 1)\}$ be a set of distinct 8 points on the projective line. Let $(x_0 : x_1, y_0 : y_1)$ be the bi-homogenious coordinates on $\mathbb{P}^1 \times \mathbb{P}^1$. Consider a smooth divisor C in $\mathbb{P}^1 \times \mathbb{P}^1$ of bidegree $(4, 2)$ given by

$$(2.1) \qquad y_0^2 \cdot \prod_{i=1}^{4}(x_0 - \lambda_i x_1) + y_1^2 \cdot \prod_{i=5}^{8}(x_0 - \lambda_i x_1) = 0.$$

Let L_0 (resp. L_1) be the divisor defined by $y_0 = 0$ (resp. $y_1 = 0$). Let ι be an involution of $\mathbb{P}^1 \times \mathbb{P}^1$ given by

$$(2.2) \qquad (x_0 : x_1, y_0 : y_1) \longrightarrow (x_0 : x_1, y_0 : -y_1)$$

which preserves C and L_0, L_1. Note that the double cover of $\mathbb{P}^1 \times \mathbb{P}^1$ branched along $C + L_0 + L_1$ has 8 rational double points of type A_1 and its minimal resolution X is a $K3$ surface. This $K3$ surface X is obtained as follows: First blow up the 8 points which are the intersection of C and $L_0 + L_1$. Then X is the double cover branched along the proper transforms of C, L_0 and L_1. We remark that the isomorphism class of X depends only on the 8 points in $\mathbb{P}^1$ (i.e. independent on the order of 8 points) because elementary transformations change the order of 8 points.

The involution ι lifts to an automorphism σ of order 4. We can easily see that $\sigma^* \omega_X = \pm \sqrt{-1} \omega_X$ where ω_X is a nowhere vanishing holomorphic 2-form on X. We denote by S_0, S_1 the inverse image of L_0, L_1 respectively. The projection

$$(x_0 : x_1, y_0 : y_1) \longrightarrow (x_0 : x_1)$$

from $\mathbb{P}^1 \times \mathbb{P}^1$ to $\mathbb{P}^1$ induces an elliptic fibration

$$\pi : X \longrightarrow \mathbb{P}^1$$

which has 8 singular fibers of type III in the sense of Kodaira [**Kod**] and two sections S_0, S_1. Let $E_i + F_i$ $(1 \le i \le 8)$ be the 8 singular fibers of π. Then we may assume that

$$E_i \cdot S_0 = F_i \cdot S_1 = 1.$$

Put

(2.3) $$H^2(X, \mathbb{Z})^+ = \{x \in H^2(X, \mathbb{Z}) \mid (\sigma^2)^*(x) = x\};$$

(2.4) $$H^2(X, \mathbb{Z})^- = \{x \in H^2(X, \mathbb{Z}) \mid (\sigma^2)^*(x) = -x\}.$$

We also denote by S_X, T_X the *Picard lattice*, the *transcendental lattice* of X respectively.

LEMMA 2.1.   (1) $H^2(X, \mathbb{Z})^+ \simeq U(2) \oplus D_4 \oplus D_4$.
(2) $H^2(X, \mathbb{Z})^- \simeq U \oplus U(2) \oplus D_4 \oplus D_4$.
(3) σ^* *acts on* $H^2(X, \mathbb{Z})^+$ *trivially*.
(4) *The following elements generate* $(H^2(X, \mathbb{Z})^+)^*/H^2(X, \mathbb{Z})^+ \simeq (\mathbb{Z}/2\mathbb{Z})^6$:

$(F_1 + F_2)/2$, $(F_1 + F_3)/2$, $(F_1 + F_4)/2$, $(F_1 + F_5)/2$, $(F_1 + F_6)/2$, $(F_1 + F_7)/2$.

(5) *Let* $U \oplus A_1^{\oplus 8}$ *be the sublattice of* $H^2(X, \mathbb{Z})^+$ *generated by the classes of a fiber,* S_0 *and* F_i $(1 \le i \le 8)$. *Then* $H^2(X, \mathbb{Z})^+$ *is obtained from* $U \oplus A_1^{\oplus 8}$ *by adding the vector* $(F_1 + \cdots + F_8)/2$.

PROOF. For the proof of the assertions (1)–(4), see [**Kon1**], Lemma 5.2. The sublattice $U \oplus A_1^{\oplus 8}$ has index 2 in $H^2(X, \mathbb{Z})^+$. The later one is obtained from the former by adding the class of S_1. Hence the last assertion follows from the fact that $S_1 = 2F + S_0 + (F_1 + \cdots + F_8)/2$ where F is a fiber of π. $\square$

It follows that $H^2(X, \mathbb{Z})^+ \subset S_X$ and $T_X \subset H^2(X, \mathbb{Z})^-$.

2.2. A quadratic form. First of all, we define:

(2.5) $$L = U^{\oplus 3} \oplus E_8^{\oplus 2}, \quad M = U(2) \oplus D_4 \oplus D_4, \quad N = U \oplus U(2) \oplus D_4 \oplus D_4.$$

Recall that $H^2(X, \mathbb{Z}) \cong L$. We consider M as a sublattice of L and N is the orthogonal complement of M in L. It follows from Theorem 1.14.4 in Nikulin [**N3**] that the embedding of M into L is unique. Let $A_N = N^*/N$ which is isomorphic to a vector space $\mathbb{F}_2^6$ of dimension 6 over $\mathbb{F}_2$ (Lemma 2.1). The *discriminant quadratic form*

(2.6) $$q_N : A_N \longrightarrow \mathbb{Q}/2\mathbb{Z}$$

is defined by $q_N(x) = \langle x, x \rangle \mod 2\mathbb{Z}$. In our situation, the image of q_N is contained in $\mathbb{Z}/2\mathbb{Z}$ and hence q_N is a quadratic form on A_N defined over $\mathbb{F}_2$ whose associated bilinear form is given by

$$b_N(x, y) = 2\langle x, y \rangle \mod 2\mathbb{Z}.$$

Let u be the hyperbolic plane defined over $\mathbb{F}_2$, that is, the quadratic form of dimension 2 defined over $\mathbb{F}_2$ corresponding to the matrix $\begin{pmatrix} 0 & 1 \\ 1 & 0 \end{pmatrix}$. The quadratic form q_N is isomorphic to the direct sum of 3 copies of u: $q_N \cong u^{\oplus 3}$. It is known that $(A_M = M^*/M, q_M) \cong (A_N, -q_N)$ ([**N3**], Corollary 1.6.2).

LEMMA 2.2. *Let* $O(q_M)$ *be the group of isomorphisms of* A_M *preserving* q_M. *Then*

(1) $O(q_M) \simeq O(q_N) \simeq O(u_2^{\oplus 3}) \simeq S_8$ *where S_8 is the symmetric group of degree 8.*

(2) *The group $O(q_M)$ naturally isomorphic to the subgroup of $O(M)$ generated by the permutations of the 8 components A_1 in $U \oplus A_1^{\oplus 8}$.*

PROOF. The assertion (1) is well known (e.g. [**Atlas**], page 22). For (2), note that the permutations of 8 components A_1 can be extended to isometries of M because they preserve $(F_1 + \cdots + F_8)/2$ (see Lemma 2.1, (5)). Now the assertion is obvious. □

2.3. A fundamental domain. Let $M = U(2) \oplus D_4 \oplus D_4$. Let $P(M)$ be a connected component of $\{x \in M \otimes \mathbb{R} : \langle x, x \rangle > 0\}$. Let $W(M)$ be the reflection group generated by (-2)-reflections

$$s_r : x \to x + \langle x, r \rangle r$$

for any $r \in M$ with $r^2 = -2$. The group $W(M)$ acts on $P(M)$ discretely. Let $C(M)$ be the finite polyhedral cone defined by the 18 (-2)-vectors which are corresponding to the 18 smooth rational curves S_0, S_1, E_i, F_i $(1 \leq i \leq 8)$ on X under an isomorphism $M \cong H^2(X, \mathbb{Z})^+$:

$$C(M) = \{x \in P(M) : \langle x, r \rangle > 0, r = S_0, S_1, E_i, F_i (1 \leq i \leq 8)\}.$$

PROPOSITION 2.3.   (1) *The group $W(M)$ is of finite index in the orthogonal group $O(M)$. Moreover the closure $\bar{C}(M)$ of $C(M)$ is a fundamental domain of $W(M)$. The symmetry group of $C(M)$ is isomorphic to $S_8 \times \mathbb{Z}/2\mathbb{Z}$.*

(2) *If $S_X \cong M$, then X contains exactly 18 smooth rational curves S_0, S_1, E_i, F_i $(1 \leq i \leq 8)$.*

PROOF. (1) The first assertion follows from Nikulin's classification of such hyperbolic 2-elementary lattices ([**N1**], Theorem 4.4.1). Moreover $C(M)$ satisfies the condition in Vinberg's theorem [**V**], Theorem 2.6, that is, any maximal extended Dynkin diagram in these 18 (-2)-vectors is either $\tilde{A}_1^{\oplus 8}$ or $\tilde{D}_4^{\oplus 2}$ both of which have the maximal rank 8. Hence $C(M)$ is of finite volume. Let $C(M)'$ be a fundamental domain of $W(M)$ with $C(M)' \subset C(M)$. Then [**V**], Lemma 2.4 implies that $C(M) = C(M)'$. The last assertion is obvious.

(2) It follows from a remark in Vinberg [**V**], p. 335 that $C(M)$ has finite volume iff the polyherdal cone $\bar{C}(M)$ is contained in the closure $\bar{P}(M)$ of $P(M)$. If there exists a smooth rational curve E different from the above 18 curves, then the intersection number of E with any one of these 18 curves is non negative, that is, the class of E is contained in $\bar{C}(M) \subset \bar{P}(M)$. This implies that $E^2 \geq 0$, which is a contradiction. □

2.4. Let X be a $K3$ surface as above. Let $P(X)$ be the component of $\{x \in S_X \otimes \mathbb{R} : \langle x, x \rangle > 0\}$ which contains an ample class. Let $\Delta(X)$ be the set of all effective divisors r with $r^2 = -2$. Let $C(X)$ be the polyhedral cone defined by:

$$C(X) = \{x \in P(X) : \langle x, r \rangle > 0, r \in \Delta(X)\}.$$

Note that the integral points in $C(X)$ are nothing but the ample classes. If $S_X \cong M$, then $C(X) = C(M)$ (Proposition 2.3, (2)).

LEMMA 2.4. *The orthogonal complement of $H^2(X, \mathbb{Z})^+$ in S_X contains no (-2)-vectors.*

PROOF. If $r \in (H^2(X, \mathbb{Z})^+)^\perp \cap S_X$ with $r^2 = -2$, then $(\sigma^*)^2(r) = -r$. On the other hand, Riemann-Roch theorem implies that r is effective. This is a contradiction. □

PROPOSITION 2.5. *Assume that $S_X = H^2(X, \mathbb{Z})^+$. Then the automorphism group of X is finite. Moreover X has exactly 18 smooth rational curves which are components of singular fibers of π and two sections.*

PROOF. Recall that $Aut(X)$ is isomorphic to $O(S_X)/W(S_X)$ up to finite groups ([**PS**]). Here $W(S_X)$ is the subgroup generated by (-2)-reflections. Hence the assertion follows from Proposition 2.3. □

2.5. The automorphism of order 4. We shall study the action of σ on $H^2(X, \mathbb{Z})^-$. Recall that

$$D_4 \cong \{(x_1, x_2, x_3, x_4) \in \mathbb{Z}^4 \mid x_1 + x_2 + x_3 + x_4 \equiv 0 \ (\text{mod } 2)\}.$$

Here we consider the standard inner product on $\mathbb{Z}^4$ with the negative sign. Let ρ_0 be the isometry of D_4 given by

$$\rho_0(x_1, x_2, x_3, x_4) = (x_2, -x_1, x_4, -x_3).$$

Obviously ρ_0 is of order 4 and fixes no non-zero vectors in D_4. Also an easy calculation shows that ρ_0 acts trivially on D_4^*/D_4. Next let e, f (resp. e', f') be a basis of U (resp. $U(2)$) with $e^2 = f^2 = 0, \langle e, f \rangle = 1$ (resp. $(e')^2 = (f')^2 = 0, \langle e', f' \rangle = 2$). Define the isometry ρ_1 of $U \oplus U(2)$ by

$$\rho_1(e) = -e - e', \ \rho_1(f) = f - f', \ \rho_1(e') = e' + 2e, \ \rho_1(f') = 2f - f'.$$

Obviously ρ_1 is of order 4, fixes no non-zero vectors in $U \oplus U(2)$ and acts trivially on the discriminant group $(U \oplus U(2))^*/U \oplus U(2)$. Thus we have an isometry $\rho = \rho_1 \oplus \rho_0 \oplus \rho_0$ of $N = U \oplus U(2) \oplus D_4 \oplus D_4$ which fixes no non-zero vectors in N and acts trivially on N^*/N. Then ρ can be extended to an isometry of the $K3$ lattice L acting trivially on M ([**N3**], Proposition 1.6.1).

LEMMA 2.6. *The isometry ρ is conjugate to σ^* under an isomorphism*

$$H^2(X, \mathbb{Z}) \cong L.$$

PROOF. Let ω be an eigenvector of ρ which is sufficiently general, that is, satisfying the condition $\omega^\perp \cap L = M$. By the surjectivity of the period map for $K3$ surfaces, there exists a $K3$ surface Y and an isometry

$$\alpha_Y : H^2(Y, \mathbb{Z}) \to L$$

with $\alpha_Y(\omega_Y) = \omega$ where ω_Y is a nowhere vanishing holomorphic 2-form on Y. By the condition $\omega^\perp \cap L = M$, we have $S_Y \cong M$. Consider the isometry $\phi = \alpha_Y^{-1} \circ \rho \circ \alpha_Y$ of $H^2(Y, \mathbb{Z})$. Since ρ acts trivially on M, ϕ preserves ample classes. Then it follows from the Torelli theorem [**PS**] that ϕ is induced from an automorphism g of Y of order 4. On the other hand, Proposition 2.5 implies that Y contains exactly 18 smooth rational curves whose dual graph is the same as that of the smooth rational curves on X. In particular, Y has an elliptic fibration with two sections and 8 singular fibers each of which is type III or I_2. Since g acts trivially on the Picard lattice M, g preserves the elliptic fibration and the class of each component of singular fibers of type III or I_2. Since the elliptic fibration has 8 reducible singular fibers, g acts trivially on the base of the elliptic fibration, and hence induces an automorphism of each fiber. Hence all singular fibers are of type III. It follows from Nikulin [**N1**], Theorem 4.2.2 that the set of fixed points of the involution g^2 is the disjoint union of two smooth rational curves R_0, R_1 and a smooth curve C of genus 3. Since g acts trivially on the base, R_0, R_1 are sections of the elliptic fibration. We can easily see that C passes through singular points of singular fibers of type III. Thus we have the same configuration of smooth rational curves on Y as that of X. By taking the quotient of Y by g^2, we can see that Y is a deformation of X. Hence we have the assertion. $\square$

2.6. Markings. Recall that $H^2(X, \mathbb{Z})^+ \cong M = U(2) \oplus D_4 \oplus D_4$ (Lemma 2.1). We fix a fundamental domain $C(M)$ (Proposition 2.3). It follows from Lemma 2.6 that there exists an isometry

$$\alpha_X : H^2(X, \mathbb{Z}) \to L$$

satisfying $\rho = \alpha_X \circ \sigma^* \circ \alpha_X^{-1}$. We call α_X a *marking* and the pair (X, α_X) a *marked* $K3$ surface. Then

PROPOSITION 2.7. *There exists a marking α_X such that*

$$\alpha_X(C(X)) \cap M \otimes \mathbb{R} \subset C(M).$$

PROOF. It follows from Lemma 2.4 that $\alpha_X(C(X)) \cap M \otimes \mathbb{R}$ is an open polyhedral cone in $M \otimes \mathbb{R}$. Hence Proposition 2.3 implies the assertion. $\square$

3. A complex ball uniformization

In this section we construct an S_8-equivariant isomorphism between the moduli space of the projective equivalence classes of the set of distinct 8 ordered points in $\mathbb{P}^1$ and an open set of the arithmetic quotient of 5-dimensional complex ball.

3.1. The period domain. Let (X, α_X) be a marked $K3$ surface and let ω_X be a nowhere vanishing holomorphic 2-form on X. Then $\alpha_X(\omega_X)$ is contained in the following domain:

$$(3.1) \qquad \mathcal{D} = \{\omega \in \mathbb{P}(N \otimes \mathbb{C}) : \langle \omega, \omega \rangle = 0, \langle \omega, \bar{\omega} \rangle > 0\}.$$

Note that $\mathcal{D}$ is a disjoint union of two copies of a bounded symmetric domain of type IV and of dimension 10. To get the period domain, we first define:

$$(3.2) \qquad V_{\pm} = \{z \in N \otimes \mathbb{C} \mid \rho(z) = \pm\sqrt{-1}z\}.$$

It follows from Nikulin [**N2**], Theorem 3.1 that $N \otimes \mathbb{C} = V_+ \oplus V_-$. Now we may assume $\sigma^*(\omega_X) = \sqrt{-1} \cdot \omega_X$. Then $\alpha_X(\omega_X)$ is, in fact, contained in $\mathcal{B}$ defined by

$$(3.3) \qquad \mathcal{B} = \{z \in \mathbb{P}(V_+) \mid \langle z, \bar{z} \rangle > 0\}.$$

If $z \in \mathcal{B}$, then

$$\langle z, z \rangle = \langle \rho(z), \rho(z) \rangle = \langle \sqrt{-1}z, \sqrt{-1}z \rangle = -\langle z, z \rangle,$$

and hence $\langle z, z \rangle = 0$. Thus we have

$$\mathcal{D} \cap \mathbb{P}(V_+) = \mathcal{B}.$$

We remark that $\mathcal{B}$ is a 5-dimensional complex ball. We call $\alpha_X(\omega_X)$ the *period* of (X, α_X). We also define two arithmetic subgroups:

(3.4) $$\Gamma = \{\gamma \in O(N) \mid \gamma \circ \rho = \rho \circ \gamma\};$$

(3.5) $$\Gamma(1 - i) = \mathrm{Ker}(\Gamma \to O(q_N)).$$

We shall see that the quotient $\mathcal{B}/\Gamma$ (resp. $\mathcal{B}/\Gamma(1 - i)$) is the coarse moduli space of distinct 8 unordered points on $\mathbb{P}^1$ (resp. distinct 8 ordered points on $\mathbb{P}^1$) (see Theorem 3.3).

3.2. Hermitian form. We consider N as a free $\mathbb{Z}[\sqrt{-1}]$-module Λ by

$$(a + b\sqrt{-1})x = ax + b\rho(x).$$

Let

$$h(x, y) = \sqrt{-1}\langle x, \rho(y)\rangle + \langle x, y\rangle.$$

Then $h(x, y)$ is a hermitian form on $\mathbb{Z}[\sqrt{-1}]$-module Λ. With respect to a $\mathbb{Z}[\sqrt{-1}]$-basis $(1, -1, 0, 0)$, $(0, 1, -1, 0)$ of D_4, the hermitian matrix of $h \mid D_4$ is given by

(3.6) $$\begin{pmatrix} -2 & 1 - \sqrt{-1} \\ 1 + \sqrt{-1} & -2 \end{pmatrix}.$$

And with respect to a $\mathbb{Z}[\sqrt{-1}]$-basis e, f of $U \oplus U(2)$, the hermitian matrix of $h \mid U \oplus U(2)$ is given by

(3.7) $$\begin{pmatrix} 0 & 1 + \sqrt{-1} \\ 1 - \sqrt{-1} & 0 \end{pmatrix}.$$

Let

$$\varphi : \Lambda \to N^*$$

be a linear map defined by $\varphi(x) = (x + \rho(x))/2$. Note that $\varphi((1 - \sqrt{-1})x) = \varphi(x - \rho(x)) = x \in N$. Hence φ induces an isomorphism

(3.8) $$\Lambda/(1 - \sqrt{-1})\Lambda \simeq N^*/N.$$

REMARK 3.1. The hermitian form h coincides with the one of Matsumoto and Yoshida in [**MY**], §6. This implies that our groups Γ, $\Gamma(1 - i)$ coincide with the ones of Matsumoto and Yoshida in [**MY**].

3.3. Reflections. For $r \in N$ with $\langle r, r \rangle = -2$, we define a *reflection*

$$s_r(x) = x + \langle r, x \rangle r$$

which is contained in $\tilde{O}(N) = \mathrm{Ker}(N \to O(q_N))$, but not in Γ. On the other hand, by considering r as in Λ, we define a *reflection*

(3.9) $$R_{r,\epsilon}(x) = x - (1 - \epsilon)\frac{h(r, x)}{h(r, r)}r$$

where $\epsilon \neq 1$ is a 4-th root of unity. We can easily see that $R_{r,-1}$ corresponds to the isometry in Γ

$$x \to x + \langle r, x \rangle r + \langle \rho(r), x \rangle \rho(r)$$

which coincides with $s_r \circ s_{\rho(r)}$. Also $R_{r,\sqrt{-1}}$ corresponds to the isometry in Γ

$$x \to x + \langle r, x \rangle (r - \rho(r))/2 + \langle \rho(r), x \rangle (r + \rho(r))/2$$

which induces a transvection of A_N defined by

$$t_\alpha(x) = x + b_N(x, \alpha)\alpha$$

where $\alpha \in A_N$ is a non-isotropic vector $(r + \rho(r))/2 \bmod N$.

3.4. Discriminant. Let $r \in N$ with $r^2 = -2$. We denote by H_r the hyperplane of $\mathcal{B}$ defined by

$$H_r = \{z \in \mathcal{B} : \langle z, r \rangle = 0\}.$$

Let $\mathcal{H}$ be the union of all hyperplanes H_r where r moves on the set of all (-2)-vectors in N. We call $\mathcal{H}$ the *discriminant locus*. By Lemma 2.4, the periods of marked $K3$ surfaces as above are contained in $\mathcal{B} \setminus \mathcal{H}$.

Conversely let $\omega \in \mathcal{B} \setminus \mathcal{H}$. Then by the surjectivity of the period map, there exists a marked $K3$ surface (X, α_X) with $\alpha_X(\omega_X) = \omega$. The condition $\omega \notin \mathcal{H}$ implies that Proposition 2.7 holds for this $K3$ surface. Hence, if necessary by replacing α_X, we may assume that the isometry $\alpha_X^{-1} \circ \rho \circ \alpha_X$ preserves the ample cone of X. It now follows from the Torelli type theorem ([**PS**]) that there exists an automorphism σ of order 4 satisfying $\alpha_X^{-1} \circ \rho \circ \alpha_X = \sigma^*$. Moreover the marking as in Proposition 2.7 defines an elliptic fibration

$$\pi : X \to \mathbb{P}^1$$

with a section s.

LEMMA 3.2.   (1) π has 8 *singular fibers of type III and two sections.*
  (2) *The set of fixed points of σ^2 is the disjoint union of two sections and a smooth curve of genus 3 which passes through 8 singular points of 8 singular fibers.*

PROOF. It is known that the set of fixed points of the involution σ^2 is the disjoint union of two smooth rational curves R_0, R_1 and a smooth curve C of genus 3 (Nikulin [**N1**], Theorem 4.2.2). Obviously the set X^σ of fixed points of σ is contained in $R_0 + R_1 + C$. Since $\sigma^* = \alpha_X^{-1} \circ \rho \circ \alpha_X$, X^σ has the Euler number 12. Since σ acts on M trivially, it preserves the section s and the class of a fiber of π. We show that σ acts trivially on the base of π. Assume otherwise, then X^σ is contained in two invariant fibers, F_1, F_2. Let l be the number of irreducible one-dimensional components of X^σ and k the number of isolated fixed points of σ. Then $2l + k = 12$. If we denote by U the sublattice generated by the classes of a fiber and the section s, then $U^\perp \cap H^2(X, \mathbb{Z})^{\sigma^*} = A_1^{\oplus 8}$. Hence the divisor $F_1 + F_2$ contains at least 10 components. Assume F_1 contains at least 5 components. Note that σ preserves the component of F_1 which meets with s. Obviously there are no singular fibers with non-trivial symmetry of order 4. Hence the involution σ^2 preserves each component of F_1. Then the sublattice generated by components of F_1 not meeting s has at least rank 4 and is isomorphic to an indecomposable root lattice R. Since σ^2 acts trivially on R, R is contained in $A_1^{\oplus 8}$ which is impossible.

Thus σ acts trivially on the base. This implies that each fiber has an automorphism of order 4. In particular, singular fibers of π are either of type III, III^* or I_0^*. Since $U^\perp \cap H^2(X, \mathbb{Z})^{\sigma^*} = A_1^{\oplus 8}$ and the singular fibers of type III^* and I_0^* have no non-trivial symmetry of order 4, every singular fiber is of type III. Since σ fixes two points on each component of a singular fiber one of which is the singular point, R_0, R_1 or C passes through these points. Now we can easily see the assertion (2).   □

THEOREM 3.3. *The period map induces an S_8-equivariant isomorphism ϕ between the moduli space $(P_1^8)^0$ of distinct ordered 8 points on the projective line and the quotient space $(\mathcal{B} \setminus \mathcal{H})/\Gamma(1-i)$.*

PROOF. As in 3.4, for each $\omega \in \mathcal{B} \setminus \mathcal{H}$, we have a marked $K3$ surface (X, α_X) with $\alpha_X(\omega_X) = \omega$. Moreover X has an automorphism σ of order 4. By Lemma 3.2, X has an elliptic fibration with two section and 8 singular fibers of type III. By taking the quotient of X by σ^2 and contracting (-1)-curves, we have the embedding of C as in 2.1. This correspondence is the inverse of the period map. □

4. Discriminant locus

In this section we shall determine the discriminant locus $\mathcal{H}$ of $\mathcal{B}$.

4.1. Let r be a (-2)-vector in N. Let $\omega \in \mathcal{B}$. Then $\langle r, \omega \rangle = \sqrt{-1} \langle \rho(r), \omega \rangle$. Hence r and $\rho(r)$ define the same hyperplane H_r in $\mathcal{B}$. Hence H_r corresponds to an embedding of the lattice $R_r \cong A_1 \oplus A_1$ generated by r and $\rho(r)$ into N. Obviously $R_r \simeq A_1 \oplus A_1$. Also every embedding of R_r into N is primitive, that is, N/R_r is torsion free.

LEMMA 4.1. *Let $R_r^\perp$ be the orthogonal complement of R_r in N. Then $R_r^\perp \simeq U \oplus U(2) \oplus D_4 \oplus A_1^2$. In particular, $(r + \rho(r))/2 \in N^*$.*

PROOF. The proof for the first assertion is similar to those of [**Kon1**], Lemmas 3.2, 3.3. Then $R_r^\perp \oplus R_r$ is a sublattice of N of index 2 and N is obtained from $R_r^\perp \oplus R_r$ by adding $(r + \rho(r))/2 + \theta$ where $\theta \in (R_r^\perp)^*$. We can see that $\langle (r+\rho(r))/2, x \rangle \in \mathbb{Z}$ for $x \in R_r \oplus R_r^\perp$ and $x = (r + \rho(r))/2 + \theta$. Hence the second assertion holds. □

4.2. Note that A_N consists of the following 64 vectors:

Type (00) : $\alpha = 0$, $\#\alpha = 1$ $(zero)$;
Type (0) : $\alpha \neq 0$, $q(\alpha) = 0$, $\#\alpha = 35$ $(non\text{-}zero\ isotropic\ vector)$;
Type (1) : $q(\alpha) = 1$, $\#\alpha = 28$ $(non\text{-}isotropic\ vector)$.

4.3. By Lemma 4.1, the vector $(r+\rho(r))/2$ is contained in N^*. In particular it defines a non-isotropic vector $(r+\rho(r))/2 \mod N$ in A_N. Conversely let δ be a (-4)-vector in N with $\delta/2 \in N^*$. Since ρ acts trivially on $A_N = N^*/N$, $\delta - \rho(\delta) \in 2N$. Put $r = (\delta - \rho(\delta))/2 \in N$. Since $\langle \delta, \rho(\delta) \rangle = 0$, $r^2 = -2$. Obviously $\delta = r + \rho(r)$. Thus we have

LEMMA 4.2. $\{\delta \in N \ : \ \delta^2 = -4, \ \delta/2 \in N^*\} = \{r + \rho(r) \ : \ r \in N, \ r^2 = -2\}$.

LEMMA 4.3. *Any non-isotropic vector in A_N is represented by $(r + \rho(r))/2$ for a suitable (-2)-vector r in N.*

PROOF. It follows from [**N3**], Theorem 1.14.2 that the natural map from $O(N)$ to $O(A_N)$ is surjective. The group $O(A_N) \cong S_8$ acts transitively on the set of non-isotropic vectors in A_N. Combining these with Lemma 4.2, we have the assertion. □

PROPOSITION 4.4.   (1) $\Gamma/\Gamma(1-i) \simeq S_8$.
(2) Γ *acts transitively on the set of cusps of $\mathcal{B}$ and on the set of ρ-invariant $R = A_1 \oplus A_1$.*
(3) $\Gamma(1-i)$-*orbits of ρ-invariant $R = A_1 \oplus A_1$ bijectively correspond to non-isotropic vectors in A_N. Also $\Gamma(1-i)$-orbits of cusps of $\mathcal{B}$ bijectively correspond to non-zero isotropic vectors in A_N.*

PROOF. Recall that the pair (N, ρ) naturally corresponds to the hermitian form h (see Remark 3.1). Hence the assertions follow from [**MY**].   □

4.4. Stable points. Next we shall construct a $K3$ surface associated to each stable point from P_1^8. Recall that 8 points is *stable* (resp. *semi-stable*) iff no four points (resp. five points) coincide (e.g. [**DO**], Chap. I, §4, Example 2 (page 31)). We denote by the symbol (11111111) for distinct 8 points in $\mathbb{P}^1$. If two points (resp. three points) coincide, then we denote it by (2111111) (resp. (311111)).

EXAMPLE 4.5. We consider the case (2111111). We use the same notation as in Section 2. We assume that $(\lambda_1 : 1)$ has multiplicity 2. Then the curve C in (2.1) degenerates to one of the following two types:

$$C_1 : \ (x_0 - \lambda_1 x_1)(y_0^2 \prod_{i=2}^{4}(x_0 - \lambda_i x_1) + y_1^2 \cdot \prod_{i=5}^{7}(x_0 - \lambda_i x_1)) = 0;$$

$$C_2 : \ y_0^2(x_0 - \lambda_1 x_1)^2 \prod_{i=2}^{3}(x_0 - \lambda_i x_1) + y_1^2 \cdot \prod_{i=4}^{7}(x_0 - \lambda_i x_1) = 0.$$

The minimal resolution Y_i of the double covering of $\mathbb{P}^1 \times \mathbb{P}^1$ branched along $C_i + L_0 + L_1$ is a $K3$ surface and has an elliptic fibration π which has 6 singular fibers of type III, one singular fiber of type I_0^* and two sections S_0, S_1 $(i = 1, 2)$. We remark that Y_1 and Y_2 are isomorphic because C_1 and C_2 are mutually transformed under elementary transformations. Thus we denote by Y instead of Y_1, Y_2. Denote by $2R_0 + R_1 + R_2 + R_3 + R_4$ the singular fiber of type I_0^*. Assume that S_0 meets R_1 and S_1 meets R_2. Then the normalization $\tilde{C}$ of C_1 meets R_3, R_4. The involution ι given in (2.2) induces an automorphism σ' of Y of order 4. Note that the restriction σ' on $\tilde{C}$ is the hyperelliptic involution of the smooth curve of genus 2. This implies that σ' switches R_3 and R_4. Let U be the sublattice generated by the classes of a fiber and S_0. Then 6 components in the fibers of type III not meeting to S_0 and $R_2, 2R_0 + R_2 + R_3 + R_4$ generate the sublattice isomorphic to $A_1^{\oplus 8}$. This gives an isometry from M into S_X such that $M^\perp \cap S_X$ contains (-2) vectors R_3, R_4. Thus the period of Y is contained in $\mathcal{H}$.

For other stable points, the process is similar. We have several types of the branch curve C depending on the order of 8 points, however, they are transformed each other under elementary transformations. Hence the corresponding $K3$ surface is determined by the isomorphism class of 8 points (independent of the order of 8 points).

If three points coincide, then the corresponding elliptic fibration has a singular fiber of type III^*. All cases except (2222), the elliptic fibration has two sections. In case of (2222), it has four sections.

The next Table 1 lists the type of 8 stable points on the projective line, type of singular fibers of the elliptic fibration, the Picard lattice and the transcendental lattice of a generic member.

4.5. Strictly semi-stable points: (44). In this case, we have the following 3 cases of curves in the quadric corresponding to the strictly semi-stable points with unique minimal closed orbit:

$$C_3 : \ (x_0 - \lambda_1 x_1)^2(x_0 - \lambda_2 x_1)^2(y_0^2 + y_1^2) = 0;$$

$$C_4 : \ (x_0 - \lambda_1 x_1)(x_0 - \lambda_2 x_1)(y_0^2(x_0 - \lambda_1 x_1)^2 + y_1^2(x_0 - \lambda_2 x_1)^2) = 0;$$

	8 points	Singular fibers	Picard lattice	Transcendental lattice
1)	(11111111)	$8III$	$U(2) \oplus D_4 \oplus D_4$	$U \oplus U(2) \oplus D_4 \oplus D_4$
2)	(2111111)	$I_0^*,\ 6III$	$U \oplus D_4 \oplus D_4 \oplus A_1^{\oplus 2}$	$U \oplus U(2) \oplus D_4 \oplus A_1^{\oplus 2}$
3)	(221111)	$2I_0^*,\ 4III$	$U \oplus D_6 \oplus D_4 \oplus A_1^{\oplus 2}$	$U \oplus U(2) \oplus A_1^{\oplus 4}$
4)	(22211)	$3I_0^*, 2\ III$	$U \oplus D_6 \oplus D_6 \oplus A_1^{\oplus 2}$	$A_1(-1)^{\oplus 2} \oplus A_1^{\oplus 4}$
5)	(2222)	$4I_0^*$	$U \oplus D_8 \oplus D_8$	$U(2) \oplus U(2)$
6)	(311111)	$III^*,\ 5III$	$U \oplus D_8 \oplus D_4$	$U \oplus U(2) \oplus D_4$
7)	(32111)	$III^*,\ I_0^*,\ III$	$U \oplus E_8 \oplus D_4 \oplus A_1^{\oplus 2}$	$U \oplus U(2) \oplus A_1^{\oplus 2}$
8)	(3221)	$III^*,\ 2I_0^*,\ III$	$U \oplus E_8 \oplus D_6 \oplus A_1^{\oplus 2}$	$A_1(-1)^{\oplus 2} \oplus A_1^{\oplus 2}$
9)	(3311)	$2III^*,\ 2III$	$U \oplus E_8 \oplus D_8$	$U \oplus U(2)$
10)	(332)	$2III^*,\ I_0^*$	$U \oplus E_8 \oplus D_{10}$	$A_1(-1)^{\oplus 2}$

TABLE 1

$$C_5 :\ y_0^2(x_0 - \lambda_1 x_1)^4 + y_1^2(x_0 - \lambda_2 x_1)^4 = 0.$$

These curves appear in the list of Shah's classification of semistable $K3$ surfaces of degree 4. See Shah [**S**], Theorem 4.8, B, Type II, (i)–(iii).

We denote by $\bar{\mathcal{B}}/\Gamma(1-i)$ the Satake-Baily-Borel compactification of $\mathcal{B}/\Gamma(1-i)$ whose boundary consists of 35 cusps. Then we conclude:

THEOREM 4.6. *The S_8-equivariant isomorphism ϕ in Theorem 3.3 can be extended to an S_8-equivariant isomorphism $\tilde{\phi}$ between P_1^8 and $\bar{\mathcal{B}}/\Gamma(1-i)$. Moreover $\tilde{\phi}$ sends strictly semistable points to cusps and stable but not distinct 8 points into $\mathcal{H}/\Gamma(1-i)$.*

PROOF. We can apply the argument of Horikawa's proof of the main theorem in [**H**]. Let $\mathcal{M}$ be the space of all 8 semi-stable points on $\mathbb{P}^1$ and $\mathcal{M}_0$ the space of all distinct 8 points on $\mathbb{P}^1$. We can easily see that $\mathcal{M} \setminus \mathcal{M}_0$ is locally contained in a divisor with normal crossing. By construction, ϕ is locally liftable to $\mathcal{B}$. It now follows from a theorem of Borel [**Bo**] that ϕ can be extended to a holomorphic map from $\mathcal{M}$ to $\bar{\mathcal{B}}/\Gamma(1-i)$ which induces a holomorphic map

$$\tilde{\phi} : P_1^8 \to \bar{\mathcal{B}}/\Gamma(1-i).$$

By using the same argument as in the proof of [**H**], Theorem 2.2, we can see that $\tilde{\phi}$ sends stable, but non-distinct 8 points to $\mathcal{H}$. More precisely we can choose a marking for $K3$ surfaces corresponding to stable, but non-distinct 8 points, and define the period for them. For each stratification as in Table 1, we can prove the similar statement as in Lemma 3.2. Then, as in the generic case (Theorem 3.3), by case by case argument according to strata, we can see that the map $\tilde{\phi}$ is injective over $\mathcal{H}$. Moreover Shah's classification [**S**] implies the image of strictly semi-stable points go to the boundaries. Hence the Zariski Main theorem implies that $\tilde{\phi}$ is an isomorphism. The S_8-equivariantness is obvious. □

5. The Weil representation

In this section we shall study the quadratic form (A_N, q_N) over $\mathbb{F}_2$ given in (2.6) and the Weil representation of $SL(2, \mathbb{Z})$ on the group ring $\mathbb{C}[A_N]$.

5.1. Let

$$(5.1) \qquad\qquad T = \begin{pmatrix} 1 & 1 \\ 0 & 1 \end{pmatrix}, \quad S = \begin{pmatrix} 0 & -1 \\ 1 & 0 \end{pmatrix},$$

We denote by $\{e_\alpha\}_{\alpha \in A_N}$ the standard basis of $\mathbb{C}[A_N]$. Let ρ be the Weil representation of $SL(2, \mathbb{Z})$ on $\mathbb{C}[A_N]$ which factors through $SL(2, \mathbb{Z}/2\mathbb{Z})$ ([**Bor**]):

$$(5.2) \qquad \rho(T)(e_\alpha) = (-1)^{q_N(\alpha)} e_\alpha; \quad \rho(S)(e_\alpha) = \frac{1}{8} \sum_\beta (-1)^{b_N(\beta,\alpha)} e_\beta.$$

Representatives of the conjugacy classes of $SL(2, \mathbb{Z}/2\mathbb{Z}) \simeq S_3$ consist of E, T, ST. A direct calculation shows that the traces of the action of E, T, ST on $\mathbb{C}[A_N]$ are

$$tr(E) = 2^6, \ tr(T) = 8, \ tr(ST) = 1.$$

Let χ_i $(1 \leq i \leq 3)$ be the characters of irreducible representations of $SL(2, \mathbb{Z}/2\mathbb{Z})$: χ_1, χ_2 or χ_3 is the trivial, alternating character or the character of 2-dimensional irreducible representation respectively. Let χ be the character of the Weil representation of $SL(2, \mathbb{Z}/2\mathbb{Z})$ on $\mathbb{C}[A_N]$ and let $\chi = \sum_i m_i \chi_i$ be its decomposition into irreducible characters. Then an elementary calculation shows that

$$(5.3) \qquad\qquad \chi = 15\chi_1 + 7\chi_2 + 21\chi_3.$$

We call a subspace I of A_N *totally isotropic* if q_N vanishes on I, and I *maximal* if it has dimension 3.

LEMMA 5.1. *For each maximal totally isotropic subspace I of A_N,*

$$\sum_{\alpha \in I} e_\alpha \in \mathbb{C}[A_N]^{SL(2,\mathbb{Z})}.$$

PROOF. The proof is the same as that of [**Kon2**], Lemma 3.2. □

Let $\alpha \in A_N$ with $q_N(\alpha) = 1$. Then

$$t_\alpha : x \longrightarrow x + b_N(x, \alpha)\alpha$$

is called a *transvection* and contained in $O(q_N)$. Note that t_α is induced from a reflection s_r associated with a (-4)-vector r in N with $r/2 \bmod N = \alpha$ and these t_α $(\alpha \in A_N$ with $q_N(\alpha) = 1)$ generate $O(q_N)$. These 28 transvections in $O(A_N)$ correspond to the 28 transpositions of S_8.

DEFINITION 5.2. Let q_s be a quadratic form on $\mathbb{F}_2^3$ given by

$$q_s(x) = \sum_{i=1}^3 x_i, \ x = (x_1, x_2, x_3) \in \mathbb{F}_2^3.$$

Note that the associated bilinear form of q_s is identically zero. Let V be a 3-dimensional subspace of A_N. We call V *maximal totally singular* if $(V, q_N \mid V)$ is isomorphic to $(\mathbb{F}_2^3, q_s)$. Obviously V has a basis consisting of 3 mutually orthogonal non-isotropic vectors $\{\alpha_1, \alpha_2, \alpha_3\}$. We remark that V consists of 4 non-isotropic vectors α_i $(1 \leq i \leq 3)$, $\alpha_1 + \alpha_2 + \alpha_3$ and 4 isotropic vectors $0, \alpha_i + \alpha_j$, $(1 \leq i < j \leq 3)$. For each maximal totally singular subspace V, we define a vector $f_V \in \mathbb{C}[A_N]^{SL(2,\mathbb{Z})}$ on which the transvection t_α $(\alpha \in V)$ acts as -1. Let I be the kernel of $q_N \mid V$.

Then I is a totally isotropic subspace of dimension 2 in A_N and there exist exactly two maximal totally isotropic subspaces I^+, I^- in A_N which contain I. We define

$$(5.4) \qquad\qquad f_V = \sum_{\alpha \in I^+} e_\alpha - \sum_{\alpha \in I^-} e_\alpha \in \mathbb{C}[A_N].$$

THEOREM 5.3. *Let V be a maximal totally singular subspace. Then f_V is contained in $\mathbb{C}[A_N]^{SL(2,\mathbb{Z})}$ satisfying the following condition : f_V is the unique vector (up to constant) in $\mathbb{C}[A_N]$ on which transvections $t_\alpha(\alpha \in V, q_N(\alpha) = 1)$ act as -1.*

PROOF. The proof is the same as that of [**Kon2**], Theorem 3.4.   □

REMARK 5.4. The group $O(q_N)(\simeq S_8)$ naturally acts on $\mathbb{C}[A_N]^{SL(2,\mathbb{Z})}$ with character $\chi_1 + \chi_{14}$, where χ_1 is the trivial character and χ_{14} is the character of an irreducible representation of S_8 of degree 14. This follows from Lemma 5.1 and [**Atlas**], page 22. Moreover it follows from Theorem 5.3 that the multiplicity of the irreducible representation of degree 14 on $\mathbb{C}[A_N]$ is one. We denote by W the subspace of dimention 14 in $\mathbb{C}[A_N]^{SL(2,\mathbb{Z})}$ with character χ_{14}.

LEMMA 5.5. *The number of maximal totally singular subspaces of A_N is equal to 105.*

PROOF. We can easily count the number of mutually orthogonal three non-isotropic vectors in A_N which is $2^3 \cdot 3^2 \cdot 5 \cdot 7$. On the other hand, the automorphism group of a maximal totally singular subspace has order $2^3 \cdot 3$. Hence the assertion follows.   □

In the Lemma 7.1, we shall give a geometric interpretation of maximal totally singular subspaces.

5.2. Heegner divisors. Let $\delta \in N$ be a (-4)-vector with $\delta/2 \in N^*$. Let D_δ be the hyperplane of $\mathcal{D}$ defined by

$$D_\delta = \delta^\perp \cap \mathcal{D}.$$

It follows from Lemma 4.2 that

$$H_r = D_\delta \cap \mathcal{B}$$

where $r = (\delta - \rho(\delta))/2$ is a (-2)-vector in N and H_r is as in 3.4. For $\alpha \in A_N$ with $q_N(\alpha) = 1$, we define *Heegner divisors* $\mathcal{D}_\alpha$ and $\mathcal{H}_\alpha$ by

$$\mathcal{D}_\alpha = \sum_\delta D_\delta, \quad \mathcal{H}_\alpha = \sum_\delta H_\delta$$

where δ varies over the set of (-4)-vectors in N with $\delta/2 \bmod N = \alpha$. Since $H_r = D_\delta \cap \mathcal{B} = D_{\rho(\delta)} \cap \mathcal{B}$, we have

$$(5.5) \qquad\qquad 2\mathcal{H}_\alpha = \mathcal{D}_\alpha \mid \mathcal{B}.$$

6. Automorphic forms

6.1. Let $\{f_\alpha\}_{\alpha \in A_N}$ be a *vector valued elliptic modular form of weight* -4 *and of type* ρ, i.e., f_α is a holomorphic function on the upper half plane satisfying

$$(6.1) \qquad f_\alpha(\tau + 1) = e^{2\pi i q(\alpha)} f_\alpha(\tau), \quad f_\alpha(-1/\tau) = \frac{\tau^{-4}}{8} \sum_{\beta \in A_N} e^{-2\pi i \langle \alpha, \beta \rangle} f_\beta.$$

Recall that there are three types of vectors in A_N denoted by type 00, 0 or 1 according to zero, non-zero isotropic or non-isotropic respectively (see 4.2) . For each $\alpha \in A_N$, we denote by m_0 or m_1 the number of vectors $\beta \in A_N$ with $b_N(\alpha, \beta) = 0$ or 1 respectively which is as in the following Table 2:

α	00	00	00	0	0	0	1	1	1
β	00	0	1	00	0	1	00	0	1
m_0	1	35	28	1	19	12	1	15	16
m_1	0	0	0	0	16	16	0	20	12

TABLE 2

We shall find a modular form h such that the components h_α are given by functions h_{00}, h_0, h_1 depending only on the type of α. Then it follows from (6.1) and Table 2 that $h = \{h_\alpha\}$ satisfies:

$$h_{00}(\tau + 1) = h_{00}(\tau), \quad h_{00}(-1/\tau) = \frac{\tau^{-4}}{8}(h_{00}(\tau) + 35h_0(\tau) + 28h_1(\tau)),$$

$$h_0(\tau + 1) = h_0(\tau), \quad h_0(-1/\tau) = \frac{\tau^{-4}}{8}(h_{00}(\tau) + 3h_0(\tau) - 4h_1(\tau)),$$

$$h_1(\tau + 1) = -h_1(\tau), \quad h_1(-1/\tau) = \frac{\tau^{-4}}{8}(h_{00}(\tau) - 5h_0(\tau) + 4h_1(\tau)).$$

LEMMA 6.1. *One solution of these equations is given as follows:*

$$h_{00}(\tau) = 56\eta(2\tau)^8/\eta(\tau)^{16} = 56 + 896q + 8064q^2 + \cdots,$$

$$h_0(\tau) = -8\eta(2\tau)^8/\eta(\tau)^{16} = -8 - 128q - 1152q^2 - \cdots,$$

$$h_1(\tau) = 8\eta(2\tau)^8/\eta(\tau)^{16} + \eta(\tau/2)^8/\eta(\tau)^{16} = q^{-1/2} + 36q^{1/2} + 402q^{3/2} + \cdots$$

where $\eta(\tau)$ *is the Dedekind eta function and* $q = e^{2\pi\sqrt{-1}\tau}$.

PROOF. The proof is the same as that of [**Kon2**], Lemma 4.3. $\square$

6.2. By applying Borcherds [**Bor**], Theorem 13.3, for the vector valued modular form h given in Lemma 6.1, we have

THEOREM 6.2. *There exists an automorphic form of weight* $28(= 56/2)$ *on* $\mathcal{D}$ *which vanishes exactly on Heegner divisors corresponding* 28 *non-isotoropic vectors in* A_N.

6.3.   On the other hand, by Borcherds [**Bor**], Theorem 14.3, we have an S_8-equivariant map

$$\varphi : W \to \mathcal{A}_4(\tilde{O}(N))$$

where W is the 14-dimensional subspace of $\mathbb{C}[A_N]^{SL(2,\mathbb{Z})}$ given in Remark 5.4,

$$\tilde{O}(N) = \mathrm{Ker}(O(N) \to O(q_N))$$

and $\mathcal{A}_4(\tilde{O}(N))$ is the space of automorphic forms on $\mathcal{D}$ of weight 4 with respect to $\tilde{O}(N)$. It follows from [**N3**], Theorem 1.14.2 that the map from $O(N)$ to $O(q_N)$ is surjective and hence S_8 ($\cong O(q_N) \cong O(N)/\tilde{O}(N)$) naturally acts on $\mathcal{A}_4(\tilde{O}(N))$. On the other hand, S_8 acts on W. With respect to these actions, φ is S_8-equivariant.

LEMMA 6.3. *The map φ is injective.*

PROOF. The proof is the same as that of [**Kon2**], Lemma 4.1. $\square$

THEOREM 6.4. *Let V be a maximal totally singular subspace of A_N. Let F_V be an automorphic form of weight 4 on $\mathcal{D}$ associated with $f_V \in \mathbb{C}[A_N]^{SL(2,\mathbb{Z})}$: $F_V = \varphi(f_V)$. Then*

$$(F_V) = \sum_{\alpha \in V, \ q(\alpha)=1} \mathcal{D}_\alpha$$

where $\mathcal{D}_\alpha$ is the Heegner divisor associated with α.

PROOF. Let Φ be the product of all F_V where V varies over all maximal totally singular subspaces. Then Φ is an automorphic form of weight 105×4 (Lemma 5.5). By Theorem 5.3 and the S_8-equivariantness of ϕ, we have

$$\sum_{\alpha \in V, \ q(\alpha)=1} \mathcal{D}_\alpha \subset (F_V).$$

Hence Φ vanishes along $\mathcal{D}_\alpha$ with vanishing order $\geq 4 \times 105/28 = 15$. On the other hand, the 15-th power of the automorphic form given in Theorem 6.2 has the same weight and vanishes on $\mathcal{D}_\alpha$ with multiplicity 15. The assertion now follows from the Koecher principle. $\square$

7. Cross ratios

7.1.   Let

$$(7.1) \qquad\qquad \tau = \begin{pmatrix} \tau_{11} & \tau_{12} \\ \tau_{21} & \tau_{22} \\ \tau_{31} & \tau_{32} \\ \tau_{41} & \tau_{42} \end{pmatrix}, \quad \tau_{ij} \in \{1, 2, ..., 8\}.$$

We call τ a *tableau*. A tableau τ is called *standard* if

$$\tau_{ij} < \tau_{ij+1}, \quad \tau_{ij} \leq \tau_{i+1j}$$

for any i, j. The number of tableaus is 105 and the number of standard tableaus is 14 (e.g. [**DO**], Chap. I). For each τ we define

$$\mu_\tau = \prod_{1 \leq i \leq 4} \det(v^{\tau_{i1}} v^{\tau_{i2}})$$

where $v^i \in \mathbb{C}^2$ is a column vector. If

(7.2) $$v^i = \begin{pmatrix} 1 \\ x^i \end{pmatrix},$$

then,

$$\mu_\tau = \prod_{1 \le i \le 4} (x^{\tau_{i2}} - x^{\tau_{i1}}).$$

These μ_τ define an S_8-equivariant map Θ from P_1^8 to $\mathbb{P}^{13}$ (e.g. [**DO**]). We identify P_1^8 with $\bar{\mathcal{B}}/\Gamma(1-i)$ under the isomorphism given in Theorem 4.6. In the following, we shall discuss a relation between Θ and the map defined by 14-dimensional space W of automorphic forms given in §6.

7.2.   We give a relation between the set of tableaus and the set of totally singular subspaces. Let $K = U \oplus A_1^{\oplus 8}$ be the sublattice given in Lemma 2.1. Then $A_K = K^*/K \cong (\mathbb{F}_2)^8$ is generated by $F_i/2$ $(1 \le i \le 8)$ corresponding to 8 points on the projective line. The discriminant quadratic form q_K of K is a map

$$q_K : (\mathbb{F}_2)^8 \to \mathbb{Q}/2\mathbb{Z}$$

defined by $q_K(x) = \langle x, x \rangle \bmod 2\mathbb{Z}$. Let $\theta = (F_1 + \cdots + F_8)/2$ which is perpendicular to all vectors in A_K. Then it is known (Nikulin [**N3**], Proposition 1.4.1) that the discriminant quadratic form of M is obtained by

$$q_M = q_K \mid \theta^\perp/\theta.$$

Finally q_N is canonically isomorphic to $-q_M$ ([**N3**], Corollary 1.6.2). Thus non-isotropic vectors with respect to q_N bijectively corresponds to the vectors $(F_i + F_j)/2$ $(i \ne j)$ in A_K. Each column (τ_{i1}, τ_{i2}) of τ in (7.1) defines a vector in $(\mathbb{F}_2)^8$ whose nonzero entries are indexed by τ_{i1}, τ_{i2}. Thus four columns of τ corresponds to mutually orthogonal 4 non-isotropic vectors which generate a maximal totally singular subspace in A_N. This implies the following:

LEMMA 7.1.   *The set of 105 tableaus τ bijectively corresponds to the set of maximal totally singular subspaces of A_N. Under this correspondence, the zero of μ_τ coincides with the zero of F_V where V is a maximal totally singular subspace corresponding to τ.*

7.3.   Consider the linear system of automorphic forms F_V of dimension 14 defined by W (see 6.3). Note that the divisor $(F_V \mid \mathcal{B})$ is given by

$$2 \sum_{\alpha \in V,\ q(\alpha)=1} \mathcal{H}_\alpha$$

(see (5.5), Theorem 6.4). Since $\mathcal{B}$ is simply connected, we can take a square root of F_V. Thus we have an automorphic form G_V of weight 2 on $\mathcal{B}$ with

(7.3) $$(G_V) = \sum_{\alpha \in V,\ q(\alpha)=1} \mathcal{H}_\alpha.$$

Then $\{G_V\}_V$ defines a map Ψ from $\bar{\mathcal{B}}/\Gamma(1-i)$ to $\mathbb{P}^{13}$.

THEOREM 7.2.   *The map Θ coincides with Ψ.*

PROOF. We shall show that $\mu_{\tau_1}/\mu_{\tau_2}$ coincides with G_{V_1}/G_{V_2} for suitable tableaux τ_1, τ_2 and the corresponding maximal totally singular subspaces V_1, V_2. We consider the following tableaux:

$$(7.4) \qquad \tau_1 = \begin{pmatrix} 1 & 2 \\ 3 & 4 \\ 5 & 6 \\ 7 & 8 \end{pmatrix}, \tau_2 = \begin{pmatrix} 1 & 2 \\ 3 & 4 \\ 5 & 7 \\ 6 & 8 \end{pmatrix}, \tau_3 = \begin{pmatrix} 1 & 2 \\ 3 & 4 \\ 5 & 8 \\ 6 & 7 \end{pmatrix}.$$

We take a decomposition of $A_N = u_1 \oplus u_2 \oplus u_3$ into three hyperbolic planes u_1, u_2, u_3 defined over $\mathbb{F}_2$. Let $\{e_i, f_i\}$ be a basis of u_i with $\langle e_i, e_i \rangle = 0$, $\langle f_i, f_i \rangle = 0$, $\langle e_i, f_i \rangle = 1$. Let $\alpha_i = e_i + f_i$ be the non-isotropic vector in u_i. We may assume that

$$V_1 = \langle \alpha_1, \alpha_2, \alpha_3 \rangle, \ V_2 = \langle \alpha_1, \alpha_2, \alpha_1 + e_3 \rangle, \ V_3 = \langle \alpha_1, \alpha_2, \alpha_1 + f_3 \rangle$$

correspond to τ_1, τ_2, τ_3 respectively. Then by (7.3), we can see that $(G_{V_1}/G_{V_2}) = (\mu_{\tau_1}/\mu_{\tau_2})$ as divisors. Note that the function $\mu_{\tau_1}/\mu_{\tau_2}$ takes the value 1 on the divisors defined by $x_5 - x_8$ and $x_6 - x_7$. On the other hand, easy calculation shows that $f_{V_1} - f_{V_2} = f_{V_3}$ and hence $F_{V_1} - F_{V_2}$ vanishes on the divisor of F_{V_3}. This implies that $G_{V_1}/G_{V_2} = 1$ on the divisors $\mathcal{H}_{a_1+f_3}$ and $\mathcal{H}_{a_2+f_3}$. Hence $G_{V_1}/G_{V_2} = \mu_{\tau_1}/\mu_{\tau_2}$. For any pair of τ_1', τ_2', the ratio $\mu_{\tau_1'}/\mu_{\tau_2'}$ can be written as the product of some $\mu_{\tau_1}/\mu_{\tau_2}$ as above type. Hence the assertion follows. $\qquad \square$

THEOREM 7.3. Ψ is an embedding from $\bar{\mathcal{B}}/\Gamma(1-i)$ into $\mathbb{P}^{13}$. The image satisfies $2^2 \cdot 3 \cdot 5 \cdot 7$ quartic relations.

PROOF. It is known that Θ is embedding (Koike [**Koi**]). The proof of the second assertion is the same as that of [**Kon2**], Theorem 7.2, that is, for each non-isotropic vector in A_N we have 15 quartic relations. Since the number of non-isotropic vectors is 28, the second assertion follows. $\qquad \square$

REMARK 7.4. In the paper [**Koi**], by using a computer, he showed that the image is the intersection of 14 quadrics.

REMARK 7.5. In the paper [**MT**], Matsumoto and Terasoma constructed an S_8-equivariant map from the moduli space of ordered 8 points on $\mathbb{P}^1$ to $\mathbb{P}^{104}$ by using the theta constants related to the curve which is the 4-fold covering of $\mathbb{P}^1$ branched at 8 points. They showed that this map coincides with the map defined by the above 105 μ_τ.

REMARK 7.6. Since 8 points on $\mathbb{P}^1$ naturally correspond to hyperelliptic curves of genus three, we can consider that our case is a degenerate one of smooth curves of genus three. The moduli space of non-hyperelliptic curves of genus three can be also described as an arithmetic quotient of a complex ball ([**Kon1**]). On the other hand, Coble constructed a map from the moduli space of non-hyperelliptic curves of genus 3 with level 2-structure to $\mathbb{P}^{14}$ by using Göpel functions (Coble [**C**], Dolgachev, Ortland [**DO**], Chap. IX). It would be interesting to extend the result in this paper to the case of curves of genus three.

References

[AF] D. Allcock, E. Freitag, *Cubic surfaces and Borcherds products*, Comm. Math. Helv., **77** (2002), 270–296.

[Atlas] J. H. Conway et al., *Atlas of finite groups*, Oxford 1985.

[Bor] R. Borcherds, *Automorphic forms with singularities on Grassmannians*, Invent. Math., **132** (1998), 491–562.

[Bo] A. Borel, *Some metric properties of arithmetic quotients of symmetric spaces and an extension theorem*, J. Diff. Geometry **6** (1972), 543–560.

[C] A. Coble, *Algebraic geometry and theta functions*, Amer. Math. Soc. Coll. Publ., **10**, Providence, R.I., 1929 (3rd ed., 1969).

[DM] P. Deligne, G. W. Mostow, *Monodromy of hypergeometric functions and non-lattice integral monodromy*, Publ. Math. IHES, **63** (1986), 5–89

[DO] I. Dolgachev, D. Ortland, *Point sets in projective spaces and theta functions*, Astérisque **165**(1988).

[H] E. Horikawa, *On the periods of Enriques surfaces. II*, Math. Ann., **235** (1978), 217–246.

[Kod] K. Kodaira, *On compact complex analytic surfaces II*, Ann. Math., **77**(1963), 563–626. III, Ann. Math., **78**(1963), 1–40.

[Koi] K. Koike, *The projective embedding of the configuration space $X(2,8)$*, preprint.

[Kon1] S. Kondō, *A complex hyperbolic structure for the moduli space of curves of genus three*, J. reine angew. Math., **525**(2000), 219–232.

[Kon2] S. Kondō, *The moduli space of Enriques surfaces and Borcherds products*, J. Algebraic Geometry **11** (2002), 601–627.

[MT] K. Matsumoto, T. Terasoma, *Theta constants associated to coverings of $\mathbb{P}^1$ branching at 8 points*, Compositio Math., **140** (2004), 1277–1301.

[MY] K. Matsumoto, M. Yoshida, *Configuration space of 8 points on the projective line and a 5-dimensional Picard modular group*, Compositio Math., **86** (1993), 265–280.

[N1] V. V. Nikulin, *Factor groups of groups of automorphisms of hyperbolic forms with respect to subgroups generated by 2-reflections*, J. Soviet Math., **22** (1983), 1401–1475.

[N2] V. V. Nikulin, *Finite automorphism groups of Kähler K3 surfaces*, Trans. Moscow Math. Soc., **38** (1980), 71–135.

[N3] V. V. Nikulin, *Integral symmetric bilinear forms and its applications*, Math. USSR Izv., **14** (1980), 103–167.

[PS] I. Piatetski-Shapiro, I. R. Shafarevich, *A Torelli theorem for algebraic surfaces of type K3*, Math. USSR Izv., **5** (1971), 547–587.

[S] J. Shah, *Degenerations of K3 surfaces of degree 4*, Trans. A. M. S., **263** (1981), 271–308.

[V] E. B. Vinberg, *Some arithmetic discrete groups in Lobachevskii spaces*, in "Discrete subgroups of Lie groups and applications to moduli", Tata-Oxford (1975), 323–348.

GRADUATE SCHOOL OF MATHEMATICS, NAGOYA UNIVERSITY, NAGOYA, 464-8602, JAPAN
E-mail address: kondo@math.nagoya-u.ac.jp

Contemporary Mathematics
Volume **422**, 2007

Invariants of quartic plane curves as automorphic forms

Eduard Looijenga

ABSTRACT. We identify the algebra of regular functions on the space of quartic polynomials in three complex variables invariant under $SL(3, \mathbb{C})$ with an algebra of meromorphic automorphic forms on the complex 6-ball. We also discuss the underlying geometry.

To Igor Dolgachev, for his 60th birthday

1. Motivation and goal

One of the most wonderful mathematical gems of the 19th century is, at least to my taste (but I expect Igor to agree), the theorem that says that the algebra of invariants of plane cubics is the algebra of modular forms. Its precise statement is as follows. Consider the vector space of cubic homogeneous forms in three complex variables, which as a $SL(3, \mathbb{C})$-representation is the third symmetric power of the dual of the defining representation $\mathbb{C}^3$. Since $SL(3, \mathbb{C})$ acts through its simple quotient $PGL(3, \mathbb{C})$, we prefer to regard this a representation of the latter. Then the theorem I am refering to says that the algebra of $PGL(3, \mathbb{C})$-invariant polynomials on that space is as a graded $\mathbb{C}$-algebra isomorphic to the algebra of $SL(2, \mathbb{Z})$-modular forms (recall that this is the polynomial algebra generated by the Eisenstein series E_4 and E_6). This theorem is in a sense the optimal algebraic form of the geometric property which says that a genus one curve is completely determined by its periods. Yet there is subtlety here which deserves to be explicated. Denote the vector space of homogeneous cubic forms on $\mathbb{C}^3$ by $\mathcal{K}$ and let $\mathcal{K}^\circ \subset \mathcal{K}$ be the open subset of those $F \in \mathcal{K}$ whose zero set $C(F)$ in $\mathbb{P}^2$ defines a smooth cubic curve (hence of genus one). Then any $F \in \mathcal{K}^\circ$ also determines a distinguished holomorphic differential $\omega(F)$ on $C(F)$, which perhaps is best described as an iterated residue:

$$\omega(F) = \mathrm{Res}_{C(F)} \mathrm{Res}_{Z(F)} \frac{dZ_0 \wedge dZ_1 \wedge dZ_2}{F}.$$

Here $Z(F)$ denotes the zero set of F in $\mathbb{C}^3$ (the affine cone over $C(F)$) and $C(F)$ is regarded as the locus where $Z(F)$ meets the hyperplane at infinity $\mathbb{P}^2$. It is easy to see that this double residue is indeed a nonzero holomorphic differential on $C(F)$. The lattice of periods of $\omega(F)$ (a lattice in $\mathbb{C}$) only depends on the $PGL(3, \mathbb{C})$-orbit

---

2000 *Mathematics Subject Classification.* Primary: 11D25, 32N15.

Key words and phrases. Quartic curve, ball quotient, meromorphic automorphic form.

of F. Now recall that the space of lattices in $\mathbb{C}$ is naturally the $\mathrm{SL}(2,\mathbb{Z})$-orbit space of the space $\mathrm{Iso}^+(\mathbb{R}^2,\mathbb{C})$ of oriented $\mathbb{R}$-isomorphisms $\mathbb{R}^2 \to \mathbb{C}$ (where $\sigma \in \mathrm{SL}(2,\mathbb{Z})$ takes $\zeta \in \mathrm{Iso}^+(\mathbb{R}^2,\mathbb{C})$ to $\zeta\sigma^{-1}$). So we have defined a map

$$\mathrm{PGL}(3,\mathbb{C})\backslash\mathcal{K}^\circ \to \mathrm{SL}(2,\mathbb{Z})\backslash \mathrm{Iso}^+(\mathbb{R}^2,\mathbb{C}).$$

It is easy to see that this map is in fact an isomorphism of (affine) algebraic varieties (as for its injectivity, observe that the $\mathrm{PGL}(3,\mathbb{C})$-stabilizer of a nonsingular cubic plane curve is the group of automorphisms of that curve which preserve its degree three polarization). Both sides come with a natural $\mathbb{C}^\times$-action: $\lambda \in \mathbb{C}^\times$ acts on $\mathcal{K}^\circ$ by multiplication with λ^{-1} and acts on $\mathrm{Iso}^+(\mathbb{R}^2,\mathbb{C})$ by composing with multiplication by λ on $\mathbb{C}$. These actions descend to the orbit spaces and make the isomorphism above $\mathbb{C}^\times$-equivariant. (These actions are no longer effective, for $-1 \in \mathbb{C}^\times$ acts as the identity.) The graded algebra $\mathbb{C}[\mathcal{K}]^{\mathrm{PGL}(3,\mathbb{C})}$ can be understood as an algebra of regular functions on the left hand side and the graded algebra of $\mathrm{SL}(2,\mathbb{Z})$-modular forms $\mathbb{C}[E_4, E_6]$ can be understood as an algebra of regular functions on right hand side. In either case the grading comes from the $\mathbb{C}^\times$-action we just described. The cited theorem says that the displayed map identifies these graded $\mathbb{C}$-algebras. This has a geometrical consequence which goes somewhat beyond the observed isomorphism: the proj construction on the left is interpreted by Geometric Invariant Theory: $\mathrm{Proj}(\mathbb{C}[\mathcal{K}]^{\mathrm{PGL}(3,\mathbb{C})})$ adds to $\mathrm{PGL}(3,\mathbb{C})\backslash\mathbb{P}(\mathcal{K}^\circ)$ a singleton which is represented by the unique closed strictly semistable $\mathrm{PGL}(3,\mathbb{C})$-orbit in $\mathbb{P}(\mathcal{K})$, namely, the orbit of three nonconcurrent lines in $\mathbb{P}^2$. The proj construction on the right compactifies the modular curve in question, that is, the j-line $\mathrm{SL}(2,\mathbb{Z})\backslash\mathbb{P}\,\mathrm{Iso}^+(\mathbb{R}^2,\mathbb{C})$, in the standard manner: it is the simplest instance of a Baily-Borel compactification.

To me this theorem serves as a model for the theory of period maps. It seems to tell us that whenever we know such a map to be an open embedding, then we should try to express that fact on a (richer) algebraic level, amounting to the identification of two graded algebras of invariants, one with respect to a reductive algebraic group, the other with respect to a discrete group. The ensueing identification of their proj's should give us then an additional piece of geometric information, which in the end is a sophisticated way to understand the period map's boundary behavior.

This phenomenon is most likely to occur when the period map takes values in a ball quotient or a locally symmetric variety of type IV. If it is an isomorphism, then one hopes for an exact analogue of the theorem above. For example, Allcock-Carlsen-Toledo [2] have essentially established this for the case of cubic surfaces, where $\mathcal{K}$ is replaced by the space of homogenenous cubic forms in 4 variables and the algebra of modular forms by the algebra of automorphic forms on a 4-ball with respect to an arithmetic group. But if the period map is not surjective, then some modifications on the automorphic side are in order and it is precisely for this purpose that I developed the geometric theory of meromorphic automorphic forms in [5] and [6] (supplemented by a joint paper [7] with Swierstra). What I want to do here is to illustrate that theory in the (other) case that logically comes after our guiding example, namely that of quartic plane curves. To be precise, let $\mathcal{Q}$ stands for the vector space of *quartic* (rather than cubic) homogeneous forms in three complex variables and regard this space as a representation of $\mathrm{SL}(3,\mathbb{C})$, then we shall interpret the algebra of invariants $\mathbb{C}[\mathcal{Q}]^{\mathrm{SL}(3,\mathbb{C})}$ (or rather $\mathbb{C}[\mathcal{Q}]^{\mu_4 \times \mathrm{SL}(3,\mathbb{C})}$ with μ_4 acting on $\mathcal{Q}$ by scalar multiplication) as an algebra of (meromorphic) automorphic forms on the complex 6-ball. This will be an algebra isomorphism which rescales the degrees by a factor three. We shall, of course, also interpret the proj construction on either side.

This example is not an isolated one. For instance, Allcock [**1**] has determined the semistable cubic threefolds and his results suggest that the situation is very much like the case of quartic curves.

It is pleasure to dedicate this paper to my longtime friend Igor Dolgachev. Igor and I share a passion for our field (which is not confined to algebraic geometry) and we have a similar mathematical taste, but I only wish I had his extensive knowledge of the classical literature of our subject. I learned a lot from him.

2. The Baily-Borel compactification of a ball quotient

Let V be a complex vector space of finite dimension, endowed with a nondegenerate Hermitian form $h : V \times V \to \mathbb{C}$ of signature $(1, n)$. So we can find a coordinate system $(z_0, \ldots, z_n)$ for V on which h takes the standard form: $h(z, z) = |z_0|^2 - |z_1|^2 - \cdots - |z_n|^2$. This shows that the unitary group $\mathrm{U}(V)$ of (V, h) is isomorphic to $\mathrm{U}(1, n)$. Let us denote by V_+ the open subset of $z \in V$ with $h(z, z) > 0$. This set is $\mathbb{C}^\times$-invariant and hence defines an open subset $\mathbb{P}(V_+)$ of $\mathbb{P}(V)$. In terms of the above coordinates, $\mathbb{P}(V_+)$ is defined by $\sum_{i=1}^{n} |z_i/z_0|^2 < 1$, which shows that $\mathbb{P}(V_+)$ is biholomorphic to the complex n-ball. The group $\mathrm{U}(V)$ acts properly and transitively on it; in fact, the stabilizer of a point is a maximal compact subgroup $\mathrm{U}(V)$, so that $\mathbb{P}(V_+)$ can be understood as the symmetric space of $\mathrm{U}(V)$.

Suppose we are also given a discrete subgroup $\Gamma \subset \mathrm{U}(V)$ of finite *covolume* (which means that $\mathrm{U}(V)/\Gamma$ has finite $\mathrm{U}(V)$-invariant volume). Since Γ is discrete, it acts properly discontinuously on $\mathbb{P}(V_+)$ (and hence also on V_+) so that the formation of the orbit spaces $\Gamma\backslash\mathbb{P}(V_+)$ and $\Gamma\backslash V_+$ takes place in the complex-analytic category (the orbit spaces will be normal). However, the Baily-Borel theory tells us that these spaces have the richer structure of quasi-projective variety. In order to state a precise result, we recall that a Γ-*automorphic form of degree* $d \in \mathbb{Z}$ is in the present setting a Γ-invariant holomorphic function $f : V_+ \to \mathbb{C}$ which is homogeneous of degree $-d$: $f(\lambda z) = \lambda^{-d} f(z)$ and which in case $n = 1$ also obeys a growth condition which we do not bother to specify. If we denote the space of such forms by A_d^Γ, then it is clear that their direct sum $A_\bullet^\Gamma$ is a graded $\mathbb{C}$-algebra. The Baily-Borel theory has the following to say about it:

THEOREM 2.1. *The graded $\mathbb{C}$-algebra $A_\bullet^\Gamma$ has finitely many generators of positive degree. This algebra separates the Γ-orbits in V_+ so that we have injective complex-analytic maps*

$$\Gamma\backslash V_+ \to \mathrm{Spec}(A_\bullet^\Gamma), \quad \Gamma\backslash\mathbb{P}(V_+) \to \mathrm{Proj}(A_\bullet^\Gamma).$$

The images of these morphisms are Zariski open-dense so that $\Gamma\backslash V_+$ resp. $\Gamma\backslash\mathbb{P}(V_+)$ acquires the structure of a quasi-affine resp. quasi-projective complex variety. Moreover, we have natural (so-called Baily-Borel*) extensions $V_+ \subset V_+^*$ and $\mathbb{P}(V_+) \subset \mathbb{P}(V_+^*)$ as topological spaces with Γ-action (which we describe below) such that these morphisms extend to homeomorphisms*

$$\Gamma\backslash V_+^* \cong \mathrm{Spec}(A_\bullet^\Gamma), \quad \Gamma\backslash\mathbb{P}(V_+^*) \cong \mathrm{Proj}(A_\bullet^\Gamma),$$

if we endow the targets with their Hausdorff topology.

In order to describe the extensions mentioned in this theorem, we introduce the following notion. Let $H \subset V$ be a linear hyperplane. If H is nondegenerate, then the U(V)-stabilizer U$(V)_H$ is simply $U(H) \times U(H^\perp)$. Suppose now that H is degenerate. Then $H^\perp \subset H$, $H/H^\perp$ is negative definite and the evident homomorphism U$(V)_H \to$ U$(H/H^\perp) \times$ GL$(H^\perp)$ is surjective with kernel a unipotent group (the unipotent radical of U$(V)_H$). This unipotent group is in fact a Heisenberg group: an extension of the complex vector group $(H/H^\perp) \otimes \overline{H^\perp}$ by a real vector group of dimension one.

DEFINITION 2.2. We call a linear hyperplane $H \subset V$ that is not negative definite Γ-*rational* if its Γ-stabilizer Γ_H maps to a subgroup of finite covolume in U(H).

(Since the U(V)-stabilizer of a negative definite hyperplane is compact, this notion would be for such hyperplanes without much interest.) Let $\mathcal{I}$ denote the collection of isotropic lines $I \subset V$ for which $I^\perp$ is Γ-rational. The Γ-rationality amounts here to requiring that Γ_I meets the unipotent radical of U$(V)_I$ in a cocompact subgroup. According to the reduction theory for such forms, Γ has only finitely many orbits in the set $\mathcal{I}$. Now we can describe the extensions mentioned in Theorem 2.1 as a set:

$$V_+^* = V_+ \sqcup \left(\coprod_{I \in \mathcal{I}} V_+/I^\perp \right) \sqcup (V_+/V), \quad \mathbb{P}(V_+^*) = \mathbb{P}(V_+) \sqcup \coprod_{I \in \mathcal{I}} \mathbb{P}(V_+/I^\perp),$$

where $V_+/I^\perp$ simply denotes the image of V_+ in $V/I^\perp$ and likewise in other cases. This notation may appear unnecessarily complicated, since $V_+/I^\perp = V/I^\perp - \{0\}$ is just a copy of $\mathbb{C}^\times$ and both V_+/V and $\mathbb{P}(V_+/I^\perp)$ are even singletons. Yet it is convenient notation, not just because we need to be able to distinguish singletons by name, but also because it is typical for the general situation that the extension is obtained by adding quotients of the very space we are extending. Still we may observe that $\mathbb{P}(V_+^*) - \mathbb{P}(V_+)$ can be identified with the discrete set $\mathcal{I}$ so that $\Gamma \backslash \mathbb{P}(V_+^*) - \Gamma \backslash \mathbb{P}(V_+)$ is identified with the finite set $\Gamma \backslash \mathcal{I}$.

We will not define the topology of these extensions; suffices to say here that this topology induces the given one on each stratum, and that the elements of both Γ and $\mathbb{C}^\times$ (acting by scalar multiplication) act as homeomorphisms. The space V_+^* has the structure of a cone with the singleton V_+/V as vertex and $\mathbb{P}(V_+^*)$ as its base.

3. Γ-arrangements and associated compactifications

We keep the situation of the previous section, but now assume that $n \geq 2$.

Γ-arrangements. Suppose that $H \subset V$ is a Γ-rational hyperplane with $H^\perp$ a negative definite line. So H has signature $(1, n-1)$, $H_+ = H \cap V_+$ and Γ_H maps to a subgroup of $U(H)$ of cofinite volume. That subgroup will be discrete and so the preceding applies: we find a quasi-affine variety $\Gamma_H \backslash H_+$ with its Baily-Borel extension $\Gamma_H \backslash H_+^*$ (a normal affine variety). The natural map $\Gamma_H \backslash H_+ \to \Gamma \backslash V_+$ is evidently complex-analytic, but since Γ-automorphic forms restrict to Γ_H-automorphic forms it is in fact an algebraic morphism. For the same reason, this map has a Baily-Borel extension Spec$(A_\bullet^{\Gamma_H}) \to$ Spec$(A_\bullet^\Gamma)$ in the algebraic category. The underlying map $\Gamma_H \backslash H_+^* \to \Gamma \backslash V_+^*$ is defined in an evident manner. In particular, the preimage of the vertex is the vertex and so this Baily-Borel extension is finite. It is also bira-tional onto its image and hence this map is a normalization of that image. Since the preimage of $\Gamma \backslash V_+$ in $\Gamma_H \backslash H_+^*$ is $\Gamma_H \backslash H_+$, it follows that $\Gamma_H \backslash H_+ \to \Gamma \backslash V_+$ is a finite morphism as well. The preimage in V_+ of the image of $\Gamma_H \backslash H_+ \to \Gamma \backslash V_+$ is of course the union of the Γ-translates of H_+. It follows that these translates form a collection of hyperplane sections of V_+ that is locally finite. This remains true if instead of a single H we take a finite number of them. This leads up to the following

DEFINITION 3.1. A Γ-*arrangement* is a collection $\mathcal{H}$ of Γ-rational hyperplanes with negative definite orthogonal complement, which, when viewed as a subset of the appropriate Grassmannian, is a union of finitely many Γ-orbits.

For the remainder of this section we fix a Γ-arrangement $\mathcal{H}$.

Meromorphic automorphic forms. The preceding discussion shows that the collection $(H_+)_{H \in \mathcal{H}}$ is locally finite on V_+ so that its union $D_{\mathcal{H}} := \cup_{H \in \mathcal{H}} H_+$ is closed in V_+. Moreover, $D_{\mathcal{H}}$ is the preimage of a hypersurface in $\Gamma \backslash V_+$. We shall denote $V_+ - D_{\mathcal{H}}$ simply by $V_{\mathcal{H}}$. Any open subset of V of this form will be refered to as a Γ-*arrangement complement*. It is clear that $\Gamma \backslash V_{\mathcal{H}}$ is the complement of a hypersurface in $\Gamma \backslash V_+$. It is our goal to state a generalization of Theorem 2.1 for $\Gamma \backslash V_{\mathcal{H}}$.

For this purpose we introduce an algebra of meromorphic Γ-automorphic forms: for any integer d, we denote by $A_{\mathcal{H},d}^{\Gamma}$ the space of Γ-invariant holomorphic functions homogenenous of degree $-d$ on $V_{\mathcal{H}}$ that are meromorphic on V_+ and have a pole of order at most d along the hyperplane sections H_+, $H \in \mathcal{H}$. It is clear that $A_{\mathcal{H},\bullet}^{\Gamma}$ is an algebra of holomorphic functions on $\Gamma \backslash V_{\mathcal{H}}$ which contains $A_{\bullet}^{\Gamma}$ as a subalgebra. In particular, this algebra separates the points of $\Gamma \backslash V_{\mathcal{H}}$. We now quote from [**5**]:

THEOREM 3.2. *The graded $\mathbb{C}$-algebra $A_{\mathcal{H},\bullet}^{\Gamma}$ has finitely many generators of positive degree and the resulting complex-analytic injections*

$$\Gamma \backslash V_{\mathcal{H}} \to \mathrm{Spec}(A_{\mathcal{H},\bullet}^{\Gamma}), \quad \Gamma \backslash \mathbb{P}(V_{\mathcal{H}}) \to \mathrm{Proj}(A_{\mathcal{H},\bullet}^{\Gamma})$$

are open embeddings onto Zariski open-dense subsets so that $\Gamma \backslash V_{\mathcal{H}}$ resp. $\Gamma \backslash \mathbb{P}(V_{\mathcal{H}})$ acquires the structure of a quasi-affine resp. quasi-projective complex variety. More-over, we have natural topological extensions $V_{\mathcal{H}} \subset V_{\mathcal{H}}^*$ and $\mathbb{P}(V_{\mathcal{H}}) \subset \mathbb{P}(V_{\mathcal{H}}^*)$ as Γ-spaces (described below) such that these embeddings extend to homeomorphisms

$$\Gamma \backslash V_{\mathcal{H}}^* \cong \mathrm{Spec}(A_{\mathcal{H},\bullet}^{\Gamma}), \quad \Gamma \backslash \mathbb{P}(V_{\mathcal{H}}^*) \cong \mathrm{Proj}(A_{\mathcal{H},\bullet}^{\Gamma})$$

when the targets are endowed with their Hausdorff topology. These topological exten-sions come as Γ-equivariant stratified spaces which result in partitions of $\mathrm{Spec}(A_{\mathcal{H},\bullet}^{\Gamma})$ and $\mathrm{Proj}(A_{\mathcal{H},\bullet}^{\Gamma})$ into subvarieties.

Small modification of the Baily-Borel extension. The Γ-extensions in the above theorem share a number of properties with the Baily-Borel and the toric extensions: the boundary material that we add comes partitioned into subvarieties ('strata'), where each stratum is given as a topological quotient of the extended space. Before we describe the extension in question, we first discuss a small modi-fication of the Baily-Borel extension.

Given $I \in \mathcal{I}$, we denote by $\mathcal{H}_I$ the collection of $H \in \mathcal{H}$ containing the isotropic line I and by $I^{\mathcal{H}}$ the intersection of $I^{\perp}$ with all the members of $\mathcal{H}_I$. So $I \subset I^{\mathcal{H}} \subset I^{\perp}$.

Our rationality assumptions imply that Γ_I has finitely many orbits in $\mathcal{H}_I$. The small modication of the Baily-Borel extension amounts to replacing in its definition $I^\perp$ by $I^\mathcal{H}$:

$$\mathbb{P}(V_+^\mathcal{H}) := \mathbb{P}(V_+) \sqcup \coprod_{I \in \mathcal{I}} \mathbb{P}(V^+/I^\mathcal{H})$$

This extension comes with a natural Γ-invariant topology which makes the evident map $\mathbb{P}(V_+^\mathcal{H}) \to \mathbb{P}(V_+^*)$ continuous and its Γ-orbit space comes with the structure of a normal complex-analytic space so that the induced map $\Gamma\backslash\mathbb{P}(V_+^\mathcal{H}) \to \Gamma\backslash\mathbb{P}(V_+^*)$ is an analytic morphism. That morphism is in fact a modification: it is proper and an isomorphism over the open-dense stratum. The lemma below describes the exceptional fibers and also helps us to understand the behavior of the collection $\mathcal{H}_I$ near I.

LEMMA 3.3. *The collection $\mathcal{H}_I$ induces on $I^\perp$ a finite arrangement (which we denote $\mathcal{H}_I|I^\perp$). Any proper intersection $L \neq I^\perp$ of members of $\mathcal{H}_I|I^\perp$ is also an intersection of members of $\mathcal{H}_I$ only; moreover, there exists an intersection $\tilde{L}$ of members of $\mathcal{H}_I$ which meets V_+ and is such that $L = I^\perp \cap \tilde{L}$.*

The fiber of $\Gamma\backslash V_+^\mathcal{H} \to \Gamma\backslash V_+^*$ over a point of the $\mathbb{C}^\times$-orbit defined by $I \in \mathcal{I}$ is the quotient of an affine space over $I^\perp/I^\mathcal{H}$ by a crystallographic group.

PROOF. We observe $V_+/I_\mathcal{H} = V/I^\mathcal{H} - I^\perp/I^\mathcal{H}$ is the complement of a linear hyperplane and so its projectivization $\mathbb{P}(V_+/I^\mathcal{H}) = \mathbb{P}(V/I^\mathcal{H}) - \mathbb{P}(I^\perp/I^\mathcal{H})$ is in a natural manner an affine space. Let us denote this affine space simply by A. The group Γ_I acts on A through a complex crystallographic group (and so the orbit space of that action is a compact complex-analytic variety (in fact, a quotient of an abelian variety by a finite group). The exceptional fibers over the image of $V_+/I^\perp$ are all isomorphic to $\Gamma_I\backslash A$ and so the last assertion of the lemma follows.

Every $H \in \mathcal{H}_I$ determines a hyperplane A_H of A and this defines a bijection between $\mathcal{H}_I$ and a collection of affine hyperplanes of A, a collection we shall denote by $\mathcal{H}|A$. By construction, the common intersection of the collection $\mathcal{H}|A$ is empty. Notice that for $H_1, H_2 \in \mathcal{H}_I$, we have $H_1 \cap I^\perp = H_2 \cap I^\perp$ if and only if A_{H_1} and A_{H_2} are parallel.

The group Γ_I has finitely many orbits in $\mathcal{H}_I$ and hence also in $\mathcal{H}_I|I^\perp$ and $\mathcal{H}_I|A$. The assertions of the lemma now follow easily. The finiteness of $\mathcal{H}_I|I^\perp$ is a consequence of the fact that Γ_I acts in $I^\perp/I$ through a finite group. This is equivalent to $\mathcal{H}|A$ decomposing into finitely many equivalence classes for the relation of parallelism. If $L \neq I^\perp$ is as in the lemma, then the collection $\mathcal{H}_L$ of $H \in \mathcal{H}$ containing L corresponds to a nonempty finite union of such equivalence classes; a minimal nonempty intersection of these is an affine subspace B of A which has L as translation space; the common intersection $\tilde{L}$ of the corresponding subset of $\mathcal{H}_L$ will have the property that $\tilde{L}$ meets V_+ and $\tilde{L} \cap I^\perp = L$. Finally, if B' is another such minimal nonempty intersection distinct from B so that $B \cap B' = \emptyset$, then the collection of $H \in \mathcal{H}_L$ with $A_H \supset B$ or $A_H \supset B'$ has L as its common intersection.   □

REMARK 3.4. The modification $\Gamma\backslash\mathbb{P}(V_+^\mathcal{H}) \to \Gamma\backslash\mathbb{P}(V_+^*)$ has a simple algebro-geometric meaning: for every $H \in \mathcal{H}$, the evident map $\Gamma_H\backslash\mathbb{P}(H_+) \to \Gamma\backslash\mathbb{P}(V_+)$ is a finite morphism which extends to the Baily-Borel compactifications: $\Gamma_H\backslash\mathbb{P}(H_+^*) \to \Gamma\backslash\mathbb{P}(V_+^*)$. The latter is also finite with image a hypersurface (in fact, it is a normalization of this image), but that hypersurface need not support a $\mathbb{Q}$-Cartier divisor.

According to Lemma 5.2 of [**5**], the morphism $\Gamma_H \backslash \mathbb{P}(H_+^*) \to \Gamma \backslash \mathbb{P}(V_+^*)$ lifts naturally to a morphism $\Gamma_H \backslash \mathbb{P}(H_+^*) \to \Gamma \backslash \mathbb{P}(V_+^{\mathcal{H}})$ whose image *does* support a $\mathbb{Q}$-Cartier divisor. The modification $\Gamma \backslash \mathbb{P}(V_+^{\mathcal{H}}) \to \Gamma \backslash \mathbb{P}(V_+^*)$ is the smallest one for which the images of all the $\Gamma_H \backslash \mathbb{P}(H_+)$ extend to hypersurfaces which are $\mathbb{Q}$-Cartier divisors.

Baily-Borel extension of an arithmetic arrangement complement. Let $\mathcal{L}_+(\mathcal{H})$ denote the collection of all the linear subspaces that arise as an intersection of members of $\mathcal{H}$ and meet V_+ (this includes V as the empty intersection) and denote by $\mathcal{L}_+(\mathcal{H}, \mathcal{I})$ the set of pairs $(L, I) \in \mathcal{L}_+(\mathcal{H}) \times \mathcal{I}$ with $L \supset I$. Then the extensions appearing in Theorem 3.2 are

$$V_{\mathcal{H}}^* = V_{\mathcal{H}} \sqcup \left(\coprod_{(L,I) \in \mathcal{L}_+(\mathcal{H},\mathcal{I})} V_{\mathcal{H}}/(L \cap I^\perp) \right) \sqcup \coprod_{L \in \mathcal{L}_+(\mathcal{H})} (V_{\mathcal{H}}/L),$$

$$\mathbb{P}(V_{\mathcal{H}}^*) = \mathbb{P}(V_{\mathcal{H}}) \sqcup \left(\coprod_{(L,I) \in \mathcal{L}_+(\mathcal{H},\mathcal{I})} \mathbb{P}(V_{\mathcal{H}}/(L \cap I^\perp)) \right) \sqcup \coprod_{V \neq L \in \mathcal{L}_+(\mathcal{H})} \mathbb{P}(V_{\mathcal{H}}/L),$$

with a topology enjoying similar properties as in the Baily-Borel case. Notice, incidentally, that we recover the latter if $\mathcal{H}$ is empty.

Structure of the strata. A valuable piece of information contained in Theorem 3.2 is the stratified structure it exhibits in the boundary $\mathrm{Proj}(A_{\mathcal{H},\bullet}^\Gamma) - \Gamma \backslash \mathbb{P}(V_+)$. Any stratum is clearly of the form $\Gamma_L \backslash \mathbb{P}(V_{\mathcal{H}}/L)$ with $L \in \mathcal{L}_+(\mathcal{H})$ or $\Gamma_{L,I} \backslash \mathbb{P}(V_{\mathcal{H}}/L \cap I^\perp)$ with $(L, I) \in \mathcal{L}_+(\mathcal{H}, \mathcal{I})$.

Let us see what we get in the first case. Denote by $\mathcal{H}_L$ the collection of members of $\mathcal{H}$ which contain L and by $(V/L)_{\mathcal{H}}$ the complement of the union of the members of $\mathcal{H}_L$ in V/L. Then $V/L \cong L^\perp$ is negative definite, $\mathcal{H}_L$ is finite and $V_{\mathcal{H}}/L = (V/L)_{\mathcal{H}}$. So $\mathbb{P}(V_{\mathcal{H}}/L) = \mathbb{P}(V/L)_{\mathcal{H}}$ is a projective arrangement complement. The stabilizer Γ_L acts on $L^\perp$ through a finite group and hence the stratum $\Gamma_L \backslash \mathbb{P}(V_{\mathcal{H}}/L)$ of $\mathrm{Proj}(A_{\mathcal{H},\bullet}^\Gamma)$ is well-understood. In particular, its codimension is the dimension of L.

In the second case, we only note that the maximal stratum associated to I is the Γ_I-orbit space of the affine arrangement complement $A_{\mathcal{H}} := A - \cup_{H \in \mathcal{H}_I} A_H$, where A is the affine space and A_H the affine hyperplane that we encountered in the proof of Lemma 3.3. (So this is an open subset of a finite quotient of an abelian variety). This discussion has an interesting corollary:

COROLLARY 3.5. *Suppose that $\mathcal{H} \neq \emptyset$. Then the codimension of the boundary that $\Gamma \backslash V_{\mathcal{H}}$ has in $\mathrm{Proj}(A_{\mathcal{H},\bullet}^\Gamma)$ is the minimal dimension of a member of $\mathcal{L}_+(\mathcal{H}) \cup \{I^{\mathcal{H}}\}_{I \in \mathcal{I}}$. In particular, if every one-dimensional intersection of members of $\mathcal{H}$ is negative definite, then the boundary of $\Gamma \backslash \mathbb{P}(V_+)$ in $\mathrm{Proj}(A_{\mathcal{H},\bullet}^\Gamma)$ is of codimension > 1 and the meromorphicity requirement in the definition of $A_{\mathcal{H},\bullet}^\Gamma$ is superfluous: $A_{\mathcal{H},\bullet}^\Gamma$ is simply the algebra of Γ-invariant holomorphic functions on $V_{\mathcal{H}}$.*

PROOF. The first statement follows immediately from Theorem 3.2 and Lemma 3.3.

Under the assumption that every one-dimensional intersection of members of $\mathcal{H}$ is negative definite, the boundary of our compactification is of codimension ≥ 2. A meromorphic Γ-invariant holomorphic functions on $V_{\mathcal{H}}$ which is homogeneous of degree $-d$ defines a holomorphic function on $\mathrm{Spec}(A_{\mathcal{H},\bullet}^\Gamma) - \Gamma \backslash V_+$. Since $\mathrm{Spec}(A_{\mathcal{H},\bullet}^\Gamma)$ is normal, such a function extends to all of $\mathrm{Spec}(A_{\mathcal{H},\bullet}^{\Gamma})$. The homegeneity assumption implies that this extension is regular, i.e., lies in $A_{\mathcal{H},d}^{\Gamma}$. $\square$

Comparison of two compactifications. There is in general no natural map $\mathbb{P}(V_{\mathcal{H}}^{*}) \to \mathbb{P}(V_{+}^{*})$ and hence no natural morphism $\mathrm{Proj}(A_{\mathcal{H},\bullet}^{\Gamma}) \to \mathrm{Proj}(A_{\bullet}^{\Gamma})$. As both $\mathrm{Proj}(A_{\mathcal{H},\bullet}^{\Gamma})$ and $\mathrm{Proj}(A_{\bullet}^{\Gamma})$ are projective compactifications of the same variety, we can only say that we have a birational map between them. In [**5**] we explicitly described the graph of the birational map $\mathrm{Proj}(A_{\mathcal{H},\bullet}^{\Gamma}) \dashrightarrow \mathrm{Proj}(A_{\bullet}^{\Gamma})$ and for its topological Γ-equivariant counterpart $\mathbb{P}(V_{\mathcal{H}}^{*}) \dashrightarrow \mathbb{P}(V_{+}^{*})$. We shall not recall this, but it is worth mentioning an interesting special case, which is relevant for the example that we will discuss below.

PROPOSITION 3.6. *If $\mathcal{H}$ has the property that any two distinct members of $\mathcal{H}$ do no intersect each other in V_{+}, then*

   (i) *we have a natural map $\pi : \mathbb{P}(V_{+}^{\mathcal{H}}) \to \mathbb{P}(V_{\mathcal{H}}^{*})$ which is continuous Γ-equivariant and such that the resulting map $\Gamma\backslash\mathbb{P}(V_{+}^{\mathcal{H}}) \to \Gamma\backslash\mathbb{P}(V_{\mathcal{H}}^{*})$ is a morphism,*

   (ii) *the dimension of the boundaries $\Gamma\backslash\mathbb{P}(V_{+}^{\mathcal{H}}) - \Gamma\backslash\mathbb{P}(V_{+})$ and $\Gamma\backslash\mathbb{P}(V_{\mathcal{H}}^{*}) - \Gamma\backslash\mathbb{P}(V_{\mathcal{H}})$ are at most one,*

   (iii) *the images of the natural morphisms $\Gamma_{H}\backslash\mathbb{P}(H_{+}^{*}) \to \Gamma\backslash\mathbb{P}(V_{+}^{\mathcal{H}})$ are disjoint if we let H run over a system of representatives of the Γ-orbits in $\mathcal{H}$, and π contracts each of these images.*

In particular, $\Gamma\backslash\mathbb{P}(V_{+}^{\mathcal{H}})$ appears as the normalization of the graph of the map of orbit spaces $\Gamma\backslash\mathbb{P}(V_{+}^{*}) \dashrightarrow \Gamma\backslash\mathbb{P}(V_{\mathcal{H}}^{*})$.

This is a special case of Theorem 5.7 of [**5**]. We confine ourselves here to describing the map π. Under the hypothesis of the proposition, the boundary strata of $\mathbb{P}(V_{\mathcal{H}}^{*})$ consist of $\mathbb{P}(V_{\mathcal{H}}/H)$ (a singletons), $H \in \mathcal{H}$ and $\mathbb{P}(V_{\mathcal{H}}/I^{\mathcal{I}})$ (a singleton in case $\mathcal{H}_{I}$ is empty and an affine line minus a discrete set otherwise), $I \in \mathcal{I}$. Now π sends $\mathbb{P}(H_{+})$ to the singleton $\mathbb{P}(V_{\mathcal{H}}/H)$; if $I \in \mathcal{I}$, then it is identity on $\mathbb{P}(V_{\mathcal{H}}/I^{\mathcal{H}})$, and if $H \in \mathcal{H}_{I}$, then the image of $\mathbb{P}(H_{+})$ in $\mathbb{P}(V_{+}/I^{\mathcal{H}})$ (a singleton) maps to the singleton $\mathbb{P}(V_{\mathcal{H}}/H)$.

4. The moduli space plane quartics

We illustrate the preceding (and especially Proposition 3.6) with the case of quartic plane curves.

Geometric invariant theory of quartics. Fix a complex vector space W of dimension three. In what follows a central role is played by the space of homogeneous quartic forms on W, $\mathrm{Sym}^{4}(W^{*})$, and so we prefer a briefer name: we abbreviate that space simply by $\mathcal{Q}$. The fundamental theorem of geometric invariant theory says that each fiber of the natural map $\mathcal{Q} \to \mathrm{Spec}\,\mathbb{C}[\mathcal{Q}]^{\mathrm{SL}(W)}$ contains a unique closed $\mathrm{SL}(W)$-orbit which lies in the closure of all other orbits in that fiber. In other words, $\mathrm{Spec}\,\mathbb{C}[\mathcal{Q}]^{\mathrm{SL}(W)}$ is the orbit space in the separated category, $\mathrm{SL}(W)\backslash\backslash\mathcal{Q}$. The preimage of 0 is called the *unstable locus* and has the origin of $\mathcal{Q}$ as its unique closed orbit. The complement of the unstable locus is by definition the *semistable locus* $\mathcal{Q}^{\mathrm{ss}}$. The *stable locus* $\mathcal{Q}^{\mathrm{s}} \subset \mathcal{Q}^{\mathrm{ss}}$ is the set of $F \in \mathcal{Q}$ for which the orbit map $\mathrm{SL}(W) \to \mathcal{Q}$, $g \mapsto gF$, is proper (so that the orbit of F is closed in $\mathcal{Q}$). Following Mumford these are precisely the F which define a reduced quartic curve with only ordinary double points or (ordinary) cusps. The closed orbits in the *strictly semistable locus* $\mathcal{Q}^{ss} - \mathcal{Q}^s$ are the $F \in \mathcal{Q}$ of the form $(Z_1 Z_2 - Z_0^2)(s Z_1 Z_2 - t Z_0^2)$ with $s \neq 0$, where Z_0, Z_1, Z_2 is a coordinate system for W. So these define quartics that are the union $C' \cup C''$ of two conics with C' smooth, C'' not a double line and for which either C' and C'' meet in two points of multiplicity two, or $C' = C''$.

The degrees appearing in $\operatorname{Spec} \mathbb{C}[\mathcal{Q}]^{\operatorname{SL}(W)}$ are divisible by 3 because the center μ_3 of $\operatorname{SL}(W)$ acts effectively on $\mathcal{Q}$.

We denote by $\mathcal{Q}^\circ \subset \mathcal{Q}^s$ the set of $F \in \mathcal{Q}$ for which $C(F)$ is smooth.

Associated K3 surface. The following construction follows S. Kondō. Fix $F \in \mathcal{Q}^s$ and denote by $C(F) \subset \mathbb{P}(W)$ the curve defined by F. The equation $T^4 = F$ defines an affine cone $Z(F)$ in $W \oplus \mathbb{C}$, whose base at infinity is a quartic surface $S(F) \subset \mathbb{P}(W \oplus \mathbb{C})$. The projection $S(F) \to \mathbb{P}(W)$ from $[0:1] \in \mathbb{P}(W \oplus \mathbb{C})$ has as its fibers the μ_4-orbits and $S(F)$ as discriminant. In this way $S(F)$ is a μ_4-covering of $\mathbb{P}(W)$ with total ramification along $C(F)$. Since the singularities of $C(F)$ are ordinary double points or ordinary cusps, those of $S(F)$ are DuVal singularities of type A_3 (local-analytic equation $w^4 = x^2 + y^2$) or E_6 (local-analytic equation $w^4 = x^3 + y^2$) and hence, if $\tilde{S}(F) \to S(F)$ resolves these in the standard (minimal) manner, then $\tilde{S}(F)$ is a nonsingular K3 surface. This resolution is unique (up to unique isomorphism, to be precise), and so the μ_4-action on $S(F)$ lifts to $\tilde{S}$. The hyperplane class $\eta(F) \in H^2(\tilde{S}(F), \mathbb{Z})$ is a semipolarization of $\tilde{S}(F)$ invariant under μ_4 and of degree 4: we have $\eta(F) \cdot \eta(F) = 4$ (the dot denotes the intersection form on the cohomology).

We now fix a generator $\mu \in \wedge^3 W^*$ and view this generator as a translation invariant 3-form on W. Then the surface $\tilde{S}(F)$ comes with a holomorphic differential defined by

$$\omega(F) := \operatorname{Res}_{S(F)} \left(\frac{\mu}{T^3} \bigg|_{Z(F)} \right).$$

Since μ_4 acts on the last factor $\mathbb{C}$ by the tautological character, it acts on the T-coordinate by $\bar\chi$ and hence on $\omega(F)$ by $\bar\chi^{-3} = \bar\chi$. We often regard $\omega(F)$ as a cohomology class in $H^2(\tilde{S}(F), \mathbb{C})_{\bar\chi}$. Since $\omega(F) \wedge \overline{\omega}(F)$ is everywhere positive (for the complex orientation), we have $\omega(F) \cdot \overline{\omega}(F) > 0$. It is clear that $\tilde{S}(F)$ is a $K3$-surface. The pull-back of the hyperplane class defines a semipolarization $\eta(F)$ of degree four for $\tilde{S}(F)$ relative to which the surface is *nonhyperelliptic*: $\tilde{S}(F)$ has no elliptic fibration such that the semipolarization is of degree 1 or 2 on the fibers. (Otherwise these fibers would map with degree 2 to their image or get contracted.)

If we carry out this construction universally, then we find a hypersurface $\mathcal{Z} \subset \mathcal{Q}^s \times (W \oplus \mathbb{C})$ (defined by the equation $F = T^4$), the projection $\mathcal{Z} \to \mathcal{Q}^s \times W$ is a μ_4-cover which ramifies (totally) over the zero set $\mathcal{C} \subset \mathcal{Q}^s \times W$ of F and at infinity we get $\mathcal{S} \subset \mathcal{Q}^s \times \mathbb{P}(W \oplus \mathbb{C})$ as a μ_4-cover of $\mathcal{Q}^s \times \mathbb{P}(W)$. If we let $\operatorname{GL}(W)$ act as the identity on the last factor $\mathbb{C}$, then this action preserves $\mathcal{Z}$ and its defining equation. We also get a section ω of the relative dualizing sheaf of $\mathcal{S}$. This section transforms under μ_4 according to the character $\det : \operatorname{GL}(W) \to \mathbb{C}^\times$.

PROPOSITION 4.1. *Let $F \in \mathcal{Q}^s$, denote by a_1 resp. a_2 the number of ordinary double points resp. ordinary cusps of C and put $d := a_1 + 2a_2$. Then μ_4 acts on $H^2(S(F), \mathbb{C})$ with character $1 + (7-d)(\chi + \chi^2 + \chi^3)$ and on $H^2(\tilde{S}(F), \mathbb{C})$ with* *character*

$$1 + 7(\chi + \chi^2 + \chi^3) + a_1(3 - \chi - \chi^2 - \chi^3) + a_2(4 - 2\chi - 2\chi^3).$$

PROOF. The fixed point set of μ_4 in $S = S(F)$ is $C = C(F)$, whereas the action of μ_4 on $S - C$ is free. We can now invoke the Lefschetz theorem which says that the (alternating) character of μ_4 on the total cohomology of S takes on $\xi \in \mu_4$ the value $e(S^\xi)$ (the Euler characteristic of the fixed point set of ξ). The latter is $e(S)$ for $\xi = 1$ and $e(C)$ otherwise. In the presence of a_1 ordinary double points and a_2 cusps, the Euler characteristic of C resp. S is $-4 + d$ resp. $24 - 3d$. It then easily follows that the character on the total cohomology is $3 + (7 - d)(\chi + \chi^2 + \chi^3)$. So the character on $H^2(S, \mathbb{C})$ is as asserted.

We use this to compute the character of μ_4 on $H^2(\tilde{S}, \mathbb{C})$: The difference between the two is accounted for by H^2 of the exceptional set. A double point resp. a cusp yields a DuVal curve D of type A_3 resp. E_6. Its reduced cohomology only lives in dimension 2 and has as basis the fundamental classes of the irreducible components of D. A generator of μ_4 leaves in the first case each irreducible component invariant (so we get character 3) and induces in the second case the only non trivial symmetry of an E_6-diagram (so we get character $4 + 2\chi^2$). Hence the character on $H^2(\tilde{S}, \mathbb{C})$ is $1 + (7 - d)(\chi + \chi^2 + \chi^3) + a_1 3 + a_2(4 + 2\chi^2) = 1 + 7(\chi + \chi^2 + \chi^3) + a_1(3 - \chi - \chi^2 - \chi^3) + a_2(4 - 2\chi - 2\chi^3)$. $\square$

It is clear that the orbit space $S'(F) := \mu_2 \backslash S(F)$ is the degree two cover of $\mathbb{P}^2$ which ramifies along $C(F)$. So $S'(F)$ is a Del Pezzo surface of degree 2 which is allowed to have DuVal singularities of type A_1 and A_2.

In Proposition 4.1, $H^2(S(F), \mathbb{C})^{\mu_4}$ is spanned by the hyperplane class and $\sqrt{-1} \in \mu_4$ acts on $S(F)$ with Lefschetz number $-4 + d$. So the following proposition provides a converse in case the latter is ≤ 0.

PROPOSITION 4.2. *Let (S, η) be a polarized K3 surface (DuVal singularities allowed) of degree 4 with μ_4-action $\rho : \mu_4 \to \mathrm{Aut}(S, \eta)$ and let ω be a generator ω of the dualizing sheaf of S such that*
   (i) $H^2(S, \mathbb{C})^{\mu_4}$ *is spanned by η,*
   (ii) *the Lefschetz number of $\rho(\sqrt{-1})$ is ≤ 0,*
   (iii) $\omega \in H^2(S, \mathbb{C})_{\bar{\chi}}$ *and*
   (iv) *if $\tilde{S} \to S$ is the minimal resolution, then for every $\varepsilon \in \mathrm{Pic}(\tilde{S})$ with $\varepsilon \cdot \varepsilon = 0$ we have $|\varepsilon \cdot \eta| > 2$.*

Then there is an embedding of S in $\mathbb{P}^3$ such that η is the hyperplane class, the image has an equation of the form $F(Z_0, Z_1, Z_2) = Z_0^4$, with F defining a stable quartic plane curve (i.e., $F \in \mathcal{Q}^s$), the μ_4-action on S is the restriction of its diagonal action on $\mathbb{P}^3$ with characters $(1, 1, 1, \chi)$ and ω is the residue of $F^{-1}dZ_0 \wedge dZ_1 \wedge dZ_2$. Moreover the $\mathrm{SL}(3, \mathbb{C})$-orbit of F is unique.

PROOF. The last assumption says that (S, η) is nonhyperelliptic, i.e., that the degree of every elliptic curve on a minimal resolution $\tilde{S}$ of S is ≥ 3. According to Mayer [8] and Saint-Donat [9], the nonhyperellipticity of $\tilde{S}$ implies that the polarization defines up to a projective transformation an embedding $S \subset \mathbb{P}^3$.

Let $S^{\mu_4} \subset S$ denote the fixed point set of the μ_4-action. The Lefschetz formula says that the Euler characteristic of S^{μ_4} is the Lefschetz number of $\sqrt{-1} \in \mu_4$, which is ≤ 0. Since S^{μ_4} is nonempty, it follows that S^{μ_4} contains an irreducible curve C of positive genus. That curve spans a subspace of $\mathbb{P}^3$ on which μ_4 acts trivially and that subspace is not a line. So μ_4 acts trivially on a plane in $\mathbb{P}^3$. It follows that we can represent the μ_4-action on $\mathbb{C}^4$ by a diagonal action of type $(1, 1, 1, \chi^{\pm 1})$. Then S will have an equation of the form $F(Z_0, Z_1, Z_2) = cT^4$ for some $c \in \mathbb{C}$. Since S has only DuVal singularities, we cannot have $c = 0$. We rescale the coordinates in such a manner that ω is the residue as above. It is now also clear that the μ_4-action is of type $(1, 1, 1, \chi)$. At the same time we see that F defines a stable quartic curve. The uniqueness of the $SL(3, \mathbb{C})$-orbit of F is left to the reader. $\square$

The period map. We briefly review the period map, again essentially following Kondō [4]. Fix an even unimodular lattice Λ of signature $(3, 19)$ and a $\eta \in \Lambda$ with $\eta \cdot \eta = 4$. As is well-known, such a pair (Λ, η) is unique up to isometry. Consider the collection of μ_4-actions $\rho : \mu_4 \to O(\Lambda)_\eta$ on Λ fixing η for which

- (i) η spans Λ^{μ_4},
- (ii) the sublattice Λ^{μ_2} is nondegenerate of signature $(1, 7)$, and
- (iii) if $\varepsilon \in \Lambda^{\mu_2}$ is such that $\varepsilon \cdot \varepsilon = 0$, then $\varepsilon \cdot \eta \neq 2$.

Notice that (i) and (ii) imply that μ_4 has character $1 + 7(\chi + \chi^2 + \chi^3)$ on Λ. The group $O(\Lambda)_\eta$ acts on this set (by composition). One can either invoke the surjectivity of the period map and the above discussion or use more intrinsically the theory of lattices to see that this action is transitive.

We therefore fix one such ρ and we write simply a' for $\rho(\sqrt{-1})a$. We let $\Lambda_\pm$ denote the set of $a \in \Lambda$ with $a'' = \pm a$ (so $\Lambda_+ = \Lambda^{\mu_2}$). According to Kondō, (Λ_+, η) is naturally isomorphic to the Picard lattice of a Del Pezzo surface of degree two with its anticanonical class, scaled by a factor two: Λ_+ admits an orthogonal basis $e_0, \ldots, e_7$ with $e_0 \cdot e_0 = 2$, $e_i \cdot e_i = -2$ for $i = 1, \ldots, 7$ and such that $\eta = 3e_0 - (e_1 + \cdots + e_7)$.

Notice that $V := (\Lambda \otimes \mathbb{C})_{\bar{\chi}}$ is a 7-dimensional complex vector space. If $u, v \in V$, then $u \cdot v = u' \cdot v' = -u \cdot v$ and so V is totally isotropic relative to the $\mathbb{C}$-bilinear extension of the form on Λ. We have $\bar{V} = (\Lambda \otimes \mathbb{C})_\chi$ and $V \oplus \bar{V}$ is real of signature $(2, 12)$. So the Hermitian extension of the bilinear form to $\Lambda \otimes \mathbb{C}$ has signature $(1, 6)$ on V. Denote half that Hermitian form by h and let $V_\mathcal{O}$ be the set of $v \in V$ with $\frac{1}{2}(v + \bar{v}) \in \Lambda_-$. So if for instance $\alpha \in \Lambda_-$ is such that $\alpha \cdot \alpha = -2$, then $\alpha \cdot \alpha' = 0$, $v := \alpha + \sqrt{-1}\alpha' \in V_\mathcal{O}$ and we have $h(v, v) = \frac{1}{2}(\alpha \cdot \alpha + \alpha' \cdot \alpha') = -2$.

Clearly, $V_\mathcal{O}$ is a Hermitian module over the Gaussian integers $\mathcal{O} := \mathbb{Z}[\sqrt{-1}]$. According to Heckman $V_\mathcal{O}$ admits a $\mathcal{O}$-basis $\{v_i\}_{i=1}^7$ which at the same time enumerates the vertices of a E_7-graph such that $h(v_i, v_j)$ equals -2 when $i = j$, 0 when v_i and v_j are not connected and $1 + \text{sign}(j - i)\sqrt{-1}$ if v_i and v_j are connected. In particular, the pair (V, h) is naturally defined over $\mathbb{Q}(\sqrt{-1})$. Denote by $\tilde{\Gamma} \subset O(\Lambda)$ the group of μ_4-automorphisms of (Λ, η) and by Γ its image in the unitary group of (Λ, η). Both groups are arithmetic and contain μ_4 as their center (which acts on V as group of scalars with character $\bar{\chi}$). This implies that Γ splits as a direct product $\mu_4 \times \Gamma_1$: since $V_\mathcal{O}$ is a free $\mathcal{O}$-module of rank 7, the kernel Γ_1 of the determinant homomorphism $\det_\mathcal{O} : \Gamma \to \mathcal{O}^\times = \mu_4$ supplements $\mu_4 \subset \Gamma$.

Given $F \in \mathcal{Q}^\circ$, then a choice of an equivariant isometry $H^2(S(F); \mathbb{Z}) \cong \Lambda$ which takes $\eta(F)$ to η will take $\omega(F)$ to a point of V_+. The latter's Γ-orbit does not change if we pick another such isometry and hence we get a well-defined element of $\Gamma \backslash V_+$. The resulting map

$$P : \mathcal{Q}^\circ \to \Gamma \backslash V_+$$

is analytic and constant on the $SL(W)$-orbits; in fact, for $g \in GL(W)$, we have $P(gF) = \det(g)P(F)$. Since $\lambda \in \mathbb{C}^\times \subset GL(W)$ takes F to $\lambda^{-4}F$ and $P(F)$ to $\lambda^3 P(F)$, it follows that P is homogeneous of degree $-3/4$ relative to scalar multiplication in $\mathcal{Q}$ and V_+. (There is no contradiction here: if we descend the $\mathbb{C}^\times$-action on V_+ given by scalar multiplication to $\Gamma \backslash V_+$, then it is no longer faithful: the kernel is $\Gamma \cap \mathbb{C}^\times = \mu_4$.)

The map P extends across $\mathbb{P}(\mathcal{Q}^s)$ (for then $S(F)$ only acquires DuVal singularities) and yields an analytic map $\mathbb{P}(\mathcal{Q}^s) \to \Gamma \backslash \mathbb{P}(V_+)$. The Torelli theorem for $K3$-surfaces implies that this map is an open embedding. But it fails to be surjective: if H is a hyperplane of V of signature $(1,5)$ that is orthogonal to some $\varepsilon \in \Lambda$ with $\varepsilon \cdot \varepsilon = 0$ and $\varepsilon \cdot \eta = 2$, then $\mathbb{P}(H \cap V_+)$ parametrizes hyperelliptic $K3$-surfaces and hence its image in $\Gamma \backslash \mathbb{P}(V_+)$ is disjoint with the image of the above period map. The following lemma describes the situation in a more precise manner:

LEMMA 4.3. *Let $\varepsilon \in \Lambda$ be such that $\varepsilon \cdot \varepsilon = 0$, $\varepsilon \cdot \eta = 2$ and the span of η and the μ_4-orbit of ε is of hyperbolic signature. Then we have the following eigenspace decomposition in $\Lambda \otimes \mathbb{C}$:*

$$2\varepsilon = \eta + \alpha + \beta = \eta + \tfrac{1}{2}(\alpha - \sqrt{-1}\alpha') + \tfrac{1}{2}(\alpha + \sqrt{-1}\alpha') + \beta,$$

with $\alpha, \beta \in \Lambda$ such that $\alpha'' = -\alpha$, $\beta' = -\beta$ and $\alpha \cdot \alpha = \beta \cdot \beta = -2$. In particular, ε is primitive and the orthogonal projection of 4ε in V, $v := \alpha + \sqrt{-1}\alpha'$, lies in $V_\mathcal{O}$ and satisfies $h(v,v) = -2$. Moreover, $\mathcal{O}v$ is an orthogonal direct summand of $V_\mathcal{O}$.

We first prove:

LEMMA 4.4. *Let $L \subset \Lambda$ be a μ_4-invariant sublattice of hyperbolic signature containing η and two distinct isotropic vectors $\varepsilon_1, \varepsilon_2$ with $\varepsilon_i \cdot \eta = 2$. Then $\varepsilon_1 \cdot \varepsilon_2 = 1$. In particular, each ε_i is primitive.*

PROOF. Since μ_4 fixes η, it will also preserve each connected component of $\{x \in L \otimes \mathbb{R} - \{0\} : x \cdot x \geq 0\}$. This implies that $\varepsilon_1 \cdot \varepsilon_2 > 0$. Now $a_i := \eta - 2\varepsilon_i$ is perpendicular to η and we have $a_i \cdot a_i = -4$ and $a_1 \cdot a_2 = -4 + 4\varepsilon_1 \cdot \varepsilon_2$. Since a_1 and a_2 span a negative definite lattice, it follows that $\varepsilon_1 \cdot \varepsilon_2 = 1$. □

PROOF OF LEMMA 4.3. We put $\alpha := \varepsilon - \varepsilon''$ and $\beta := \varepsilon + \varepsilon'' - \eta$ and so that we have the orthogonal decomposition $2\varepsilon = \alpha + \beta + \eta$ with $\alpha'' = -\alpha$ and $\beta' = -\beta$. If we apply the preceding lemma to the pairs $\varepsilon, \varepsilon'$ and $\varepsilon, \varepsilon''$, we find that $\alpha \cdot \alpha = \beta \cdot \beta = -2$. Clearly $\alpha + \sqrt{-1}\alpha'$ is an eigenvector of our automorphism with eigenvalue $-\sqrt{-1}$.

To prove the last assertion, let $z \in \Lambda_-$. Then we have $\alpha \cdot z = (2\varepsilon - \eta - \beta) \cdot z = 2\varepsilon \cdot z \in 2\mathbb{Z}$ and likewise $\alpha' \cdot z \in 2\mathbb{Z}$. Hence z can be written as $z_1 + \tfrac{1}{2}(\alpha \cdot z)\alpha + \tfrac{1}{2}(\alpha' \cdot z)\alpha'$ with $z_1 \in \Lambda_-$ perpendicular to α and α'. This proves that $\mathbb{Z}\alpha + \mathbb{Z}\alpha'$ is an orthogonal direct summand of Λ_-. Hence $\mathcal{O}v$ is an orthogonal direct summand of $V_\mathcal{O}$. □

It is clear that in this situation the orthogonal complement H of v in V is also the intersection of V with the orthogonal complement of ε in $\Lambda_\mathbb{C}$. Lemma 4.3 tells us that $V_\mathcal{O}$ is the orthogonal direct sum of $H \cap V_\mathcal{O}$ and $\mathcal{O}v$, in particular, H is defined over $\mathbb{Q}(\sqrt{-1})$. Kondō shows that the underlying $\mathbb{Z}$-lattice, the orthogonal complement of $\mathbb{Z}\alpha + \mathbb{Z}\alpha'$, is isometric to $U(2) \perp U(2) \perp D_8(-1)$, where U denotes the hyperbolic plane, D_8 the root lattice of type D_8, and the number between parenthesis indicates the scaling factor of the form. The signature of H is $(1,5)$ and so H is Γ-rational. We shall refer to a ε as in this lemma as a *hyperelliptic vector* and we call the associated $H \subset V$ a *hyperelliptic hyperplane*. We denote the collection of the latter by $\mathcal{H}_h$. This is clearly a Γ-arrangement. In particular, we have defined $V^*_{\mathcal{H}_h}$. We verify that this arrangement satisfies the hypotheses of Proposition 3.6.

LEMMA 4.5. *Two distinct members of $\mathcal{H}_h$ do not meet inside V_+.*

PROOF. Suppose we have two distinct hyperelliptic hyperplanes H_1, H_2 which meet inside V_+. This means that the corresponding elliptic vectors $\varepsilon_1, \varepsilon_2$ will be contained in a μ_4-invariant sublattice $L \subset \Lambda$ of hyperbolic signature. According to Lemma 4.4 we then have $\varepsilon_1^{(k)} \cdot \varepsilon_2^{(l)} = 1$ for all k, l. It follows that $\alpha_i = \varepsilon_i - \varepsilon_i''$ satisfy $\alpha_1^{(k)} \cdot \alpha_2^{(l)} = 0$ so that by Lemma 4.3 we have an orthogonal μ_4-invariant decomposition

$$\Lambda_- = (\mathbb{Z}\alpha_1 \perp \mathbb{Z}\alpha_1') \perp (\mathbb{Z}\alpha_2 \perp \mathbb{Z}\alpha_2') \perp K$$

This implies that $M := U(2) \perp U(2) \perp D_8(-1)$ has an orthogonal direct summand (of rank one) spanned by a vector α with $\alpha \cdot \alpha = -2$. This, in turn, implies that the discriminant quadratic form of M, $M^*/M \to \mathbb{Q}/\mathbb{Z}$ (the reduction of $x \in M^* \mapsto \frac{1}{2}x \cdot x$), represents $-\frac{1}{4} \in \mathbb{Q}/\mathbb{Z}$ (namely its value on $\frac{1}{2}\alpha$). But a straightforward calculation shows that this is not the case. This proves the first assertion. $\square$

The following statements are proved by Kondō or are implicit in his discussion [**4**] (a more detailed discussion can be found in the thesis by M. Artebani [**3**]):

(i) Γ acts transitively on the collection $\mathcal{I}$ of degenerate hyperplanes defined over $\mathbb{Q}(\sqrt{-1})$ and

(ii) $\tilde{\Gamma}$ acts transitively on the collection of hyperelliptic vectors and hence Γ acts transitively on the collection $\mathcal{H}_h$ of hyperelliptic hyperplanes.

According to Corollary 3.5, the algebra of Γ-invariant holomorphic functions on $V_{\mathcal{H}_h}$ is a $\mathbb{C}$-graded algebra admitting a finite set of homogeneous generators of positive degree. Since μ_4 acts faithfully as a group of scalars on V, $A^\Gamma_{\mathcal{H}_h, k}$ is zero when k is not a multiple of 4. It follows from the two properties above that the projective compactification $\mathbb{P}(\Gamma \backslash V_{\mathcal{H}_h}) \subset \mathbb{P}(\Gamma \backslash V^*_{\mathcal{H}_h}) = \mathrm{Proj}(A^\Gamma_{\mathcal{H}_h})$ adds to $\Gamma \backslash \mathbb{P}(V_{\mathcal{H}_h})$ just two strata: a singleton $\{\infty_h\}$ (corresponding to a member of $\mathcal{H}_h$) and an affine curve C_h (corresponding to some $I^{\mathcal{H}_h}$). The closure of C_h is the union of these two: $\overline{C_h} = C_h \cup \{\infty_h\}$.

THEOREM 4.6. *Let μ_4 acts on $\mathcal{Q}$ by scalar multiplication. Then the period map defines an isomorphism $(\mu_4 \times \mathrm{SL}(W)) \backslash \mathcal{Q}^s \to \Gamma \backslash V_{\mathcal{H}_h}$ and induces an isomorphism of $\mathbb{C}$-algebras*

$$A^\Gamma_{\mathcal{H}_h} \xrightarrow{\cong} \mathbb{C}[\mathcal{Q}]^{\mu_4 \times \mathrm{SL}(W)}$$

which multiplies degrees by 3 and gives rise to an isomorphism $(\mu_4 \times \mathrm{SL}(W)) \backslash\backslash \mathcal{Q} \cong \Gamma \backslash V^*_{\mathcal{H}_h}$. The points ∞_h corresponds to the strictly semistable orbit defined by $(Z_1 Z_2 - Z_0^2)^2$ (double conic) and the curve C_h corresponds to the curve of strictly semistable orbits defined by $(Z_1 Z_2 - Z_0^2)(Z_1 Z_2 - t Z_0^2)$ with $t \neq 1, \infty$.

PROOF. The Torelli theorem for $K3$-surfaces implies that $\mathrm{SL}(W) \backslash \mathcal{Q}^s \to \Gamma \backslash V_{\mathcal{H}_h}$ is an isomorphism after projectivization. We noticed that it is homogeneous of degree -3 relative to the $\mathbb{C}^\times$-actions on W and V_+. In $\Gamma \backslash V_+$ the $\mathbb{C}^\times$-action is not effective, but has kernel $\mathbb{C}^\times \cap \Gamma = \mu_4$. It follows that this map drops to an isomorphism $(\mu_4 \times \mathrm{SL}(W)) \backslash \mathcal{Q}^s \to \Gamma \backslash V_{\mathcal{H}_h}$.

Since $Q - Q^s$ is of codimension > 1 in Q, $\mathbb{C}[Q]^{\mu_4 \times \mathrm{SL}(W)}$ is the algebra of regular functions on $(\mu_4 \times \mathrm{SL}(W))\backslash Q^s$. Similarly, since $\Gamma\backslash V^*_{\mathcal{H}_h} - \Gamma\backslash V_{\mathcal{H}_h}$ is of codimension > 1 in $\Gamma\backslash V^*_{\mathcal{H}_h}$, $A^{\Gamma}_{\mathcal{H}_h}$ is the algebra of regular functions on $\Gamma\backslash V_{\mathcal{H}_h}$. It follows that the isomorphism $(\mu_4 \times \mathrm{SL}(W))\backslash Q^s \cong \Gamma\backslash V_{\mathcal{H}_h}$ induces an isomorphism $A^{\Gamma}_{\mathcal{H}_h} \cong \mathbb{C}[Q]^{\mu_4 \times \mathrm{SL}(W)}$ which multiplies the degrees by 3. $\qquad\square$

QUESTION 4.7. It seems likely that the same statements hold for $\mathrm{SL}(W)$ in relation to Γ_1, so that for instance $A^{\Gamma_1}_{\mathcal{H}_h}$ gets identified with $\mathbb{C}[Q]^{\mathrm{SL}(W)}$. This raises the following question: given $F \in Q^\circ$, does the naturally defined $\mathcal{O}$-lattice inside $H^2(S(F); \mathbb{C})_{\bar{\chi}}$ (of rank 7) have a canonical generator for its top exterior power?

REMARK 4.8. It is clear that the Baily-Borel compactification $\Gamma\backslash\mathbb{P}(V^*_+)$ of $\Gamma\backslash\mathbb{P}(V_+)$ has a unique cusp and so is a one-point compactification. According to Proposition 3.6, the small modification $\Gamma\backslash\mathbb{P}(V^{\mathcal{H}_h})$ replaces the cusp by a curve and there is a natural contraction $\Gamma\backslash\mathbb{P}(V^{\mathcal{H}_h}) \to \Gamma\backslash\mathbb{P}(V^*_{\mathcal{H}_h})$ whose exceptional divisor is the image of a natural morphism $\Gamma_H\backslash\mathbb{P}(H^*_+) \to \Gamma\backslash\mathbb{P}(V^{\mathcal{H}_h})$ (for any $H \in \mathcal{H}_h$). The intersection of this exceptional divisor with $\Gamma\backslash\mathbb{P}(V_+)$ parametrizes the hyperelliptic curves of genus three.

The small blow-up has a counterpart in the geometric invariant theory of quartic curves: Artebani [3] shows in her thesis that it is a GIT quotient of a blow-up of the orbit of the double conic in $\mathbb{P}(Q^{ss})$ and that we thus obtain a compactification of the moduli space of genus three curves.

References

[1] D. Allcock: *The moduli space of cubic threefolds*, J. Algebraic Geom. 12, 201–223 (2003).

[2] D. Allcock, J.A. Carlson, D. Toledo: *The complex hyperbolic geometry for moduli of cubic surfaces*, J. Algebraic Geom. 11, 659–724 (2002), see also arXiv:AG/9709016.

[3] M. Artebani: *The moduli space of genus three curves as a period domain of K3 surfaces*, Thesis submitted at the university of Genova (to appear). See also arXiv:AG/0511031.

[4] S. Kondō: *A complex hyperbolic structure for the moduli space of curves of genus three*, J. Reine Angew. Math. 525, 219–232 (2000).

[5] E. Looijenga: *Compactifications defined by arrangements I: the ball quotient case*, Duke Math. J. 118 (2003), pp. 157–181, see also arXiv:AG/0106228.

[6] E. Looijenga: *Compactifications defined by arrangements II: locally symmetric varieties of type IV*, Duke Math. J. 119 (2003), pp. 527–588, see also arXiv:AG/0201218.

[7] E. Looijenga, R. Swierstra: *On period maps that are open embeddings*, arXiv:AG/0512489.

[8] A.L. Mayer: *Families of K3 surfaces*, Nagoya Math. J. 48 (1972), 1–17.

[9] B. Saint-Donat: *Projective models of $K - 3$ surfaces*, Am. J. Math. 96 (1974), 602–639.

MATHEMATISCH INSTITUUT, UNIVERSITEIT UTRECHT, P.O. BOX 80.010, NL-3508 TA UTRECHT, NEDERLAND

E-mail address: looijeng@math.uu.nl

Contemporary Mathematics
Volume **422**, 2007

On correspondences of a K3 surface with itself. II

Viacheslav V. Nikulin

To 60th Birthday of Igor Dolgachev

ABSTRACT. Let X be a K3 surface with a polarization H of degree $H^2 = 2rs$, r, $s \geq 1$, and the isotropic Mukai vector $v = (r, H, s)$ is primitive. The moduli space of sheaves over X with the Mukai vector $v = (r, H, s)$ is again a K3 surface, Y.

We prove that $Y \cong X$ if the Picard lattice $N(X)$ has an element h_1 with $(h_1)^2 = f(v)$ and minor additional congruence conditions modulo $N_i(v)$. All these conditions are exactly written, very efficient, and they are necessary if X is general with rk $N(X) \leq 2$.

Existence of such kind a criterion is very surprising, and it also gives some geometric interpretation of elements in $N(X)$ with a negative square. Moreover, we describe all irreducible divisorial conditions on moduli of (X, H) which imply $Y \cong X$, and we prove that their number is always infinite.

Thus, we treat in general problems considered in [**MN1**], [**MN2**] and [**N4**], where the additional condition $H \cdot N(X) = \mathbb{Z}$ had been imposed.

0. Introduction

Let X be a K3 surface with a polarization H of degree $H^2 = 2rs$ where r, $s \in \mathbb{N}$. Assume that the isotropic Mukai vector $v = (r, H, s)$ is primitive.

Let Y be the moduli space of sheaves (coherent and semi-stable with respect to H) over X with the isotropic Mukai vector $v = (r, H, s)$. The Y (or, in special cases, its minimal resolution of singularities which we denote by the same letter Y) is again a K3 surface which is equipped with a natural *nef* element h with $h^2 = 2ab$ where we denote $c = \text{g.c.d}(r, s)$ and $a = r/c$, $b = s/c$ (see Sect. 2.1 below). The surface Y is isogenous to X in the sense of Mukai. The second Chern class of the corresponding quasi-universal sheaf gives then a 2-dimensional algebraic cycle $Z \subset X \times Y$ and an algebraic correspondence between X and Y. See Mukai [**Mu1**]—[**Mu5**] and also Abe [**A**] about these results.

1991 *Mathematics Subject Classification.* Primary 14J28, 14J60, 14C25; Secondary 14C30.

Key words and phrases. moduli space, vector bundle, K3 surface, correspondence, algebraic cycle.

Supported by EPSRC grant EP/D061997/1.

Let H be divisible by $d \in \mathbb{N}$ where $\widetilde{H} = H/d$ is primitive in the Picard lattice $N(X)$ of X. Primitivity of $v = (r, H, s)$ means that g.c.d$(r, d, s) =$ g.c.d$(c, d) = 1$. We have $d^2 | ab$. Let $\gamma = \gamma(\widetilde{H})$ is defined by $\widetilde{H} \cdot N(X) = \gamma \mathbb{Z}$, i.e. $H \cdot N(X) = \gamma d$. Clearly, $\gamma | (2rs/d^2) = \widetilde{H}^2$.

We denote
$$n(v) = \text{g.c.d}(r, s, d\gamma).$$
By Mukai, [**Mu2**], [**Mu3**], $T(X) \subset T(Y)$, and $n(v) = [T(Y) : T(X)]$ where $T(X)$ and $T(Y)$ are transcendental lattices of X and Y. We assume that
$$n(v) = \text{g.c.d}(r, s, d\gamma) = \text{g.c.d}(c, d\gamma) = 1. \qquad (0.1)$$
Since g.c.d$(c, d) = 1$, this is equivalent to g.c.d$(c, \gamma) = 1$. By Mukai [**Mu2**], [**Mu3**], the transcendental periods $(T(X), H^{2,0}(X))$ and $(T(Y), H^{2,0}(Y))$ are isomorphic in this case. We can expect that sometimes the surfaces X and Y are also isomorphic, and we then get a cycle $Z \subset X \times X$, and a correspondence of X with itself. Thus, an interesting for us question is

QUESTION 1. *When is Y isomorphic to X?*

We want to answer this question in terms of Picard lattices $N(X)$ and $N(Y)$ of X and Y. Then our question can be reformulated as follows:

QUESTION 2. *Assume that N is a hyperbolic lattice, $H_1 \in N$ an element with square $2rs$. What are conditions on N and H_1 such that for any K3 surface X with Picard lattice $N(X)$ and s polarization $H \in N(X)$ the corresponding K3 surface Y is isomorphic to X, if the the pairs $(N(X), H)$ and (N, H_1) are isomorphic as abstract lattices with fixed elements?*

In other words, what are conditions on $(N(X), H)$ as an abstract lattice with an element H which are sufficient for Y to be isomorphic to X, and they are necessary, if X is a general K3 surface with the Picard lattice $N(X)$?

We answered this question in [**MN1**], [**MN2**] and [**N4**] under the condition $d = \gamma = 1$ (equivalently, $H \cdot N(X) = \mathbb{Z}$): in [**MN1**] for $r = s = 2$; in [**MN2**] for $r = s$; in [**N4**] for arbitrary r and s.

The main surprising result of [**MN1**], [**MN2**] and [**N4**] was that $Y \cong X$ if the Picard lattice $N(X)$ has an element h_1 with some prescribed square $(h_1)^2$ and some minor additional conditions. Moreover, these conditions are necessary to have $Y \cong X$ for a general K3 surface X with $\rho(X) = \text{rk } N(X) = 2$. Thus, sometimes, elements of Picard lattice $N(X)$ deliver important 2-dimensional algebraic cycles on $X \times X$. Moreover, here $(h_1)^2$ can be negative, and this gives geometric meaning for elements of the Picard lattice with negative square (it is well-known only for $h_1^2 = -2$; then $\pm h_1$ is effective).

Here we prove similar results in general.

We assume (0.1). Then $d^2 | ab$ and $\gamma | 2ab/d^2$. Moreover, g.c.d$(a, b) = 1$. We have $d = d_a d_b$ where $d_a = $ g.c.d(d, a) and $d_b = $ g.c.d(d, b). We define
$$a_1 = \frac{a}{d_a^2}, \quad b_1 = \frac{b}{d_b^2}.$$
We have $\gamma = \gamma_2 \gamma_a \gamma_b$ where $\gamma_a = $ g.c.d(a_1, γ), $\gamma_b = $ g.c.d(b_1, γ), $\gamma_2 = \gamma/(\gamma_a \gamma_b) | 2$. We define
$$a_2 = \frac{a_1}{\gamma_a}, \quad b_2 = \frac{b_1}{\gamma_b}, \quad e_2 = \frac{2}{\gamma_2}.$$

Due to Mukai [**Mu3**] (see also Examples 2.3.4 and 2.3.5 below), one has the following result for $\rho(X) = 1$. *If $\rho(X) = 1$ and X is a general K3 surface with its Picard lattice (i. e. $Aut(T(X), H^{2,0}(X)) = \pm 1$), then $Y \cong X$ if and only if $c = 1$ and either $a_1 = 1$ or $b_1 = 1$. In particular (e. g. see Lemma 2.1.1 below), for a primitive Mukai vector (r, H, s) one always has $Y \cong X$ if and only if $c = 1$ and either $a_1 = 1$ or $b_1 = 1$.*

We prove (see Theorem 4.4) the following our main result for $\rho(X) = 2$. We denote as $\mathbb{Z}f(\widetilde{H})$ the orthogonal complement to $\widetilde{H}$ in the 2-dimensional lattice N and use the invariants $\gamma, \delta \in \mathbb{N}$ and $\mu \in \left(\mathbb{Z}/(2a_1b_1c^2/\gamma)\right)^*$ of the pair $\widetilde{H} \in N$ (see Proposition 3.1.1). Here $\widetilde{H} \cdot N = \gamma\mathbb{Z}$, $\det N = -\gamma\delta$, $N = [\widetilde{H}, f(\widetilde{H}), (\mu\widetilde{H} + f(\widetilde{H}))/(2a_1b_1c^2/\gamma)]$. One always has $\delta \equiv \gamma\mu^2 \mod 4a_1b_1c^2/\gamma$. Moreover, below $\widetilde{h}_1 = (p_1\widetilde{H} + q_1f(\widetilde{H}))/((2/\gamma_2)(a_1/\gamma_a)c)$ for a-series, and $\widetilde{h}_1 = (p_1\widetilde{H} + q_1f(\widetilde{H}))/((2/\gamma_2)(b_1/\gamma_b)c)$ for b-series. Also we denote by $n^{(l)}$ the l-component of a natural number n for a prime l, i. e. $n^{(l)} = l^k | n$ and g.c.d$(n, n/n^{(l)}) = 1$.

THEOREM 0.1. *Let X be a K3 surface and H a polarization of X of degree $H^2 = 2rs$ where $r, s \in \mathbb{N}$. Assume that the Mukai vector (r, H, s) is primitive. Let Y be the moduli space of sheaves on X with the isotropic Mukai vector $v = (r, H, s)$. Let $\widetilde{H} = H/d$ be the corresponding primitive polarization.*

We have $Y \cong X$ if there exists $\widetilde{h}_1 \in N(X)$ such that $\widetilde{H}$, $\widetilde{h}_1$ belong to a 2-dimensional primitive sublattice $N \subset N(X)$ such that $\widetilde{H} \cdot N = \gamma\mathbb{Z}$, $\gamma > 0$, and

$$g.c.d.(c, d\gamma) = 1,$$

moreover, for one of $\epsilon = \pm 1$ the element $\widetilde{h}_1$ belongs to the a-series or to the b-series described below:

$\widetilde{h}_1$ belongs to the a-series if

$$\widetilde{h}_1^2 = \epsilon 2b_1c \text{ and } \widetilde{H} \cdot \widetilde{h}_1 \equiv 0 \mod \gamma(b_1/\gamma_b)c,$$

$$\widetilde{H} \cdot \widetilde{h}_1 \not\equiv 0 \mod \gamma(b_1/\gamma_b)cl_1, \ \widetilde{h}_1/l_2 \notin N(X)$$

for any prime l_1 such that $l_1^2 | a_1$ and g.c.d$(l_1, \gamma) = 1$, and any prime l_2 such that $l_2^2 | b_1$ and g.c.d$(l_2, \gamma) = 1$, and

$$p_1 = \frac{\widetilde{H} \cdot \widetilde{h}_1}{\gamma(b_1/\gamma_b)c}, \quad q_1 = -\frac{f(\widetilde{H}) \cdot \widetilde{h}_1}{\delta(b_1/\gamma_b)c}$$

satisfy the singular condition (AS) of a-series:

if odd prime $l|\gamma$ and $l^2|a_1$, then $q_1 \not\equiv 0 \mod l$ and

either $\delta \not\equiv 0 \mod l$ or $(\delta - \gamma\mu^2) \not\equiv 0 \mod (a_1^{(l)}/\gamma_a^{(l)})l$;

if odd prime $l|\gamma$ and $l|b_1$, then $q_1 \equiv 0 \mod \gamma_b^{(l)}$;

if odd prime $l|\gamma$ and $l^2|b_1$, then $p_1 \not\equiv 0 \mod l$;

if $2|\gamma$, $\gamma_2 = 1$ and $2|a_1$, then $p_1 \equiv 1 \mod 2$;

if $2|\gamma$, $\gamma_2 = 1$ and $4|a_1$, then $\delta - \gamma\mu^2 \not\equiv 0 \mod (8a_1b_1c^2/\gamma)$;

if $2|\gamma$, $\gamma_2 = 1$, and $2|b_1$, then $p_1 - \mu q_1 \not\equiv 0 \mod 4$ and $q_1 \equiv 0 \mod \gamma_b^{(2)}$;

if $2|\gamma$, $\gamma_2 = 2$ and $2|b_1$, then $p_1 \equiv 1 \mod 2$ and $q_1 \equiv 0 \mod \gamma^{(2)}/2$.

$\widetilde{h}_1$ *belongs to the b-series if*

$$\widetilde{h}_1^2 = \epsilon 2a_1 c \text{ and } \widetilde{H} \cdot \widetilde{h}_1 \equiv 0 \mod \gamma(a_1/\gamma_a)c,$$

$$\widetilde{H} \cdot \widetilde{h}_1 \not\equiv 0 \mod \gamma(a_1/\gamma_a)cl_1, \ \widetilde{h}_1/l_2 \notin N(X)$$

for any prime l_1 such that $l_1^2|b_1$ and $g.c.d(l_1,\gamma) = 1$ and any prime l_2 such that $l_2^2|a_1$ and $g.c.d(l_2,\gamma) = 1$, and

$$p_1 = \frac{\widetilde{H} \cdot \widetilde{h}_1}{\gamma(a_1/\gamma_a)c}, \quad q_1 = -\frac{f(\widetilde{H}) \cdot \widetilde{h}_1}{\delta\,(a_1/\gamma_a)c}$$

satisfy the singular condition (BS) of b-series:

  if odd prime $l|\gamma$ and $l|a_1$, then $q_1 \equiv 0 \mod \gamma_a^{(l)}$;

  if odd prime $l|\gamma$ and $l^2|a_1$, then $p_1 \not\equiv 0 \mod l$;

  if odd prime $l|\gamma$ and $l^2|b_1$, then $q_1 \not\equiv 0 \mod l$ and

  either $\delta \not\equiv 0 \mod l$ or $(\delta - \gamma\mu^2) \not\equiv 0 \mod (b_1^{(l)}/\gamma_b^{(l)})l$;

  if $2|\gamma$, $\gamma_2 = 1$, and $2|a_1$, then $p_1 - \mu q_1 \not\equiv 0 \mod 4$ and $q_1 \equiv 0 \mod \gamma_a^{(2)}$;

  if $2|\gamma$, $\gamma_2 = 1$ and $2|b_1$, then $p_1 \equiv 1 \mod 2$;

  if $2|\gamma$, $\gamma_2 = 1$ and $4|b_1$, then $\delta - \gamma\mu^2 \not\equiv 0 \mod (8a_1b_1c^2/\gamma)$;

  if $2|\gamma$, $\gamma_2 = 2$ and $2|a_1$, then $p_1 \equiv 1 \mod 2$ and $q_1 \equiv 0 \mod \gamma^{(2)}/2$.

Moreover, one has formulae (4.23) and (4.24) in terms of X for the canonical primitive nef element $\widetilde{h}$ of Y defined by $(-a, 0, b) \mod \mathbb{Z}v$.

These conditions are necessary to have $Y \cong X$ if $\rho(X) \leq 2$ and X is a general K3 surface with its Picard lattice, i. e. the automorphism group of the transcendental periods $(T(X), H^{2,0}(X))$ is ± 1.

As concrete examples, in Sect. 6 we specialise the theorem for $\gamma = 1$ and $\gamma = 2$. The same can be done for any γ.

For the Mukai case when $c = 1$ and either $a_1 = 1$ or $b_1 = 1$, one satisfies conditions of Theorem 0.1 for $\widetilde{h}_1 = \widetilde{H}$.

It seems many (if not all) known examples when $\rho(X) \geq 2$ and $Y \cong X$ follow from this Theorem. E.g. see [C], [T1]—[T3] and [V].

Like in [MN1], [MN2] and [N4] we also describe all irreducible divisorial conditions on moduli of polarized K3 surfaces (X, H) which imply $\widetilde{H} \cdot N(X) = \gamma\mathbb{Z}$ and $Y \cong X$. We show that they are labelled by pairs $(\pm\mu, \delta)$ where $\pm\mu \in \left(\mathbb{Z}/(2a_1b_1c^2/\gamma)\right)^*$, $\delta \in \mathbb{N}$ and $\delta \equiv \mu^2\gamma \mod 4a_1b_1c^2/\gamma$, moreover the pair belongs to the a-series or to the b-series. It belongs to the a-series if at least for one $\epsilon = \pm 1$ the equation

$$\gamma p_1^2 - \delta q_1^2 = \epsilon 2(2/\gamma_2)(a_1/\gamma_a)\gamma_b c$$

has an integral solution (p_1, q_1) where (p_1, q_1) satisfy conditions (A) of a-series (3.3.54)—(3.3.57). Similarly one can consider b-series changing a and b places. See Sect. 4.

In Sect. 5, as an application, we prove that the number of the irreducible divisorial conditions is infinite if non-empty. If $\gamma = 1$, the same considerations as for $d = \gamma = 1$ in [N4] show that for any type of a primitive isotropic Mukai vector (r, H, s) the number of divisorial conditions on moduli of K3 which imply that $Y \cong X$ and $\gamma = 1$ is always non-empty and infinite. In particular, for any type of a primitive isotropic Mukai vector the number of divisorial conditions on moduli of K3 which imply that $Y \cong X$ is always non-empty and infinite.

This paper generalises to the general case results of [N4] (see also [MN1] and [MN2]) where a particular case $d = \gamma = 1$ had been considered.

As in [MN1], [MN2] and [N4], the fundamental tools to get the results above is the Global Torelli Theorem for K3 surfaces proved by Piatetsky-Shapiro and Shafarevich in [PS], and results of Mukai [Mu2], [Mu3]. By results of [Mu2], [Mu3], we can calculate periods of Y using periods of X; comparing the periods, by the Global Torelli Theorem for K3 surfaces [PS], we can find out if Y is isomorphic to X.

These paper treats in general problems considered in [MN1], [MN2] and [N4] where the additional condition $\gamma = d = 1$ had been imposed. It makes results of this paper much more complicated. For instance, in these paper we don't consider the question of non-emptiness of the divisorial conditions on moduli for $\gamma > 1$. It is more difficult in the general setting of this paper. We hope to consider this problem later.

1. Preliminary notations and results about lattices and K3 surfaces

1.1. Some notations about lattices. We use notations and terminology from [N2] about lattices, their discriminant groups and forms. A *lattice* L is a non-degenerate integral symmetric bilinear form. I. e. L is a free $\mathbb{Z}$-module equipped with a symmetric pairing $x \cdot y \in \mathbb{Z}$ for $x, y \in L$, and this pairing should be non-degenerate. We denote $x^2 = x \cdot x$. The *signature* of L is the signature of the corresponding real form $L \otimes \mathbb{R}$. The lattice L is called *even* if x^2 is even for any $x \in L$. Otherwise, L is called *odd*. The *determinant* of L is defined to be $\det L = \det(e_i \cdot e_j)$ where $\{e_i\}$ is some basis of L. The lattice L is *unimodular* if $\det L = \pm 1$. The *dual lattice* of L is $L^* = Hom(L, \mathbb{Z}) \subset L \otimes \mathbb{Q}$. The *discriminant group* of L is $A_L = L^*/L$. It has the order $|\det L|$. The group A_L is equipped with the *discriminant bilinear form* $b_L : A_L \times A_L \to \mathbb{Q}/\mathbb{Z}$ and the *discriminant quadratic form* $q_L : A_L \to \mathbb{Q}/2\mathbb{Z}$ if L is even. To get this forms, one should extend the form of L to the form on the dual lattice L^* with values in $\mathbb{Q}$.

For $x \in L$, we shall consider the invariant $\gamma(x) \geq 0$ where

$$x \cdot L = \gamma(x)\mathbb{Z}. \tag{1.1.1}$$

Clearly, $\gamma(x)|x^2$ if $x \neq 0$.

We denote by $L(k)$ the lattice obtained from a lattice L by multiplication of the form of L by $k \in \mathbb{Q}$. The orthogonal sum of lattices L_1 and L_2 is denoted by $L_1 \oplus L_2$. For a symmetric integral matrix A, we denote by $\langle A \rangle$ a lattice which is given by the matrix A in some bases. We denote

$$U = \begin{pmatrix} 0 & 1 \\ 1 & 0 \end{pmatrix}. \tag{1.1.2}$$

Any even unimodular lattice of the signature $(1,1)$ is isomorphic to U.

An embedding $L_1 \subset L_2$ of lattices is called *primitive* if L_2/L_1 has no torsion. We denote by $O(L)$, $O(b_L)$ and $O(q_L)$ the automorphism groups of the corresponding forms. Any $\delta \in L$ with $\delta^2 = -2$ defines a reflection $s_\delta \in O(L)$ which is given by the formula

$$x \to x + (x \cdot \delta)\delta,$$

$x \in L$. All such reflections generate the *2-reflection group* $W^{(-2)}(L) \subset O(L)$.

1.2. Some notations about K3 surfaces. Here we remind some basic notions and results about K3 surfaces, e. g. see [**PS**], [**S-D**], [**Sh**]. A K3 surface X is a non-singular projective algebraic surface over $\mathbb{C}$ such that its canonical class K_X is zero and the irregularity $q_X = 0$. We denote by $N(X)$ the *Picard lattice* of X which is a hyperbolic lattice with the intersection pairing $x \cdot y$ for $x, y \in N(S)$. Since the canonical class $K_X = 0$, the space $H^{2,0}(X)$ of 2-dimensional holomorphic differential forms on X has dimension one over $\mathbb{C}$, and

$$N(X) = \{x \in H^2(X, \mathbb{Z}) \mid x \cdot H^{2,0}(X) = 0\} \tag{1.2.1}$$

where $H^2(X, \mathbb{Z})$ with the intersection pairing is a 22-dimensional even unimodular lattice of signature $(3, 19)$. The orthogonal lattice $T(X)$ to $N(X)$ in $H^2(X, \mathbb{Z})$ is called the *transcendental lattice of X*. We have $H^{2,0}(X) \subset T(X) \otimes \mathbb{C}$. The pair $(T(X), H^{2,0}(X))$ is called the *transcendental periods of X*. The *Picard number* of X is $\rho(X) = \mathrm{rk}\, N(X)$. A non-zero element $x \in N(X) \otimes \mathbb{R}$ is called *nef* if $x \neq 0$ and $x \cdot C \geq 0$ for any effective curve $C \subset X$. It is known that an element $x \in N(X)$ is ample (i. e. it defines a polarization) if $x^2 > 0$, x is *nef*, and the orthogonal complement $x^\perp$ to x in $N(X)$ has no elements with square -2. For any non-zero element $x \in N(X)$ with $x^2 \geq 0$, there exists a reflection $w \in W^{(-2)}(N(X))$ such that the element $\pm w(x)$ is nef; it then is ample if $x^2 > 0$ and $x^\perp$ had no elements with square -2 in $N(X)$. The *nef* element $\pm w(x)$ is defined canonically by x. It is called *the canonical nef element of x*.

We denote by $V^+(X)$ the light cone of X, which is the half-cone of

$$V(X) = \{x \in N(X) \otimes \mathbb{R} \mid x^2 > 0\} \tag{1.2.2}$$

containing a polarization of X. In particular, all *nef* elements x of X belong to $\overline{V^+(X)}$: one has $x \cdot V^+(X) > 0$ for them.

The reflection group $W^{(-2)}(N(X))$ acts in $V^+(X)$ discretely, and its fundamental chamber is the closure $\overline{\mathcal{K}(X)}$ of the Kähler cone $\mathcal{K}(X)$ of X. It is the same as the set of all *nef* elements of X. Its faces are orthogonal to the set $\mathrm{Exc}(X)$ of all exceptional curves r on X which are non-singular rational curves r on X with $r^2 = -2$. Thus, we have

$$\overline{\mathcal{K}(X)} = \{0 \neq x \in \overline{V^+(X)} \mid x \cdot \mathrm{Exc}(X) \geq 0\}. \tag{1.2.3}$$

2. Condition of $Y \cong X$ for a general K3 surface X with a given Picard lattice

2.1. The correspondence. Let X be a smooth complex projective K3 surface with a polarization H of degree $2rs$ where $r, s \in \mathbb{N}$.

Assume that H is divisible by $d \in \mathbb{N}$ and $\widetilde{H} = H/d$ is primitive in $N(X)$. Then $\widetilde{H}^2 = 2rs/d^2$ and $d^2 | rs$. We denote

$$c = \mathrm{g.c.d}(r, s), \quad a = r/c, \quad b = s/c. \tag{2.1.1}$$

We assume that the Mukai vector (r, H, s) is primitive, i. e.

$$\mathrm{g.c.d}(r, s, d) = \mathrm{g.c.d}(c, d) = 1. \tag{2.1.2}$$

Let Y be the moduli space of sheaves $\mathcal{E}$ (coherent and semi-stable with respect to H) on X with the primitive isotropic Mukai vector $v = (r, H, s)$. Then rk $\mathcal{E} = r$, $\chi(\mathcal{E}) = r + s$ and $c_1(\mathcal{E}) = H$. The Y (or, in special cases, its minimal resolution of singularities which we denote by the same letter Y) is again a K3 surface. See [**Mu1**]—[**Mu5**] and also [**A**] about these results.

Let
$$H^*(X, \mathbb{Z}) = H^0(X, \mathbb{Z}) \oplus H^2(X, \mathbb{Z}) \oplus H^4(X, \mathbb{Z}) \tag{2.1.3}$$
be the full cohomology lattice of X equipped with the Mukai pairing
$$(u, v) = -(u_0 \cdot v_2 + u_2 \cdot v_0) + u_1 \cdot v_1 \tag{2.1.4}$$
for $u_0, v_0 \in H^0(X, \mathbb{Z})$, $u_1, v_1 \in H^2(X, \mathbb{Z})$, $u_2, v_2 \in H^4(X, \mathbb{Z})$. We naturally identify $H^0(X, \mathbb{Z})$ and $H^4(X, \mathbb{Z})$ with $\mathbb{Z}$. Then the Mukai pairing is
$$(u, v) = -(u_0 v_2 + u_2 v_0) + u_1 \cdot v_1. \tag{2.1.5}$$
The element
$$v = (r, H, s) = (r, H, \chi - r) \in H^*(X, \mathbb{Z}) \tag{2.1.6}$$
is isotropic, i.e. $v^2 = 0$, since $H^2 = 2rs$. In this case (for a primitive v), Mukai [**Mu2**]—[**Mu5**] (see also Abe [**A**]) showed that Y is a K3 surface, and one has the natural identification
$$H^2(Y, \mathbb{Z}) \cong (v^\perp / \mathbb{Z}v) \tag{2.1.7}$$
which also gives the isomorphism of the Hodge structures of X and Y, i. e. $H^{2,0}(Y)$ will be identified with the image of $H^{2,0}(X)$. The Y has the canonical *nef* element h defined by $(-a, 0, b) \mod \mathbb{Z}v$ with $h^2 = 2ab$ (see Sect. 1.2).

In particular, (2.1.7) gives the embedding
$$T(X) \subset T(Y) \tag{2.1.8}$$
of the transcendental lattices of the index
$$[T(Y) : T(X)] = n(v) = \min |v \cdot x| \tag{2.1.9}$$
where $x \in H^0(X, \mathbb{Z}) \oplus N(X) \oplus H^4(X, \mathbb{Z})$ and $v \cdot x \neq 0$ (see [**Mu2**], [**Mu3**]). In this paper, we are interested in the case when $Y \cong X$. By (2.1.9), it may happen if $n(v) = 1$ only.

We can introduce the invariant $\gamma = \gamma(\widetilde{H}) \in \mathbb{N}$ which is defined by
$$\widetilde{H} \cdot N(X) = \gamma \mathbb{Z}, \tag{2.1.10}$$
equivalently, $H \cdot N(X) = \gamma d\mathbb{Z}$. Clearly, $\gamma | 2rs/d^2 = \widetilde{H}^2$, and
$$n(v) = \text{g.c.d}(r, s, \gamma d) = \text{g.c.d}(c, \gamma d). \tag{2.1.11}$$
Thus, $n(v) = 1$, and it is possible to have $Y \cong X$ only if
$$\text{g.c.d}(r, s, \gamma d) = \text{g.c.d}(c, \gamma) = \text{g.c.d}(c, d) = 1. \tag{2.1.12}$$
This is exactly the case when, according to Mukai, the transcendental periods
$$(T(X), H^{2,0}(X)) \cong (T(Y), H^{2,0}(Y)) \tag{2.1.13}$$
are isomorphic.

From (2.1.7), we obtain the following *specialisation principle.*

We say that a K3 surface X is *general (for its Picard lattice)* if the automorphism group of the transcendental periods $\text{Aut}(T(X), H^{2,0}(X)) = \pm 1$. We have

LEMMA 2.1.1. *(The specialisation principle.)* *Assume that for a general K3 surface X with $N = N(X)$ and a primitive isotropic Mukai vector $v = (r, H, s)$ where $H \in N$ is a polarization of X, one has $Y \cong X$*

Then the same is valid for any K3 surface X' such that $H \in N \subset N(X')$ if $N \subset N(X')$ is a primitive sublattice in $N(X')$ and H is a polarization of X.

PROOF. Since $Y \cong X$ and X is general, there exist only two isomorphisms of transcendental periods $(T(X), H^{2,0}(X)) \cong (T(Y), H^{2,0}(Y))$ which are ± 1 of the identification of the transcendental periods $(T(X), H^{2,0}(X)) = (T(Y), H^{2,0}(Y))$ which is defined by Mukai identification (2.1.7). Since $Y \cong X$, there exists an extension of this identification to the isomorphism $H^2(X, \mathbb{Z}) \cong H^2(Y, \mathbb{Z})$ of the cohomology lattices.

Now assume that $H \in N \subset N(X')$. Let Y' be the corresponding moduli space of sheaves on X' with the same Mukai vector v.

By local epimorphicity of the period map for K3, we can assume that $N = N(X)$ is the Picard lattice of a K3 surface X with the polarization H, X is general and the embedding $N(X) \subset N(X')$ extends to the identification of the cohomology lattices $H^2(X, \mathbb{Z}) = H^2(X', \mathbb{Z})$. Then $T(X) \supset T(X')$ is a primitive sublattice. By the Mukai identification (2.1.7), the identification (2.1.7) $(T(X'), H^{2,0}(X')) = (T(Y'), H^{2,0}(Y'))$ is extending to the identification of the transcendental periods $(T(X), H^{2,0}(X)) = (T(Y), H^{2,0}(Y))$ and to the identification of the cohomology lattices $H^2(Y, \mathbb{Z}) = H^2(Y', \mathbb{Z})$. Since X is general, $Y \cong X$, and the identification above of their transcendental periods extends to an isomorphism of the lattices $H^2(X, \mathbb{Z}) \cong H^2(Y, \mathbb{Z})$. This gives the isomorphism $H^2(X', \mathbb{Z}) = H^2(X, \mathbb{Z}) \cong H^2(Y, \mathbb{Z}) = H^2(Y', \mathbb{Z})$ which extends the above isomorphism $(T(X'), H^{2,0}(X')) \cong (T(Y'), H^{2,0}(Y'))$.

By global Torelli Theorem for K3 surfaces [**PS**], this defines an isomorphism $Y' \cong X'$. This finishes the proof.

2.2. The characteristic map of a primitive element of a lattice. Let S be an even lattice and $P \in S$ its primitive element with $P^2 = 2m \neq 0$ and $\gamma(P) = \gamma | 2m$ (in S), i. e. $P \cdot S = \gamma \mathbb{Z}$.

We want to calculate the discriminant quadratic form of S. Consider

$$K(P) = P_S^\perp \qquad (2.2.1)$$

the orthogonal complement to P in S. Put $P^* = P/2m$. Then any element $x \in S$ can be written as

$$x = n\gamma P^* + k^* \qquad (2.2.2)$$

where $n \in \mathbb{Z}$ and $k^* \in K(P)^*$, because

$$\mathbb{Z}P \oplus K(P) \subset S \subset S^* \subset \mathbb{Z}P^* \oplus K(P)^*$$

and $P \cdot S = \gamma \mathbb{Z}$. Since $\gamma(P) = \gamma$, the map $n\gamma P^* + [P] \to k^* + K(P)$ gives an isomorphism of the groups $\mathbb{Z}/\frac{2m}{\gamma} \cong [\gamma P^*]/[P] \cong [u^*(P) + K(P)]/K(P)$ where $u^*(P) + K(P)$ has order $2m/\gamma$ in $A_{K(P)} = K(P)^*/K(P)$. Clarify in [**N2**]. It follows,

$$S = [\mathbb{Z}P, K(P), \gamma P^* + u^*(P)]. \qquad (2.2.3)$$

The element $u^*(P)$ is defined canonically mod $K(P)$ by the condition that $\gamma P^* + u^* \in S$. The element

$$u^*(P) + K(P) \in K(P)^*/K(P) \qquad (2.2.4)$$

is called *the canonical element of P*. Since $\gamma P^* + u^*(P)$ belongs to the even lattice S, it follows

$$(\gamma P^* + u^*(P))^2 = \frac{\gamma^2}{2m} + u^*(P)^2 \equiv 0 \mod 2. \tag{2.2.5}$$

For $n \in \mathbb{Z}$ and $k^* \in K(P)^*$, we have $x = nP^* + k^* \in S^*$ if and only if

$$(nP^* + k^*) \cdot (\gamma P^* + u^*(P)) = \frac{n\gamma}{2m} + k^* \cdot u^*(P) \in \mathbb{Z}.$$

It follows,

$$S^* = \{nP^* + k^* \mid n \in \mathbb{Z},\ k^* \in K(P)^*,\ n \equiv -\frac{2m}{\gamma}\ (k^* \cdot u^*(P)) \mod \frac{2m}{\gamma}\} \subset$$
$$\subset \mathbb{Z}P^* + K(P)^*.$$
$$\tag{2.2.6}$$

It gives the calculation of the discriminant group $A_S = S^*/S$ where S is given by (2.2.3) and S^* is given by (2.2.6).

We define *the canonical submodule* $\widetilde{K}(P)^* \subset \mathbb{Z} \oplus K(P)^*$ by the condition

$$\widetilde{K}(P)^* = \{n \in \mathbb{Z},\ k^* \in K(P)^* \mid n \equiv -\frac{2m}{\gamma}\ (k^* \cdot u^*(P)) \mod \frac{2m}{\gamma}\}. \tag{2.2.7}$$

Now we define the *characteristic map*

$$\kappa(P) : \widetilde{K}(P)^* \to A_S, \tag{2.2.8}$$

by the condition

$$\kappa(P)(n, k^*) = nP^* + k^* \mod S. \tag{2.2.9}$$

Obviously, the characteristic map is epimorphic. Its kernel is

$$\widetilde{K}(P)_0^* = [(2m\mathbb{Z}, K), \mathbb{Z}(\gamma, u^*(P))] \cong S. \tag{2.2.10}$$

Thus, we correspond to a primitive $P \in S$ with $P^2 = 2m$ and $\gamma(P) = \gamma$ the canonical triplet

$$(K(P), u^*(P) + K(P), \kappa(P)). \tag{2.2.11}$$

This triplet is important because of the trivial but very important for us

LEMMA 2.2.1. *Let $P_1 \in S$ and $P_2 \in S$ are two primitive elements of an even lattice S with $P_1^2 \neq 0$ and $P_2^2 \neq 0$.*

There exists an automorphism $f \in O(S)$ such that $f(P_1) = P_2$ and f gives ± 1 on the discriminant group $A_S = S^*/S$ if and only if $P_1^2 = P_2^2$, $\gamma(P_1) = \gamma(P_2)$, and there exists an isomorphism of lattices $\phi : K(P_1) \to K(P_2)$ such that $\phi^*(u^*(P_2) + K(P_2)) = u^*(P_1) + K(P_1)$ and $\widetilde{\phi}^* = (id, \phi^*) : \widetilde{K}(P_2)^* \to \widetilde{K}(P_1)^*$ is $\pm$ commuting with the characteristic maps $\kappa(P_1)$ and $\kappa(P_2)$, i. e.

$$\kappa(P_1)\widetilde{\phi^*} = \pm\kappa(P_2).$$

PROOF. Trivial.

We also mention that

$$\det(S) = \gamma^2 \det K(P)/2m. \tag{2.2.12}$$

because $[S : \mathbb{Z}P \oplus K(P)] = 2m/\gamma$

2.3. Relation between periods of X and Y. Here we consider the case and notations of Sect. 2.1. Thus, for the primitive isotropic Mukai vector $v = (r, H, s)$, $r, s \geq 1$, $H^2 = 2rs$, we assume that for $d \in \mathbb{N}$, the element $\widetilde{H} = H/d$ is primitive in $N(X)$, $\gamma(\widetilde{H}) = \gamma$ (in $N(X)$). We remind that $c = \mathrm{g.c.d}(r, s)$, $a = r/c$, $b = s/c$, and $n(v) = \mathrm{g.c.d}(r, s, d\gamma) = (c, d\gamma) = 1$. It follows, $d^2 | ab$ and $\gamma d^2 | 2ab$. Thus, our data are defined by

$$a, b, c, d, \gamma \in \mathbb{N}, \text{ such that } (a, b) = (d, c) = (d, \gamma) = 1, \ d^2 | ab, \ \gamma d^2 | 2ab, \quad (2.3.1)$$

and by a primitive polarization

$$\widetilde{H} \in N(X) \text{ such that } \widetilde{H}^2 = 2abc^2/d^2, \ \gamma(\widetilde{H}) = \gamma. \quad (2.3.2)$$

Then, the Mukai vector $v = (r, H, s) = (ac, d\widetilde{H}, bc)$.

Let us denote by e_1 the canonical generator of $H^0(X, \mathbb{Z})$ and by e_2 the canonical generator of $H^4(X, \mathbb{Z})$. They generate the sublattice U in $H^*(X, \mathbb{Z})$ with the Gram matrix U. Consider Mukai vector $v = (re_1 + se_2 + H)$. We have

$$N(Y) = v^{\perp}_{U \oplus N(X)}/\mathbb{Z}v. \quad (2.3.3)$$

Let us calculate $N(Y)$. Let $K(H) = K(\widetilde{H}) = (H)^{\perp}_{N(X)}$. We denote $H^* = H/(2rs) \in (\mathbb{Z}H)^* = \mathbb{Z}H^*$, and $\widetilde{H}^* = \widetilde{H}/(2rs/d^2) = dH^* \in (\mathbb{Z}\widetilde{H})^* = \mathbb{Z}\widetilde{H}^*$. Then we have an embedding of lattices of finite index

$$\mathbb{Z}\widetilde{H} \oplus K(H) \subset N(X) \subset N(X)^* \subset \mathbb{Z}\widetilde{H}^* \oplus K(H)^* \quad (2.3.4)$$

We have the orthogonal decomposition up to finite index

$$U \oplus \mathbb{Z}\widetilde{H} \oplus K(H) \subset U \oplus N(X) \subset U \oplus \mathbb{Z}\widetilde{H}^* \oplus K(H)^*. \quad (2.3.5)$$

Let $f = x_1 e_1 + x_2 e_2 + yH^* + z^* \in v^{\perp}_{U \oplus N(X)}$, $z^* \in K(H)^*$. Then $-sx_1 - rx_2 + y = 0$ since $f \in v^{\perp}$ and hence $(f, v) = 0$. Thus, $y = (sx_1 + rx_2)$ and

$$f = x_1 e_1 + x_2 e_2 + (sx_1 + rx_2)H^* + z^*. \quad (2.3.6)$$

Here $f \in U \oplus N(X)$ if and only if

$$x_1, x_2 \in \mathbb{Z}, \quad sx_1 + rx_2 \equiv 0 \mod d, \quad \frac{sx_1 + rx_2}{d}\widetilde{H}^* + z^* \in N(X). \quad (2.3.7)$$

Equivalently, by Sect. 2.2, we have

$$sx_1 + rx_2 \equiv 0 \mod d\gamma, \quad z^* = \frac{sx_1 + rx_2}{d\gamma}u^*(H) \mod K(H). \quad (2.3.8)$$

We denote

$$h' = (-a, b) \oplus 0 \in U \oplus N(X).$$

Clearly, $h' \in v^{\perp}$ and $h = h' \mod \mathbb{Z}v \in N(Y)$. Thus, the orthogonal complement contains

$$[\mathbb{Z}v, \mathbb{Z}h', K(H)] \quad (2.3.9)$$

where $h' = -ae_1 + be_2$, and (2.3.9) is a sublattice of finite index in $(v^{\perp})_{U \oplus N(X)}$. The generators v, h' and generators of $K(H)$ are free, and we can rewrite f above using these generators with rational coefficients. We have

$$e_1 = \frac{v - ch' - H}{2r}, \quad e_2 = \frac{v + ch' - H}{2s}. \quad (2.3.10)$$

It follows,

$$f = \frac{sx_1 + rx_2}{2rs}v + \frac{c(-sx_1 + rx_2)}{2rs}h' + z^* \qquad (2.3.11)$$

where x_1, x_2, z^* satisfy (2.3.8). Considering $\mod \mathbb{Z}v$, we finally get

$$N(Y) = \frac{c(-sx_1 + rx_2)}{2rs}h + \frac{sx_1 + rx_2}{d\gamma}u^*(\widetilde{H}) + K(H), \text{ where } sx_1 + rx_2 \equiv 0 \mod d\gamma.$$
$$(2.3.12)$$

Let us calculate the lattice $K(h) = h_{N(Y)}^\perp$. It is equal to $(sx_1 + rx_2)/(d\gamma)u^*(\widetilde{H}) + K(H)$ where $sx_1 + rx_2 \equiv 0 \mod d\gamma$ and $-sx_1 + rx_2 = 0$. It follows, $x_1 = ax$, $x_2 = bx$, $x \in \mathbb{Z}$, and $sx_1 + rx_2 = 2abcx \equiv 0 \mod d\gamma$, which is always true. Thus, $K(h) = [2abc/(d\gamma)u^*(\widetilde{H}) + K(H)]$. By Sect. 2.2, $u^*(\widetilde{H}) + K(H)$ has the order $2abc^2/(d^2\gamma)$. We have $(2abc/(d\gamma), 2abc^2/(d^2\gamma)) = 2abc/(d^2\gamma)(d,c)$ where $2abc/(d^2\gamma) \in \mathbb{N}$ and $(d,c) = 1$. It follows, $K(h) = [K(H), 2abc/(d^2\gamma)u^*(\widetilde{H})]$, and the index $[K(h) : K(H)] = c$.

Let us show that $d|h$ in $N(Y)$ and h/d is primitive in $N(Y)$. By (2.3.12), the primitive submodule in $N(Y)$, generated by h, is $c(-sx_1 + rx_2)/(2rs)h$ where $(sx_1 + rx_2)/(d\gamma)u^*(\widetilde{H}) \in K(h)$ and $sx_1 + rx_2 \equiv 0 \mod d\gamma$. From calculation of $K(h)$ above, we then get $(sx_1 + rx_2)/(d\gamma) \equiv 0 \mod 2abc/(d^2\gamma)$. It follows, $bx_1 + ax_2 \equiv 0 \mod 2ab/d$. Let $d = d_a d_b$ where $d_a|a$ and $d_b|b$. Then $(a/d_a)|x_1$ and $(b/d_b)|x_2$. It follows, $x_1 = (a/d_a)\widetilde{x}_1$ and $x_2 = (b/d_b)\widetilde{x}_2$ where $\widetilde{x}_1, \widetilde{x}_2 \in \mathbb{Z}$ and $d_b\widetilde{x}_1 + d_a\widetilde{x}_2 \equiv 0 \mod 2$. It follows that the module $c(-sx_1 + rx_2)/(2rs)h = (-d_b\widetilde{x}_1 + d_a\widetilde{x}_2)/(2d_a d_b)$ where $d_b\widetilde{x}_1 + d_a\widetilde{x}_2 \equiv 0 \mod 2$. It follows that this module is $\mathbb{Z}(h/d)$. It proves the statement. We denote

$$\widetilde{h} = \frac{h}{d}. \qquad (2.3.13)$$

We had proved that $\widetilde{h}$ is primitive in $N(Y)$ and $h = d\widetilde{h}$. We have $\widetilde{h}^2 = 2ab/d^2$.

Let us show that $\gamma(\widetilde{h}) = \gamma(H) = \gamma$. We remind that $\gamma(\widetilde{h})\mathbb{Z} = \widetilde{h} \cdot N(Y)$. By Sect. 2.2 and (2.3.12), $\widetilde{h}^2/\gamma(\widetilde{h})$ is equal to the index $[[(sx_1 + rx_2)/(d\gamma)u^*(\widetilde{H}) + K(H)] : K(h)]$ where $sx_1 + rx_2 \equiv 0 \mod d\gamma$. Such elements x_1, x_2 give $cd\gamma\mathbb{Z}$. Thus, $[(sx_1 + rx_2)/(d\gamma)u^*(\widetilde{H}) + K(H)] = [cu^*(\widetilde{H}) + K(H)]$. We had proved that $K(h) = [2abc/(d^2\gamma)u^*(\widetilde{H}) + K(H)]$. Thus, the index is equal to $2ab/d^2\gamma$. Then we get $\widetilde{h}^2/\gamma(\widetilde{h}) = 2ab/(d^2\gamma(\widetilde{h})) = 2ab/(d^2\gamma)$. It follows $\gamma(\widetilde{h}) = \gamma$. We had also proved that $u^*(\widetilde{h}) = mcu^*(H) + K(h)$ where m is defined $\mod 2ab/(d^2\gamma)$. We shall calculate m below.

Now we can rewrite (2.3.12) in the form (2.2.2). We denote $\widetilde{h}^* = \widetilde{h}/\widetilde{h}^2 = \widetilde{h}/(2ab/d^2)$. We have

$$N(Y) = \frac{-bx_1 + ax_2}{d}\widetilde{h}^* + \frac{(bx_1 + ax_2)}{d\gamma}cu^*(\widetilde{H}) + K(H) \qquad (2.3.14)$$

where $bx_1 + ax_2 \equiv 0 \mod d\gamma$. We have proved that the elements $(-bx_1 + ax_2)/d$ give $\mathbb{Z}\gamma$.

Let us write $d = d_a d_b$ where $d_a|a$ and $d_b|b$. Since $\gamma|2ab/d^2$ and g.c.d$(a,b) = 1$, we can write $\gamma = \gamma_2\gamma_a\gamma_b$ where $\gamma_a = $ g.c.d$(\gamma, a/d_a^2)$, $\gamma_b = $ g.c.d$(\gamma, b/d_b^2)$ and $\gamma_2 = \gamma/(\gamma_a\gamma_b)$. Clearly, $\gamma_2|2$.

We define $m = m(a, b, d, \gamma) \mod 2ab/(d^2\gamma)$ by the conditions

$$m \equiv -1 \mod 2a/(d_a^2\gamma_a\gamma_2) \text{ and } m \equiv 1 \mod 2b/(d_b^2\gamma_b\gamma_2). \qquad (2.3.15)$$

The congruences (2.3.15) define m uniquely $\mod 2ab/(d^2\gamma)$. Really, assume that $x \equiv 0 \mod 2b/(d_b^2\gamma_b\gamma_2)$ and $x \equiv 0 \mod 2a/(d_a^2\gamma_a\gamma_2)$. Then $x/(2/\gamma_2) \equiv 0 \mod b/(d_b^2\gamma_b)$ and $x/(2/\gamma_2) \equiv 0 \mod a/(d_a^2\gamma_a)$. It follows that $x/(2/\gamma_2) \equiv 0 \mod ab/(d^2\gamma_a\gamma_b)$. Thus, $x \equiv 0 \mod 2ab/(d^2\gamma_a\gamma_b\gamma_2)$ where $2ab/(d^2\gamma_a\gamma_b\gamma_2) = 2ab/(d^2\gamma)$. Thus $x \equiv 0 \mod 2ab/(d^2\gamma)$. Clearly, there exists a unique $m(a,b) \mod 2ab$ defined by the condition

$$m(a,b) \equiv -1 \mod 2a \text{ and } m(a,b) \equiv 1 \mod 2b. \qquad (2.3.16)$$

Then

$$m(a,b,d,\gamma) \equiv m(a,b) \mod \frac{2ab}{d^2\gamma}, \qquad (2.3.17)$$

and one can take (2.3.16) and (2.3.17) as definition of $m(a,b,d,\gamma)$.

Let us prove that $u^*(\widetilde{h}) + K(h) = m(a,b,d,\gamma)cu^*(\widetilde{H}) + K(h)$. We had proved that $u^*(\widetilde{h}) + K(h) = mcu^*(\widetilde{H}) + K(h)$ where m is defined $\mod 2ab/(d^2\gamma)$. To find $m \mod 2ab/(d^2\gamma)$, one should put $(-bx_1 + ax_2)/d = \gamma$ in (2.3.14) or

$$-bx_1 + ax_2 = d\gamma. \qquad (2.3.18)$$

From (2.3.18) we have $d_a\gamma_a|x_1$ and $d_b\gamma_b|x_2$. From (2.3.14) and (2.3.18), we get

$$m \equiv \frac{ax_2 + bx_1}{d\gamma} \equiv 1 + \frac{2bx_1}{d\gamma} \mod 2ab/(d^2\gamma), \qquad (2.3.19)$$

and

$$m \equiv -1 + \frac{2ax_2}{d\gamma} \mod 2ab/(d^2\gamma). \qquad (2.3.20)$$

Since $d_a\gamma_a|x_1$, we get from (2.3.19) that $m \equiv 1 \mod 2b/(d_b^2\gamma_b\gamma_2)$, and from (2.3.20) that $m \equiv -1 \mod 2a/(d_a^2\gamma_a\gamma_2)$. Thus, $m = m(a,b,d,\gamma) \mod 2ab/(d^2\gamma)$. It proves the statement.

Since $h^2 = 2ab$ and $H^2 = 2abc^2$, we can formally put $h = H/c$. By our construction, $K(H) \subset K(h)$ is a sublattice. Thus, we can consider $N(X)$ and $N(Y)$ as extensions of finite index of a common sublattice $\mathbb{Z}\widetilde{H} + K(H)$.

Finally, we get the very important for us

PROPOSITION 2.3.1. *The Picard lattice of X is*

$$N(X) = [\widetilde{H}, K(H), \gamma\widetilde{H}^* + u^*(\widetilde{H})] \qquad (2.3.21)$$

where $\widetilde{H} = H/d$ is primitive in $N(X)$ with $H^2 = 2abc^2$ and $\widetilde{H}^* = d^2\widetilde{H}/(2abc^2)$, the lattice $K(H) = H^\perp_{N(X)}$, and $u^*(\widetilde{H}) + K(H)$ has the order $2abc^2/(d^2\gamma)$ in $K(H)^*/K(H)$.

The Picard lattice of Y is

$$N(Y) = [\widetilde{h} = h/d, K(h), \gamma\widetilde{h}^* + u^*(\widetilde{h})], \qquad (2.3.22)$$

where the element $h = (-a,0,b) \mod \mathbb{Z}v$, $h^2 = 2ab$ and $\widetilde{h} = h/d$ is primitive in $N(Y)$, the element $\widetilde{h}^* = d^2\widetilde{h}/(2ab)$, and $u^*(\widetilde{h}) + K(h)$ has the order $2ab/(d^2\gamma)$ in $K(h)^*/K(h)$.

They are related as follows:

$$K(h) = h^\perp_{N(Y)} = [K(H), \frac{2ab}{d^2\gamma}cu^*(\widetilde{H})], \qquad (2.3.23)$$

and

$$u^*(\tilde{h}) + K(h) = m(a,b)cu^*(\tilde{H}) + K(h) \qquad (2.3.24)$$

where $m(a,b)$ mod $2ab$ is defined by $m(a,b) \equiv -1$ mod $2a$ and $m(a,b) \equiv 1$ mod $2b$. To define $u^*(\tilde{h}) + K(h)$ above, it is enough to consider $m(a,b)$ mod $2ab/(d^2\gamma)$.

We can formally put $h = H/c$, equivalently $\tilde{h} = \tilde{H}/c$. Then $N(X)$ and $N(Y)$ become the extensions of a common sublattice:

$$N(X) \supset [\tilde{H} = c\,\tilde{h},\, K(H)] \subset N(Y). \qquad (2.3.25)$$

Since $n(v) = 1$, the transcendental lattices $T(X) = T(Y)$ are canonically identified in $v^\perp$. It follows that we have the canonical identifications

$$N(X)^*/N(X) = T(X)^*/T(X) = T(Y)^*/T(Y) = N(Y)^*/N(Y). \qquad (2.3.26)$$

Here we use that discriminant groups of orthogonal complements in a unimodular lattice are canonically isomorphic. For example, here the identification $N(X)^*/N(X) = T(X)^*/T(X)$ is given by $n^* + N(X) \to t^* + T(X)$, if $n^* + t^* \in H^2(X, \mathbb{Z})$.

Let us calculate the identification

$$N(X)^*/N(X) = N(Y)^*/N(Y). \qquad (2.3.27)$$

Obviously (from the description above), it is given by the canonical maps

$$N(X)^* \leftarrow (U \oplus N(X))^* \supset (U \oplus N(X))^{*\perp}_v \to (v^\perp)^*_0 \leftarrow (v^\perp/\mathbb{Z}v)^*. \qquad (2.3.28)$$

Here $(U \oplus N(X))^{*\perp}_v = \{x \in (U \oplus N(X))^* | x \cdot v = 0\}$, $(v^\perp) = \{x \in U \oplus N(X) | x \cdot v = 0\}$ and $(v^\perp)^*_0 = \{x \in (v^\perp)^* | x \cdot v = 0\}$.

Let $f^* = n\tilde{H}^* + k^* \in N(X)^*$ where $n \in \mathbb{Z}$ and $k^* \in K(H)^*$. The element $\tilde{f}^* = x_1 e_1 + x_2 e_2 + n\tilde{H}^* + k^*$ is its lift to $(U \oplus N(X))^*$ where $x_1, x_2 \in \mathbb{Z}$. We have $\tilde{f}^* \in (U \oplus N(X))^{*\perp}_v$ if $-sx_1 - rx_2 + dn = 0$. It follows that $n = c(bx_1 + ax_2)/d$ where $bx_1 + ax_2 \equiv 0$ mod d, and $\tilde{f}^* = x_1 e_1 + x_2 e_2 + (c(bx_1 + ax_2)/d)\tilde{H}^* + k^*$. It follows that

$$f^* = \frac{c(bx_1 + ax_2)}{d}\tilde{H}^* + k^*, \quad bx_1 + ax_2 \equiv 0 \mod d. \qquad (2.3.29)$$

Like in (2.3.11), (2.3.12) and (2.3.14), we finally get that the corresponding to $f^* + N(X)$ element in $N(Y)^*/N(Y)$ is $\epsilon(f^*) + N(Y)$ where

$$\epsilon(f^*) = \frac{-bx_1 + ax_2}{d}\tilde{h}^* + k^* \in N(Y)^*. \qquad (2.3.30)$$

Thus, the identification (2.3.27) is given by $f^* = c\left((bx_1 + ax_2)/d\right)\tilde{H}^* + k^* + N(X) \to \epsilon(f^*) = (-bx_1 + ax_2)\tilde{h}^* + k^* + N(Y)$ where $bx_1 + ax_1 \equiv 0$ mod d. We have $x_1 = d_a\tilde{x}_1$ and $x_2 = d_b\tilde{x}_2$ where $\tilde{x}_1, \tilde{x}_2 \in \mathbb{Z}$. Then

$$f^* = c\left((b/d_b)\tilde{x}_1 + (a/d_a)\tilde{x}_2\right)\tilde{H}^* + k^* \to \epsilon(f^*) = \left(-(b/d_b)\tilde{x}_1 + (a/d_a)\tilde{x}_2\right)\tilde{h}^* + k^*.$$

Let us denote $n_1 = (b/d_b)\tilde{x}_1 + (a/d_a)\tilde{x}_2$ and $n_2 = -(b/d_b)\tilde{x}_1 + (a/d_a)\tilde{x}_2$. We have $n_1 \equiv m(a,b)n_2$ mod $2ab/d^2$. Really, $n_1 - m(a,b)n_2 \equiv (2a/d_a)\tilde{x}_2 \equiv 0$ mod $2a/d_a^2$ because $m(a,b) \equiv -1$ mod $2a/d_a^2$. We have $n_1 - m(a,b)n_2 \equiv$

$(2b/d_b)\widetilde{x}_1 \equiv 0 \mod 2b/d_b^2$ because $m(a,b) \equiv 1 \mod 2b/d_b^2$. It follows the statement. Here we consider $m(a,b) \mod 2ab/d^2$.

The expression $(b/d_b)\widetilde{x}_1 + (a/d_a)\widetilde{x}_2$ gives all integers $n \in \mathbb{Z}$. Thus, we have proved that if $f^* = cn\widetilde{H}^* + k^* \in N(X)^*$ where $n \in \mathbb{Z}$, $k^* \in K(H)^*$, then $\epsilon(f^*) = m(a,b)\widetilde{h}^* + k^* \in N(Y)$. Thus, we have proved

PROPOSITION 2.3.2. *In notations of Proposition 2.3.1, the canonical identification $\epsilon : N(X)^*/N(X) \cong N(Y)^*/N(Y)$ of the discriminant groups given by periods:*

$$\epsilon : N(X)^*/N(X) = T(X)^*/T(X) = T(Y)^*/T(Y) = N(Y)^*/N(Y) \qquad (2.3.31)$$

is given by

$$\epsilon : cn\widetilde{H}^* + k^* + N(X) \mapsto m(a,b)n\widetilde{h}^* + k^* \qquad (2.3.32)$$

where $n \in \mathbb{Z}$, $k^* \in K(H)^*$. Here one should consider $m(a,b) \mod 2ab/d^2$.

Equivalently, the characteristic maps $\kappa(\widetilde{H}) : \widetilde{K}(H) \to N(X)^*/N(X)$ and $\kappa(\widetilde{h}) : \widetilde{K}(h) \to N(Y)^*/N(Y)$ are related as follows:

$$\epsilon(\kappa(\widetilde{H})((cn, k^*) + \widetilde{K}(H)_0) = \kappa(\widetilde{h})(m(a,b)n, k^*) + \widetilde{K}(h)_0 \qquad (2.3.33)$$

(here one should consider $m(a,b) \mod 2ab/d^2$). Here we set $\widetilde{K}(H) = \widetilde{K}(\widetilde{H})$ and $\widetilde{K}(h) = \widetilde{K}(\widetilde{h})$ (see (2.2.7)).

This finishes the calculation of the periods of $N(Y)$ in terms of the periods of $N(X)$.

Applying Lemma 2.2.1 and Propositions 2.3.1 and 2.3.2, by Global Torelli Theorem for K3 surfaces [**PS**], we get

THEOREM 2.3.3. *Assume that X is a K3 surface with a polarization H with $H^2 = 2rs$, $r, s \geq 1$, and a primitive Mukai vector $v = (r, H, s)$ with the invariants*

$$(a, b, c, d, \gamma) \qquad (2.3.34)$$

introduced above, i.e. $\widetilde{H} = H/d$ is primitive. Let Y be the moduli space of coherent sheaves on X with the Mukai vector $v = (r, H, s)$. We denote by $(K(\widetilde{H}), u^*(\widetilde{H}), \kappa(\widetilde{H}))$ the invariants (2.2.11) of $\widetilde{H} \in N(X)$.

The transcendental periods $(T(X), H^{2,0}(X))$ and $(T(Y), H^{2,0}(Y))$ are isomorphic if and only if

$$n(v) = g.c.d(c, d\gamma) = 1 \qquad (2.3.35)$$

(this is Mukai's result). We denote by $(K(H) = K(\widetilde{H}), u^*(\widetilde{H}), \kappa(\widetilde{H}))$ the invariants (2.2.11) of $\widetilde{H} \in N(X)$.

Assume that (2.3.35) is valid. Then $Y \cong X$, if the following conditions (a), (b) and (c) are valid:

(a) there exists a primitive $\widetilde{h} \in N(X)$ with $\widetilde{h}^2 = 2ab/d^2$ and $\gamma(\widetilde{h}) = \gamma$.

(b) There exists an embedding $\phi : K(H) \subset K(\widetilde{h})$ of lattices such that $\phi^*(K(\widetilde{h})) = [K(H), 2abc/(d^2\gamma)u^*(\widetilde{H})]$ and $\phi^*(u^*(\widetilde{h})) + \phi^*(K(\widetilde{h})) = m(a,b)cu^*(\widetilde{H}) + \phi^*(K(\widetilde{h}))$. Here $m(a,b) \mod 2ab/(d^2\gamma)$ is considered.

(c) There exists a choice of $\pm$ such that $\kappa(\widetilde{h})(m(a,b)n, z^*) = \pm\kappa(H)(cn, \phi^*(z^*))$ if $(cn, \phi^*(z^*)) \in \widetilde{K}(H)^* = \widetilde{K}(\widetilde{H})^*$. Here $m(a,b) \mod 2ab/d^2$ is considered.

The conditions (a), (b) and (c) are necessary for a K3 surface X with $\rho(X) \leq$
19 which is general for its Picard lattice $N(X)$ in the following sense: the automor-
phism group of the transcendental periods $(T(X), H^{2,0}(X))$ is ± 1. If $\rho(X) = 20$,
then always $Y \cong X$ if (2.3.35) holds.

EXAMPLE 2.3.4. Let us assume that $\rho(X) = 1$. Thus, $N(X) = \mathbb{Z}\widetilde{H}$. Assume
conditions (a), (b) and (c) of Theorem 2.3.3 satisfy. Then $\widetilde{h} = \pm \widetilde{H}$. It follows $c = 1$.
The lattices $K(H)$ and $K(\widetilde{h})$ are zero, and then (b) is valid. The discriminant
group $N(X) = \mathbb{Z}\widetilde{H}^*/\mathbb{Z}\widetilde{H} \equiv \mathbb{Z}/(2ab/d^2)\mathbb{Z}$. The condition (c) is valid if and only if
$m(a, b) \equiv \pm 1 \mod 2ab/d^2$. This is true if and only if $(a/d_a^2) = 1$ or $(b/d_b^2) = 1$.

Thus, $X \cong Y$ if $c = 1$ and either $a_1 = a/d_a^2 = 1$ or $b_1 = b/d_b^2 = 1$. These
conditions are necessary to have $Y \cong X$, if X is a general K3 surface with $\rho(X) = 1$.
We recover the result of Mukai from [**Mu3**].

EXAMPLE 2.3.5. Following Example 2.3.5, let us consider the case when we
can satisfy conditions of Theorem 2.3.3 taking $\widetilde{h} = \pm H$ and $\phi = \pm id$. Again
we get $c = 1$. (b) satisfies, if and only if $m(a, b) \equiv \pm 1 \mod 2ab/(d^2\gamma)$. This is
equivalent to either $2a/(d_a^2 \gamma_a \gamma_2) \leq 2$ or $2b/(d_b^2 \gamma_b \gamma_2) \leq 2$. (c) satisfies, if and only
if $m(a, b) \equiv \pm 1 \mod 2ab/d^2$. Thus either $a/d_a^2 = 1$ or $b/d_b^2 = 1$. Thus, always
$Y \cong X$, if $c = 1$ and either $a_1 = a/d_a^2 = 1$ or $b_1 = b/d_b^2 = 1$. This is just a
specialisation (see Lemma 2.1.1) of the $\rho = 1$ case above. This result is also due to
Mukai [**Mu3**]

In Sect. 3 we consider $\rho = 2$. We shall analyse when we can satisfy conditions
of Theorem 2.3.3 in this case. By specialisation (see Lemma 2.1.1) of these cases,
we shall get results about K3 surfaces with any Picard number $\rho \geq 2$.

3. Conditions of $Y \cong X$ for a general K3 surface X with $\rho = 2$

3.1. Main results for $\rho(X) = 2$. Here we apply results of Sect. 2 to X and
Y with Picard number 2. Thus, we assume that $\rho(X) = \mathrm{rk}\, N(X) = 2$.

We start with some preliminary considerations on a primitive element $P \in S$
of an even hyperbolic lattice S of $\mathrm{rk}\, S = 2$. We assume that $P^2 = 2n$, $n \in \mathbb{N}$, and
$\gamma(P) = \gamma|2n$.

Let

$$K(P) = P_S^\perp = \mathbb{Z}f(P) \tag{3.1.1}$$

and $f(P)^2 = -t$ where $t > 0$ is even. Then $\pm f(P) \in S$ is defined uniquely by P.
Below we set $f = f(P)$.

By elementary considerations, we have

$$S = [\mathbb{Z}P, \mathbb{Z}f, \frac{\gamma(\mu P + f)}{2n}] \tag{3.1.2}$$

where

$$\mathrm{g.c.d}(\mu, \frac{2n}{\gamma}) = 1. \tag{3.1.3}$$

The element

$$\pm \mu \mod \frac{2n}{\gamma} \in (\mathbb{Z}/\frac{2n}{\gamma})^* \tag{3.1.4}$$

is *the invariant of the pair $P \in S$* up to isomorphisms of lattices with a primitive
vector P of $P^2 = 2n$ and $\gamma(P) = \gamma$. If f changes to $-f$, then $\mu \mod 2n/\gamma$ changes
to $-\mu \mod 2n/\gamma$.

We have $(\gamma(\mu P + f)/(2n))^2 = \gamma^2(\mu^2 - t/2n)/2n \equiv 0 \mod 2$. It follows $2n\mu^2 - t \equiv 0 \mod 8n^2/\gamma^2$. It follows that for some $\delta \in \mathbb{N}$ we have

$$f^2 = -\frac{2n\delta}{\gamma}, \quad n\delta \equiv 0 \mod \gamma, \quad \text{and} \quad \delta \equiv \mu^2\gamma \mod \frac{4n}{\gamma}. \tag{3.1.5}$$

We have

$$\det S = -\gamma\delta. \tag{3.1.6}$$

Any element $z \in S$ can be written as $z = \gamma(xP + yf)/2n$ where $x \equiv \mu y$ mod $2n/\gamma$. We have

$$z^2 = \frac{\gamma x^2 - \delta y^2}{(2n/\gamma)} \tag{3.1.7}$$

It is convenient to put

$$\widetilde{n} = \frac{2n}{\gamma}. \tag{3.1.8}$$

Thus, the considered above case of a primitive $P \in S$ where S is an even hyperbolic lattice of rk $S = 2$ is described by the invariants

$$\widetilde{n}, \ \gamma, \ \delta, \ \pm\mu \in (\mathbb{Z}/\widetilde{n})^*, \tag{3.1.9}$$

where $\widetilde{n}, \gamma, \delta \in \mathbb{N}$. The invariants (3.1.9) must satisfy

$$\widetilde{n}\gamma \equiv \widetilde{n}\delta \equiv 0 \mod 2, \quad \delta \equiv \mu^2\gamma \mod 2\widetilde{n}. \tag{3.1.10}$$

Then $P^2 = \widetilde{n}\gamma$, $f^2 = -\widetilde{n}\delta$, $P \perp f$, and

$$S = \{\frac{xP + yf}{\widetilde{n}} \mid x, y \in \mathbb{Z}, \ x \equiv \mu y \mod \widetilde{n}\}. \tag{3.1.11}$$

We have

$$z^2 = \frac{\gamma x^2 - \delta y^2}{\widetilde{n}}. \tag{3.1.12}$$

Moreover,

$$\det S = -\delta\gamma. \tag{3.1.13}$$

We denote

$$P^* = \frac{P}{\widetilde{n}\gamma}, \quad f^* = \frac{f}{\widetilde{n}\delta}. \tag{3.1.14}$$

Then

$$S^* = \{vP^* + wf^* \mid \mu v - w \equiv 0 \mod \widetilde{n}\} \tag{3.1.15}$$

and

$$S = \{vP^* + wf^* \mid v \equiv 0 \mod \gamma, \ w \equiv 0 \mod \delta, \ \frac{v}{\gamma} \equiv \frac{\mu w}{\delta} \mod \widetilde{n}\}. \tag{3.1.16}$$

Here $v, w \in \mathbb{Z}$. From (3.1.15), $w = \mu v + \widetilde{n}t$, and

$$S^* = v(P^* + \mu f^*) + t\widetilde{n}f^*, \quad v, t \in \mathbb{Z}, \tag{3.1.17}$$

and $v(P^* + \mu f^*) + t\widetilde{n}f^* \in S$ if and only if

$$v \equiv 0 \mod \gamma, \quad \mu v + \widetilde{n}t \equiv 0 \mod \delta, \quad \delta v \equiv \gamma\mu(\mu v + \widetilde{n}t) \mod \delta\gamma\widetilde{n}. \tag{3.1.18}$$

We have

$$u^*(P) = \frac{\mu^{-1}f}{\widetilde{n}} + \mathbb{Z}f = \mu^{-1}\delta f^* + \mathbb{Z}f. \tag{3.1.19}$$

We remind notations we have used in Sect. 2:

$$a_1 = \frac{a}{d_a^2}, \ b_1 = \frac{b}{d_b^2} \qquad (3.1.20)$$

where $d_a = \mathrm{g.c.d}(d, a)$ and $d_b = \mathrm{g.c.d}(d, b)$. We put

$$a_2 = \frac{a_1}{\gamma_a}, \ b_2 = \frac{b_1}{\gamma_b}, \ e_2 = \frac{2}{\gamma_2}. \qquad (3.1.21)$$

where $\gamma_a = \mathrm{g.c.d}(a_1, \gamma)$, $\gamma_b = \mathrm{g.c.d}(b_1, \gamma)$, $\gamma_2 = \gamma/(\gamma_a \gamma_b)$. Then $\gamma_2 | 2$.

Applying calculations above to $S = N(X)$ and primitive $P = \widetilde{H}$ with $n = 2a_1 b_1 c^2$, $\tilde{n} = 2a_1 b_1 c^2/\gamma = e_2 a_2 b_2 c^2$, and $\gamma(\widetilde{H}) = \gamma$, we get

PROPOSITION 3.1.1. *Let X be a K3 surface with Picard number $\rho = 2$ equipped with a primitive polarization (or vector) $\widetilde{H} \in N(X)$ of degree $\widetilde{H}^2 = 2a_1 b_1 c^2$ and $\gamma(\widetilde{H}) = \gamma | 2a_1 b_1$.*

Let $K(\widetilde{H}) = (\widetilde{H})^\perp = \mathbb{Z}f(\widetilde{H})$. We have $f(\widetilde{H})^2 = -2a_1 b_1 c^2 \delta/\gamma$ where $\det N(X) = -\delta\gamma$.

For some $\mu \in (\mathbb{Z}/(2a_1 b_1 c^2/\gamma))^*$ (the $\pm\mu$ is the invariant of the pair $\widetilde{H} \in N(X)$) where $\delta \equiv \mu^2 \gamma \mod 4a_1 b_1 c^2/\gamma$, one has

$$N(X) = [\widetilde{H}, f(\widetilde{H}), \frac{(\mu\widetilde{H} + f(\widetilde{H}))}{2a_1 b_1 c^2/\gamma}], \qquad (3.1.22)$$

$$N(X) = \{z = \frac{x\widetilde{H} + yf(\widetilde{H})}{2a_1 b_1 c^2/\gamma} \mid x, y \in \mathbb{Z} \text{ and } x \equiv \mu y \mod \frac{2a_1 b_1 c^2}{\gamma}\}. \qquad (3.1.23)$$

We have

$$z^2 = \frac{\gamma x^2 - \delta y^2}{2a_1 b_1 c^2/\gamma}. \qquad (3.1.24)$$

For any primitive element $P \in N(X)$ with $P^2 = 2a_1 b_1 c^2$, $\gamma(P) = \gamma$ and the same invariant $\pm\mu$, there exists an automorphism $\phi \in O(N(X))$ such that $\phi(\widetilde{H}) = P$.

Applying calculations above to $S = N(Y)$ and primitive $P = \tilde{h} \in N(Y)$ with $n = 2a_1 b_1$, $\tilde{n} = 2a_1 b_1/\gamma = e_2 a_2 b_2$ and $\gamma(\tilde{h}) = \gamma$, we get

PROPOSITION 3.1.2. *Let Y be a K3 surface with Picard number $\rho = 2$ equipped with a primitive polarization (or vector) $\tilde{h} \in N(Y)$ of degree $\tilde{h}^2 = 2a_1 b_1$ and $\gamma(\tilde{h}) = \gamma | 2a_1 b_1$.*

Let $K(\tilde{h}) = (\tilde{h})^\perp = \mathbb{Z}f(\tilde{h})$. We have $f(\tilde{h})^2 = -2a_1 b_1 \delta/\gamma$ where $\det N(Y) = -\delta\gamma$.

For some $\nu \in (\mathbb{Z}/(2a_1 b_1/\gamma))^*$ (the $\pm\nu$ is the invariant of the pair $\tilde{h} \in N(Y)$) where $\delta \equiv \nu^2 \gamma \mod 4a_1 b_1/\gamma$, one has

$$N(Y) = [\tilde{h}, f(\tilde{h}), \frac{(\nu\tilde{h} + f(\tilde{h}))}{2a_1 b_1/\gamma}], \qquad (3.1.25)$$

$$N(Y) = \{z = \frac{x\tilde{h} + yf(\tilde{h})}{2a_1 b_1/\gamma} \mid x, y \in \mathbb{Z} \text{ and } x \equiv \nu y \mod \frac{2a_1 b_1}{\gamma}\}. \qquad (3.1.26)$$

We have

$$z^2 = \frac{\gamma x^2 - \delta y^2}{2a_1 b_1/\gamma}. \qquad (3.1.27)$$

For any primitive element $P \in N(Y)$ with $P^2 = 2a_1b_1$, $\gamma(P) = \gamma$ and the same invariant $\pm\nu$, there exists an automorphism $\phi \in O(N(Y))$ such that $\phi(\widetilde{h}) = P$.

The crucial statement is

THEOREM 3.1.3. *Let X be a K3 surface, $\rho(X) = 2$ and H a polarization of X of degree $H^2 = 2rs$, $r, s \geq 1$, and Mukai vector (r, H, s) is primitive. Let Y be the moduli space of sheaves on X with the isotropic Mukai vector $v = (r, H, s)$ and the canonical nef element $h = (-a, 0, b) \mod \mathbb{Z}v$. We assume that*

$$g.c.d(c, d\gamma) = 1.$$

With notations of Propositions 3.1.1, all elements

$$\widetilde{h} = \frac{x\widetilde{H} + yf(\widetilde{H})}{2a_1b_1c^2/\gamma} \in N(X)$$

with square $\widetilde{h}^2 = 2a_1b_1$ satisfying Theorem 2.2.3 are in one to one correspondence with integral solutions (x, y) of the equation

$$\gamma x^2 - \delta y^2 = 4a_1^2b_1^2c^2/\gamma \tag{3.1.28}$$

which satisfy conditions (i) — (v) below:
  (i)

$$x \equiv \mu y \mod 2a_1b_1c^2/\gamma, \tag{3.1.29}$$

$$\mu\gamma x \equiv \delta y \mod 2a_1b_1c^2; \tag{3.1.30}$$

  (ii) (x, y) belongs to one of a-series (the sign +) or b-series (the sign −) of solutions defined below:

$$\pm m(a, b)\mu x + (\delta y/\gamma) \equiv 0 \mod 2a_1b_1/\gamma, \tag{3.1.31}$$

$$x \pm m(a, b)\mu y \equiv 0 \mod 2a_1b_1/\gamma, \tag{3.1.32}$$

$$\pm m(a, b)\mu x + (\delta y/\gamma) \equiv \mu(x \pm m(a, b)\mu y) \mod (2a_1b_1c^2/\gamma)(2a_1b_1/\gamma); \tag{3.1.33}$$

  (iii) there exists a choice of $\beta = \pm 1$ such that

$$\begin{cases} \left(\frac{m(a,b)\gamma x \pm \mu\gamma y}{2a_1b_1} - \beta c\right) \equiv 0 \mod \gamma \\[2mm] \left(\frac{\delta m(a,b)y \pm \mu\gamma x}{2a_1b_1} - \beta\mu c\right) \equiv 0 \mod \delta \\[2mm] \delta\left(\frac{m(a,b)\gamma x \pm \mu\gamma y}{2a_1b_1} - \beta c\right) \equiv \mu\gamma\left(\frac{\delta m(a,b)y \pm \mu\gamma x}{2a_1b_1} - \beta\mu c\right) \mod 2a_1b_1c^2\delta \end{cases} \tag{3.1.34}$$

and

$$\begin{cases} c\delta y \equiv 0 \mod \gamma \\[2mm] cx - (\pm\beta)\frac{2a_1b_1c^2}{\gamma} \equiv 0 \mod \delta \\[2mm] \delta y \equiv \mu\left(\gamma x - (\pm\beta)2a_1b_1c\right) \mod 2a_1b_1c\delta \end{cases} \tag{3.1.35}$$

where + is taken for a-series, and − is taken for b-series.
  (iv) the pair (x, y) is μ-primitive:

$$g.c.d\left(x, y, \frac{x - \mu y}{2a_1b_1c^2/\gamma}\right) = 1; \tag{3.1.36}$$

  (v) $\gamma(\widetilde{h}) = \gamma$, equivalently

$$g.c.d\left(\gamma x, \delta y, \frac{\mu\gamma x - \delta y}{2a_1b_1c^2/\gamma}\right) = \gamma. \tag{3.1.37}$$

In particular (by Theorem 2.3.3), for a general X with $\rho(X) = 2$ we have $Y \cong X$ if and only if the equation $\gamma x^2 - \delta y^2 = 4a_1^2 b_1^2 c^2/\gamma$ has an integral solution (x, y) satisfying conditions (i)—(v) above. Moreover, a nef primitive element $P = (x\widetilde{H} + f(\widetilde{H}))/(2a_1 b_1 c^2/\gamma)$ with $P^2 = 2a_1 b_1$ and $\gamma(P) = \gamma$ defines a pair (X, P) which is isomorphic to the $(Y, \widetilde{h})$ if and only if (x, y) satisfies the conditions (ii) and (iii) (it satisfies conditions (i), (iv) and (v) since it corresponds to a primitive element of $N(X)$ with $\gamma(P) = \gamma$).

PROOF. We denote

$$H^* = \frac{\widetilde{H}}{2a_1 b_1 c^2}, \quad f(\widetilde{H})^* = \frac{\gamma f(\widetilde{H})}{2a_1 b_1 c^2 \delta} \tag{3.1.38}$$

where $K(H) = \mathbb{Z}f(\widetilde{H}) = H^\perp$ in $N(X)$. By (3.1.19), we have

$$u^*(\widetilde{H}) = \frac{\mu^{-1}\gamma f(\widetilde{H})}{2a_1 b_1 c^2} + \mathbb{Z}f(\widetilde{H}) = \mu^{-1}\delta f(\widetilde{H})^* + \mathbb{Z}f(\widetilde{H}). \tag{3.1.39}$$

Let

$$\widetilde{h} = \frac{x\widetilde{H} + yf(\widetilde{H})}{2a_1 b_1 c^2/\gamma} \in N(X) \tag{3.1.40}$$

satisfies conditions of Theorem 2.2.1. Then $x, y \in \mathbb{Z}$ and $x \equiv \mu y \mod 2a_1 b_1 c^2/\gamma$. We get (3.1.29) in (i). Moreover, $\widetilde{h}$ is primitive which is equivalent to (iv). We also have $\gamma(\widetilde{h}) = \gamma$. It is equivalent to

$$\text{g.c.d}\left(\widetilde{H} \cdot \widetilde{h},\, f(\widetilde{H}) \cdot \widetilde{h},\, (\mu\widetilde{H} + f(\widetilde{H}))/(2a_1 b_1 c^2/\gamma) \cdot \widetilde{h}\right) = \gamma.$$

It follows (3.1.30) in (i), and (v). We have $\widetilde{h}^2 = 2a_1 b_1$. This is equivalent to $\gamma x^2 - \delta y^2 = 4a_1^2 b_1^2 c^2/\gamma$.

Consider $K(\widetilde{h}) = \widetilde{h}^\perp$ in $N(X)$. Let us denote

$$f(\widetilde{h}) = \frac{(\delta y/\gamma)\widetilde{H} + xf(\widetilde{H})}{2a_1 b_1 c^2/\gamma}. \tag{3.1.41}$$

The element $f(\widetilde{h}) \in N(X)$ because of (i). We have $f(\widetilde{h}) \perp \widetilde{h}$ and $f(\widetilde{h})^2 = -2a_1 b_1 \delta/\gamma$. Since $\gamma(\widetilde{h}) = \gamma$, it follows that $K(\widetilde{h}) = \mathbb{Z}f(\widetilde{h})$ and $f(\widetilde{h})$ is primitive. We have

$$N(X) = \left[\widetilde{h}, f(\widetilde{h}), \frac{\nu\widetilde{h} + f(\widetilde{h})}{2a_1 b_1/\gamma}\right] \tag{3.1.42}$$

where $\nu \in (\mathbb{Z}/(2a_1 b_1/\gamma))^*$ (according to Proposition 3.1.2). We denote

$$\widetilde{h}^* = \frac{\widetilde{h}}{2a_1 b_1}, \quad f(\widetilde{h})^* = \frac{\gamma f(\widetilde{h})}{2a_1 b_1 \delta}. \tag{3.1.43}$$

By (3.1.19), we have

$$u^*(\widetilde{h}) = \frac{\nu^{-1}f(\widetilde{h})}{2a_1 b_1/\gamma} + \mathbb{Z}f(\widetilde{h}) = \nu^{-1}\delta f(\widetilde{h})^* + \ddot{\mathbb{Z}}f(\widetilde{h}). \tag{3.1.44}$$

There exists a unique (up to ± 1) embedding

$$\phi : K(\widetilde{H}) = \mathbb{Z}f(\widetilde{H}) \to K(\widetilde{h}) = \mathbb{Z}f(\widetilde{h}), \quad \phi(f(\widetilde{H})) = \pm cf(\widetilde{h}). \tag{3.1.45}$$

of one-dimensional lattices. Its dual is defined by $\phi^*(f(\widetilde{h})^*) = \pm cf(\widetilde{H})^*$.

We have
$$\phi^*(K(\widetilde{h})) = \mathbb{Z}\phi^*(f(\widetilde{h})) = \mathbb{Z}f(\widetilde{H})/c =$$
$$[K(\widetilde{H}), (2a_1b_1c/\gamma)u^*(\widetilde{H})] = [K(\widetilde{H}), (2abc/(d^2\gamma))u^*(\widetilde{H})]$$

because of (3.1.39). This gives the first part of (b) in Theorem 2.3.3.

We have
$$\phi^*(u^*(\widetilde{h})) + \phi^*(K(\widetilde{h})) = \nu^{-1}\delta\phi^*(f(\widetilde{h})^*) + \mathbb{Z}\phi^*(f(\widetilde{h})) = \pm\nu^{-1}\delta cf(\widetilde{H})^* + \mathbb{Z}f(\widetilde{H})/c =$$
$$\pm\nu^{-1}\delta cf(\widetilde{H})^* + \mathbb{Z}(2a_1b_1c\delta/\gamma)f(\widetilde{H})^*.$$

On the other hand, by (3.1.39)
$$m(a,b)cu^*(\widetilde{H}) + \mathbb{Z}(2a_1b_1c\delta/\gamma)f(\widetilde{H})^* = m(a,b)\mu^{-1}\delta cf(\widetilde{H})^* + \mathbb{Z}(2a_1b_1c\delta/\gamma)f(\widetilde{H})^*.$$

Thus, by (3.1.44), second part of (b) in Theorem 2.3.3 is $\pm\nu^{-1} \equiv m(a,b)\mu^{-1}$ mod $2a_1b_1/\gamma$. Equivalently,
$$\nu \equiv \pm m(a,b)\mu \quad \mathrm{mod} \ \frac{2a_1b_1}{\gamma}. \tag{3.1.46}$$

Thus, for ν given by (3.1.46) one has (this is the definition of ν)
$$\frac{\nu\widetilde{h} + f(\widetilde{h})}{2a_1b_1/\gamma} = \left(\frac{\nu x + (\delta y/\gamma)}{2a_1b_1/\gamma}\widetilde{H} + \frac{x + \nu y}{2a_1b_1/\gamma}f(\widetilde{H})\right)/(2a_1b_1c^2/\gamma) \in N(X). \tag{3.1.47}$$

This is equivalent for $\nu \equiv m(a,b)\mu$ mod $2a_1b_1/\gamma$ to
$$\begin{cases} \pm m(a,b)\mu x + (\delta y/\gamma) \equiv 0 \ \mathrm{mod} \ 2a_1b_1/\gamma \\ x \pm m(a,b)\mu y \equiv 0 \ \mathrm{mod} \ 2a_1b_1/\gamma \\ \pm m(a,b)\mu x + (\delta y/\gamma) \equiv \mu(x \pm m(a,b)\mu y) \ \mathrm{mod} \ (2a_1b_1c^2/\gamma)(2a_1b_1/\gamma) \end{cases}. \tag{3.1.48}$$

This gives (ii).

Let us consider the condition (c) of Theorem 2.3.3. For a choice of $\beta = \pm 1$, one has
$$m(a,b)n\widetilde{h}^* + z^* \equiv \beta(cn\widetilde{H}^* + \phi^*(z^*)) \quad \mathrm{mod} \ N(X), \quad n \in \mathbb{Z}, \ z \in K(\widetilde{h})^* \tag{3.1.49}$$

if
$$cn\widetilde{H}^* + \phi^*(z^*) \in N(X)^*. \tag{3.1.50}$$

Let $z^* = kf(\widetilde{h})^*$, $k \in \mathbb{Z}$, then $\phi^*(kf(\widetilde{h})^*) = \pm kcf(\widetilde{H})^*$ and $cn\widetilde{H}^* + \phi^*(z^*) = cn\widetilde{H}^* \pm kcf(\widetilde{H})^*$. By (3.1.15), $cn\widetilde{H}^* \pm kcf(\widetilde{H})^* \in N(X)^*$ if and only if $\mu cn \mp kc \equiv 0$ mod $2a_1b_1c^2/\gamma$. This is equivalent to $k \equiv \pm\mu n$ mod $2a_1b_1c/\gamma$. Like in (3.1.17), we get $k = \pm\mu n + (2a_1b_1c/\gamma)t$ where $n, t \in \mathbb{Z}$. We get
$$cn\widetilde{H}^* \pm kcf(\widetilde{H})^* = cn(\widetilde{H}^* + \mu f(\widetilde{H})^*) \pm t(2a_1b_1c^2/\gamma)f(\widetilde{H})^*. \tag{3.1.51}$$

Thus, it is enough to check
$$m(a,b)n\widetilde{h}^* + kf(\widetilde{h})^* \equiv \beta(cn\widetilde{H}^* \pm kcf(\widetilde{H})^*) \quad \mathrm{mod} \ N(X), \tag{3.1.52}$$

where $(n,k) = (1, \pm\mu)$ or $(n,k) = (0, 2a_1b_1c/\gamma)$. Thus, one should check for one of $\beta = \pm 1$ that
$$m(a,b)\widetilde{h}^* \pm \mu f(\widetilde{h})^* \equiv \beta(c\widetilde{H}^* + c\mu f(\widetilde{H})^*) \quad \mathrm{mod} \ N(X) \tag{3.1.53}$$

and

$$\frac{2a_1b_1cf(\widetilde{h})^*}{\gamma} \equiv \pm\beta\frac{2a_1b_1c^2f(\widetilde{H})^*}{\gamma} \quad \mod N(X). \tag{3.1.54}$$

We have

$$m(a,b)n\widetilde{h}^* + kf(\widetilde{h})^* = m(a,b)n\frac{\widetilde{h}}{2a_1b_1} + k\frac{\gamma f(\widetilde{h})}{2a_1b_1\delta} =$$

$$\frac{1}{2a_1b_1\delta}\left(\delta m(a,b)n\widetilde{h} + \gamma kf(\widetilde{h})\right) =$$

$$\frac{1}{2a_1b_1\delta}\left(\delta m(a,b)n\frac{x\widetilde{H} + yf(\widetilde{H})}{2a_1b_1c^2/\gamma} + \gamma k\frac{(\delta y/\gamma)\widetilde{H} + xf(\widetilde{H})}{2a_1b_1c^2/\gamma}\right) =$$

$$\frac{1}{2a_1b_1\delta}\left(\delta m(a,b)n(\gamma x\widetilde{H}^* + \delta yf(\widetilde{H})^*) + \gamma k(\delta y\widetilde{H}^* + \delta xf(\widetilde{H})^*)\right) =$$

$$\frac{m(a,b)n\gamma x + \gamma ky}{2a_1b_1}\widetilde{H}^* + \frac{\delta m(a,b)ny + \gamma kx}{2a_1b_1}f(\widetilde{H})^*. \tag{3.1.55}$$

Thus, (3.1.53) is

$$\left(\frac{m(a,b)\gamma x \perp \mu\gamma y}{2a_1b_1} - \beta c\right)\widetilde{H}^* + \left(\frac{\delta m(a,b)y \pm \mu\gamma x}{2a_1b_1} - \beta\mu c\right)f(\widetilde{H})^* \in N(X) \tag{3.1.56}$$

and (3.1.54) is

$$cy\widetilde{H}^* + \left(cx - (\pm\beta)\frac{2a_1b_1c^2}{\gamma}\right)f(\widetilde{H})^* \in N(X). \tag{3.1.57}$$

Here one should take $+$ if (x,y) belongs to a-series, and one should take $-$ if (x,y) belongs to b-series.

By (3.1.16), we can reformulate (3.1.56) as (3.1.34) in (iii), and we can reformulate (3.1.57) as (3.1.35) in (iii).

This finishes the proof.

Now we analyse conditions of Theorem 3.1.3. The most important are congruences $\mod \delta$ since δ is not bounded by a constant depending on (r,s).

Let us consider the congruence $cx - (\pm\beta)(2a_1b_1c^2/\gamma) \equiv 0 \mod \delta$ in (3.1.35). We know that $\delta \equiv \gamma\mu^2 \mod (2a_1b_1/\gamma)c^2$ where $\mu \mod (2a_1b_1/\gamma)c^2$ is invertible, $\gamma|2a_1b_1$ and g.c.d$(\gamma,c) = 1$. It follows that g.c.d$(c,\delta) = 1$. Thus, the considered congruence is equivalent to

$$x \equiv \pm\frac{2a_1b_1c}{\gamma} \quad \mod \delta \tag{3.1.58}$$

since $\beta = \pm 1$. Later we shall see that all congruences $\mod \delta$ which follow from conditions of Theorem 3.1.3 are consequences of (3.1.58). Thus, (3.1.58) is the most important condition of Theorem 3.1.3.

Let us consider all integral (x,y) which satisfy (3.1.28) and (3.1.58), i. e.

$$\gamma x^2 - \delta y^2 = \frac{4a_1^2b_1^2c^2}{\gamma} \quad \text{and} \quad x \equiv \pm\frac{2a_1b_1c}{\gamma} \quad \mod \delta. \tag{3.1.59}$$

We apply the main trick used in [**MN1**], [**MN2**] and [**N4**]. Considering $\pm(x, y)$, we can assume that $x \equiv 2a_1b_1c/\gamma \mod \delta$, i. e. $x = 2a_1b_1c/\gamma - k\delta$ where $k \in \mathbb{Z}$. We have $\gamma^2 x^2 = 4a_1^2 b_1^2 c^2 - 4a_1 b_1 ck\delta\gamma + \gamma^2 k^2 \delta^2 = \gamma\delta y^2 + 4a_1^2 b_1^2 c^2$. It follows

$$\delta = \frac{y^2 + 4a_1b_1ck}{\gamma k^2}. \tag{3.1.60}$$

Consider a prime l such that $l \nmid 2a_1b_1c$. Assume that $l^{2t+1}|k$, but $l^{2t+2} \nmid k$. We have $k|y^2$. Then $l^{2t+1}|y^2$. Then $l^{2t+2}|y^2$. Since $k^2|y^2 + 4a_1b_1ck$, it follows that $l^{2t+2}|y^2 + 4a_1b_1ck$ We then get $l^{2t+2}|k$. We get a contradiction. It follows that $k = -\alpha q^2$ where $q \in \mathbb{Z}$ (we can additionally assume that $q \geq 0$), $\alpha|2a_1b_1c$ and α is square-free. Remark that α can be negative.

From (3.1.60), we get

$$\delta = \frac{y^2 - 4a_1b_1c\alpha q^2}{\gamma\alpha^2 q^4}. \tag{3.1.61}$$

It follows $\alpha q|y$ and $y = \alpha pq$ where $p \in \mathbb{Z}$. From (3.1.61),

$$\delta = \frac{p^2 - 4a_1b_1c/\alpha}{\gamma q^2}. \tag{3.1.62}$$

Equivalently,

$$p^2 - \gamma\delta q^2 = \frac{4a_1b_1c}{\alpha}. \tag{3.1.63}$$

If integral p, q satisfy (3.1.63), we have

$$(x, y) = \pm \left(\frac{2a_1b_1c}{\gamma} + \alpha\delta q^2, \alpha pq \right). \tag{3.1.64}$$

satisfy (3.1.59). We call solutions (3.1.64) of (3.1.59) as associated solutions. Thus, we get the very important for us

THEOREM 3.1.4. *All integral solutions (x, y) of*

$$\gamma x^2 - \delta y^2 = \frac{4a_1^2 b_1^2 c^2}{\gamma} \quad and \quad x \equiv \pm\frac{2a_1b_1c}{\gamma} \quad \mod \delta \tag{3.1.65}$$

are associated solutions

$$(x, y) = \pm \left(\frac{2a_1b_1c}{\gamma} + \alpha\delta q^2, \alpha pq \right) \tag{3.1.66}$$

to integral solutions α, (p, q) of

$$\alpha|2a_1b_1c \text{ where } \alpha \text{ is square-free, and } p^2 - \gamma\delta q^2 = \frac{4a_1b_1c}{\alpha}. \tag{3.1.67}$$

Any solution (x, y) of (3.1.65) can be written in the form (3.1.66) where α, (p, q) is solution of (3.1.67). Any solution of (3.1.67) gives a solution (3.1.66) of (3.1.65).

Solutions of (3.1.65) and (3.1.67) are in one-to-one correspondence if we additionally assume that $q \geq 0$.

Now we can write (x, y) of Theorem 3.1.3 in the form (3.1.66) as associated solutions to (3.1.67). Putting that (x, y) to relations (i)—(v) of Theorem 3.1.3, we get some relations on α and (p, q). They are a finite number of congruences mod N_i where N_i depend only on (r, s) (or (a, b, c)). All N_i are bounded by functions depending only on (r, s). These congruences have a lot of relations between them and with (3.1.67). All together they give many very strong restrictions on α and (p, q). We analyse them below.

We fix

$$\mu \in \left(\mathbb{Z}/(2a_1b_1c^2/\gamma)\right)^* \tag{3.1.68}$$

and consider $\delta \in \mathbb{N}$ such that

$$\delta \equiv \mu^2\gamma \quad \mathrm{mod}\ \frac{4a_1b_1c^2}{\gamma} \tag{3.1.69}$$

The relation (3.1.29) in (i) of Theorem 3.1.3 is equivalent to

$$2a_1b_1c + \alpha\gamma\delta q^2 \equiv \gamma\mu\alpha pq \quad \mathrm{mod}\ 2a_1b_1c^2. \tag{3.1.70}$$

By (3.1.67), we have

$$-4a_1b_1c + \alpha p^2 - \alpha\gamma\delta q^2 = 0. \tag{3.1.71}$$

Taking sum, we get

$$2a_1b_1c \equiv \alpha p(p - \mu\gamma q) \quad \mathrm{mod}\ 2a_1b_1c^2. \tag{3.1.72}$$

The relation (3.1.72) is equivalent to (3.1.29). Taking 2(3.1.70)+(3.1.71), we get

$$\alpha p^2 + \alpha\gamma\delta q^2 \equiv 2\gamma\mu\alpha pq \quad \mathrm{mod}\ 4a_1b_1c^2 \tag{3.1.73}$$

which is also equivalent to (3.1.29). From (3.1.69), we get

$$\alpha p^2 + \alpha\mu^2\gamma^2 q^2 \equiv 2\gamma\mu\alpha pq \quad \mathrm{mod}\ 4a_1b_1c^2 \tag{3.1.74}$$

and

$$\alpha(p - \mu\gamma q)^2 \equiv 0 \quad \mathrm{mod}\ 4a_1b_1c^2. \tag{3.1.75}$$

This is equivalent to (3.1.73). It follows $\alpha(p - \mu\gamma q)^2 \equiv 0\ \mathrm{mod}\ 4c^2$. Since α is square-free, it follows

$$2c|(p - \mu\gamma q). \tag{3.1.76}$$

From (3.1.72), we get

$$2a_1b_1 \equiv \alpha p\frac{p - \mu\gamma q}{c} \quad \mathrm{mod}\ 2a_1b_1c \tag{3.1.77}$$

where $(p - \mu\gamma q)/c$ is an integer. It follows, $\alpha|2a_1b_1$. Thus, we get

PROPOSITION 3.1.5. *The condition (3.1.29) of Theorem 3.1.3 is equivalent to*

$$\alpha(p - \mu\gamma q)^2 \equiv 0 \quad \mathrm{mod}\ 4a_1b_1c^2. \tag{3.1.78}$$

We have

$$\alpha|2a_1b_1. \tag{3.1.79}$$

The condition (3.1.78) is also equivalent to

$$2a_1b_1c \equiv \alpha p(p - \mu\gamma q) \quad \mathrm{mod}\ 2a_1b_1c^2. \tag{3.1.80}$$

Considering the condition (3.1.30) of Theorem 3.1.3, we similarly get

PROPOSITION 3.1.6. *The condition (3.1.30) of Theorem 3.1.3 is equivalent to*

$$\mu\alpha p^2 + \mu\alpha\gamma\delta q^2 \equiv 2\alpha\delta pq \quad \mod 4a_1b_1c^2. \tag{3.1.81}$$

Congruences (3.1.78) and (3.1.81) are equivalent $\mod (4a_1b_1c^2/\gamma)$, *and (3.1.81) is equivalent to (3.1.78) together with*

$$2\alpha pq(\delta - \gamma\mu^2) \equiv 0 \quad \mod 4a_1b_1c^2 \tag{3.1.82}$$

where $\delta - \gamma\mu^2 \equiv 0 \mod 4a_1b_1c^2/\gamma$.

PROOF. To get (3.1.82), consider μ (3.1.78) - (3.1.81).

Now consider (3.1.31) of Theorem 3.1.3. Let us consider the a-series (the sign $+$). Since $m(a,b) \equiv -1 \mod 2a_1/(\gamma_2\gamma_a)$, we get $-\mu x + (\delta y/\gamma) \equiv 0 \mod 2a_1/(\gamma_2\gamma_a)$. It is satisfied because of (3.1.30). Since $m(a,b) \equiv 1 \mod 2b_1/(\gamma_2\gamma_b)$, we get $\mu x + (\delta y/\gamma) \equiv 0 \mod 2b_1/(\gamma_2\gamma_b)$. By (3.1.30), we get $\mu x - (\delta y/\gamma) \equiv 0 \mod 2b_1/(\gamma_2\gamma_b)$. Thus, we get $2\mu x \equiv 0 \mod 2b_1/(\gamma_2\gamma_b)$. If $\gamma_2 = 1$, this is equivalent to $\mu x \equiv 0 \mod b_1/\gamma_b$ and $x \equiv 0 \mod b_1/\gamma_b$. If $\gamma_2 = 2$, then b_1/γ_b is odd, and we get $2\mu x \equiv 0 \mod (b_1/\gamma_b)$ which is equivalent to $x \equiv 0 \mod b_1/\gamma_b$. Thus, at any case we get $(b_1/\gamma_b)|x$, equivalently, $x \equiv 0 \mod (b_1/\gamma_b)$. By (3.1.66), this is equivalent to $\alpha\delta q^2 \equiv 0 \mod (b_1/\gamma_b)$. We have $\delta \equiv \mu^2\gamma \mod (b_1/\gamma_b)$ and $\mu \mod (b_1/\gamma_b)$ is invertible. Thus, we get $\alpha\gamma q^2 \equiv 0 \mod (b_1/\gamma_b)$ which is equivalent to $\alpha(\gamma_b q)^2 \equiv 0 \mod b_1$.

Let us consider (3.1.32) of Theorem 3.1.3. We consider the a-series. We get $x - \mu y \equiv 0 \mod 2a_1/(\gamma_2\gamma_a)$. It satisfies because of (3.1.29). We get $x + \mu y \equiv 0 \mod 2b_1/(\gamma_2\gamma_b)$. Using (3.1.29), we similarly get that this is equivalent to $(b_1/\gamma_b)|x$.

Thus, we finally get

PROPOSITION 3.1.7. *The conditions (3.1.31) and (3.1.32) in (ii) of Theorem 3.1.3 are equivalent to*

$$x \equiv 0 \quad \mod \frac{b_1}{\gamma_b} \tag{3.1.83}$$

or to

$$\alpha(\gamma_b q)^2 \equiv 0 \quad \mod b_1 \tag{3.1.84}$$

for a-series (the sign $+$), and it is equivalent to

$$x \equiv 0 \quad \mod \frac{a_1}{\gamma_a} \tag{3.1.85}$$

or to

$$\alpha(\gamma_a q)^2 \equiv 0 \quad \mod a_1 \tag{3.1.86}$$

for b-series (the sign $-$).

Consider (3.1.33) in Theorem 3.1.3. We consider the a-series. We get mod $2a_1/(\gamma_2\gamma_a)$ that

$$2\mu\gamma x \equiv (2\delta + (\mu^2\gamma - \delta))y \quad \mod (2a_1b_1c^2)(2a_1/(\gamma_2\gamma_a)).$$

We get $\mod 2b_1/(\gamma_2\gamma_b)$ that

$$(\delta - \mu^2\gamma)y \equiv 0 \quad \mod (2a_1b_1c^2)(2b_1/\gamma_2\gamma_b).$$

Using (3.1.66) (and (3.1.71) to make the relations homogeneous), we finally get

PROPOSITION 3.1.8. *The condition (3.1.33) of Theorem 3.1.3 is equivalent to*

$$2\mu\gamma x \equiv (\delta + \mu^2\gamma)y \quad \mod (2a_1b_1c^2)\,(2a_1/(\gamma_2\gamma_a)) \tag{3.1.87}$$

and

$$(\delta - \mu^2\gamma)y \equiv 0 \quad \mod (2a_1b_1c^2)\,(2b_1/(\gamma_2\gamma_b)) \tag{3.1.88}$$

or

$$\mu\alpha p^2 - \alpha(\delta + \mu^2\gamma)pq + \mu\alpha\gamma\delta q^2 \equiv 0 \quad \mod (2a_1b_1c^2)\,(2a_1/(\gamma_2\gamma_a)) \tag{3.1.89}$$

and

$$\alpha(\delta - \mu^2\gamma)pq \equiv 0 \quad \mod (2a_1b_1c^2)\,(2b_1/(\gamma_2\gamma_b)) \tag{3.1.90}$$

for the a-series (the sign +), and it is equivalent to

$$2\mu\gamma x \equiv (\delta + \mu^2\gamma)y \quad \mod (2a_1b_1c^2)\,(2b_1/(\gamma_2\gamma_b)) \tag{3.1.91}$$

and

$$(\delta - \mu^2\gamma)y \equiv 0 \quad \mod (2a_1b_1c^2)\,(2a_1/(\gamma_2\gamma_a)) \tag{3.1.92}$$

or

$$\mu\alpha p^2 - \alpha(\delta + \mu^2\gamma)pq + \mu\alpha\gamma\delta q^2 \equiv 0 \quad \mod (2a_1b_1c^2)\,(2b_1/(\gamma_2\gamma_b)) \tag{3.1.93}$$

and

$$\alpha(\delta - \mu^2\gamma)pq \equiv 0 \quad \mod (2a_1b_1c^2)\,(2a_1/(\gamma_2\gamma_a)) \tag{3.1.94}$$

for the b-series (the sign −).

Consider the condition (3.1.35) of Theorem 3.1.3. Consider the a-series (the sign +). Second condition in (3.1.35) gives $x \equiv \beta(2a_1b_1c/\gamma) \mod \delta$. By (3.1.66), we get $(x, y) = \beta(2a_1b_1c/\gamma + \alpha\delta q^2, \alpha pq)$. Then third condition in (3.1.35) is equivalent to

$$\beta\delta\alpha pq \equiv \mu(\beta 2a_1b_1c + \beta\alpha\gamma\delta q^2 - \beta 2a_1b_1c) \quad \mod 2a_1b_1c\delta.$$

This is equivalent to $\alpha q(p - \mu\gamma q) \equiv 0 \mod 2a_1b_1c$. Since $\gamma | 2a_1b_1c$, it follows the first condition in (3.1.35) which is $c\delta\beta\alpha pq \equiv 0 \mod \gamma$.

For b-series we get the same. Thus, we have

PROPOSITION 3.1.9. *The condition (3.1.35) of Theorem 3.1.3 is equivalent to*

$$\alpha q(p - \mu\gamma q) \equiv 0 \quad \mod 2a_1b_1c. \tag{3.1.95}$$

Consider (3.1.34) of Theorem 3.1.3. We consider the a-series. Then

$$(x, y) = \beta\left(\frac{2a_1b_1c}{\gamma} + \alpha\delta q^2,\ \alpha pq\right) \tag{3.1.96}$$

The first relation of (3.1.34) gives

$$m(a, b)\gamma x + \mu\gamma y \equiv 2a_1b_1\beta c \quad \mod 2a_1b_1\gamma.$$

Using (3.1.96), we get

$$m(a, b)\alpha\gamma\delta q^2 + \mu\alpha\gamma pq \equiv 2a_1b_1c(1 - m(a, b)) \quad \mod 2a_1b_1\gamma.$$

This is equivalent to

$$-\alpha\delta q^2 + \mu\alpha pq \equiv \frac{4a_1b_1c}{\gamma} \quad \mod 2a_1$$

and

$$\alpha\delta q^2 + \mu\alpha pq \equiv 0 \mod 2b_1.$$

Using (3.1.71), we can rewrite the first relation in the homogeneous form

$$\alpha p(p - \mu\gamma q) \equiv 0 \mod 2a_1\gamma.$$

The second relation of (3.1.34) gives

$$\delta m(a,b)y + \mu\gamma x - 2a_1 b_1 \beta\mu c \equiv 0 \mod 2a_1 b_1 \delta.$$

By (3.1.96) it is equivalent to

$$m(a,b)\alpha pq + \alpha\mu\gamma q^2 \equiv 0 \mod 2a_1 b_1.$$

This is equivalent to

$$-\alpha pq + \alpha\mu\gamma q^2 \equiv 0 \mod 2a_1$$

and

$$\alpha pq + \alpha\mu\gamma q^2 \equiv 0 \mod 2b_1.$$

The third relation in (3.1.34) is

$$\delta\left(m(a,b)\gamma x + \mu\gamma y - 2\beta a_1 b_1 c\right) \equiv$$
$$\mu\gamma\left(\delta m(a,b)y + \mu\gamma x - 2\mu\beta a_1 b_1 c\right) \mod (2a_1 b_1 c^2 \delta)(2a_1 b_1).$$

Using (3.1.96), one can calculate that it is equivalent to

$$\mu\alpha\gamma pq\left(m(a,b) - 1\right) + \alpha\gamma q^2\left(\gamma\mu^2 - m(a,b)\delta\right)$$
$$\equiv 2a_1 b_1 c\left(m(a,b) - 1\right) \mod (2a_1 b_1 c^2)(2a_1 b_1).$$

This is equivalent to

$$4a_1 b_1 c \equiv \alpha\gamma q\left(2\mu p - (\gamma\mu^2 + \delta)q\right) \mod (2a_1 b_1 c^2)(2a_1)$$

and

$$\alpha\gamma q^2(\delta - \gamma\mu^2) \equiv 0 \mod (2a_1 b_1 c^2)(2b_1).$$

Using (3.1.71), we can rewrite the first relation in the homogeneous form

$$\alpha(p - \mu\gamma q)^2 \equiv 0 \mod (2a_1 b_1 c^2)(2a_1).$$

Thus, we get

PROPOSITION 3.1.10. *The condition* (3.1.34) *of Theorem 3.1.3 is equivalent to the system of congruences:*
For the a-series (the sign +)

$$\alpha p(p - \mu\gamma q) \equiv 0 \mod 2a_1\gamma, \tag{3.1.97}$$

$$\alpha\delta q^2 + \alpha\mu pq \equiv 0 \mod 2b_1, \tag{3.1.98}$$

$$-\alpha pq + \alpha\mu\gamma q^2 \equiv 0 \mod 2a_1, \tag{3.1.99}$$

$$\alpha pq + \alpha\mu\gamma q^2 \equiv 0 \mod 2b_1, \tag{3.1.100}$$

$$\alpha(p - \mu\gamma q)^2 \equiv 0 \mod (2a_1 b_1 c^2)(2a_1), \tag{3.1.101}$$

$$\alpha\gamma q^2(\delta - \gamma\mu^2) \equiv 0 \mod (2a_1 b_1 c^2)(2b_1). \tag{3.1.102}$$

For the b-series (the sign −):

$$\alpha p(p - \mu\gamma q) \equiv 0 \mod 2b_1\gamma, \tag{3.1.103}$$

$$\alpha\delta q^2 + \alpha\mu pq \equiv 0 \mod 2a_1, \tag{3.1.104}$$

$$-\alpha pq + \alpha\mu\gamma q^2 \equiv 0 \mod 2b_1, \tag{3.1.105}$$

$$\alpha pq + \alpha\mu\gamma q^2 \equiv 0 \mod 2a_1, \tag{3.1.106}$$

$$\alpha(p - \mu\gamma q)^2 \equiv 0 \mod (2a_1b_1c^2)(2b_1), \tag{3.1.107}$$

$$\alpha\gamma q^2(\delta - \gamma\mu^2) \equiv 0 \mod (2a_1b_1c^2)(2a_1). \tag{3.1.108}$$

Consider the condition (iv) of Theorem 3.1.3. It means that the corresponding element $\widetilde{h}$ is primitive. Since $\widetilde{h}^2 = 2a_1b_1$ and the lattice $N(X)$ is even, it is not valid only if $\widetilde{h}/l \in N(X)$ for some prime l such that $l^2|a_1b_1$. Thus, (3.1.36) is not valid if and only if

$$x \equiv y \equiv \frac{x - \mu y}{2a_1b_1c^2/\gamma} \equiv 0 \mod l$$

for some prime l such that $l^2|a_1b_1$. Using (3.1.66) and (3.1.71), we get

PROPOSITION 3.1.11. *The condition (iv) of Theorem 3.1.3 is equivalent to the non-existence of a prime l such that $l^2|a_1b_1$ and*

$$x \equiv y \equiv \frac{x - \mu y}{2a_1b_1c^2/\gamma} \equiv 0 \mod l. \tag{3.1.109}$$

Equivalently, the system of congruences

$$\begin{cases} \alpha p^2 + \alpha\gamma\delta q^2 \equiv 0 \mod 2\gamma l \\ \alpha pq \equiv 0 \mod l \\ \alpha p^2 + \alpha\gamma\delta q^2 \equiv 2\alpha\gamma\mu pq \mod 4a_1b_1c^2l \end{cases} \tag{3.1.110}$$

is not satisfied for any prime l such that $l^2|a_1b_1$.

Consider the condition (v) of Theorem 3.1.3. This is equivalent to $\gamma(\widetilde{h}) = \gamma$ where $\widetilde{h}\cdot N(X) = \gamma(\widetilde{h})\mathbb{Z}$. All other conditions of Theorem 3.1.3 give that $\gamma|\widetilde{h}\cdot N(X)$, and $\gamma|\gamma(\widetilde{h})$. Since $\widetilde{h}^2 = 2a_1b_1$, it follows that $(\gamma(\widetilde{h})/\gamma)\,|\,2a_1b_1/\gamma$. Equivalently, (v) is not satisfied if and only if for some prime $l|2a_1b_1/\gamma$ one has

$$x \equiv \frac{\delta y}{\gamma} \equiv \frac{\mu\gamma x - \delta y}{2a_1b_1c^2} \equiv 0 \mod l.$$

Using (3.1.66) and (3.1.71), we get

PROPOSITION 3.1.12. *The condition (v) of Theorem 3.1.3 is equivalent to the non-existence of a prime $l\,|\,(2a_1b_1/\gamma)$ such that*

$$x \equiv \frac{\delta y}{\gamma} \equiv \frac{\mu\gamma x - \delta y}{2a_1b_1c^2} \equiv 0 \mod l. \tag{3.1.111}$$

Equivalently, the system of congruences

$$\begin{cases} \alpha p^2 + \alpha\gamma\delta q^2 \equiv 0 \mod 2\gamma l \\ \delta\alpha pq \equiv 0 \mod \gamma l \\ \mu\alpha p^2 + \mu\alpha\gamma\delta q^2 \equiv 2\alpha\delta pq \mod 4a_1b_1c^2l \end{cases} \tag{3.1.112}$$

is not satisfied for any prime $l\,|\,(2a_1b_1/\gamma)$.

Now we collect analysed conditions of Theorem 3.1.3 all together. We divide them in *general conditions:* (G') which are valid for both a and b-series, *conditions* (A') which are valid for the a-series, and *conditions* (B') which are valid for the b-series.

(G'): **General conditions**

$$\alpha(p - \mu\gamma q)^2 \equiv 0 \mod 4a_1 b_1 c^2, \tag{3.1.113}$$

$$2\alpha pq(\delta - \gamma\mu^2) \equiv 0 \mod 4a_1 b_1 c^2, \tag{3.1.114}$$

$$\alpha q(p - \mu\gamma q) \equiv 0 \mod 2a_1 b_1 c, \tag{3.1.115}$$

$$\begin{cases} \alpha p^2 + \alpha\gamma\delta q^2 \equiv 0 \mod 2\gamma l \\ \alpha pq \equiv 0 \mod l \\ \alpha p^2 + \alpha\gamma\delta q^2 \equiv 2\alpha\gamma\mu pq \mod 4a_1 b_1 c^2 l \end{cases} \tag{3.1.116}$$

is not satisfied for any prime l such that $l^2 | a_1 b_1$,

$$\begin{cases} \alpha p^2 + \alpha\gamma\delta q^2 \equiv 0 \mod 2\gamma l \\ \delta\alpha pq \equiv 0 \mod \gamma l \\ \mu\alpha p^2 + \mu\alpha\gamma\delta q^2 \equiv 2\alpha\delta pq \mod 4a_1 b_1 c^2 l \end{cases} \tag{3.1.117}$$

is not satisfied for any prime $l \mid (2a_1 b_1/\gamma)$.

(A'): **Conditions of the a-series**

$$\alpha(\gamma_b\, q)^2 \equiv 0 \mod b_1, \tag{3.1.118}$$

$$\mu\alpha p^2 - \alpha(\delta + \mu^2\gamma)pq + \mu\alpha\gamma\delta q^2 \equiv 0 \mod (2a_1 b_1 c^2)\,(2a_1/(\gamma_2\gamma_a)) \tag{3.1.119}$$

$$\alpha(\delta - \mu^2\gamma)pq \equiv 0 \mod (2a_1 b_1 c^2)\,(2b_1/(\gamma_2\gamma_b)), \tag{3.1.120}$$

$$\alpha p(p - \mu\gamma q) \equiv 0 \mod 2a_1\gamma, \tag{3.1.121}$$

$$\alpha\delta q^2 + \alpha\mu pq \equiv 0 \mod 2b_1, \tag{3.1.122}$$

$$-\alpha pq + \alpha\mu\gamma q^2 \equiv 0 \mod 2a_1, \tag{3.1.123}$$

$$\alpha pq + \alpha\mu\gamma q^2 \equiv 0 \mod 2b_1, \tag{3.1.124}$$

$$\alpha(p - \mu\gamma q)^2 \equiv 0 \mod (2a_1 b_1 c^2)(2a_1), \tag{3.1.125}$$

$$\alpha\gamma q^2(\delta - \gamma\mu^2) \equiv 0 \mod (2a_1 b_1 c^2)(2b_1). \tag{3.1.126}$$

(B'): **Conditions of the b-series**

$$\alpha(\gamma_a\, q)^2 \equiv 0 \mod a_1, \tag{3.1.127}$$

$$\mu\alpha p^2 - \alpha(\delta + \mu^2\gamma)pq + \mu\alpha\gamma\delta q^2 \equiv 0 \mod (2a_1 b_1 c^2)\,(2b_1/(\gamma_2\gamma_b)) \tag{3.1.128}$$

$$\alpha(\delta - \mu^2\gamma)pq \equiv 0 \mod (2a_1 b_1 c^2)\,(2a_1/(\gamma_2\gamma_a)), \tag{3.1.129}$$

$$\alpha p(p - \mu\gamma q) \equiv 0 \mod 2b_1\gamma, \tag{3.1.130}$$

$$\alpha\delta q^2 + \alpha\mu pq \equiv 0 \mod 2a_1, \tag{3.1.131}$$

$$-\alpha pq + \alpha\mu\gamma q^2 \equiv 0 \mod 2b_1, \tag{3.1.132}$$

$$\alpha pq + \alpha\mu\gamma q^2 \equiv 0 \mod 2a_1, \tag{3.1.133}$$

$$\alpha(p - \mu\gamma q)^2 \equiv 0 \mod (2a_1 b_1 c^2)(2b_1), \tag{3.1.134}$$

$$\alpha\gamma q^2(\delta - \gamma\mu^2) \equiv 0 \mod (2a_1 b_1 c^2)(2a_1). \tag{3.1.135}$$

3.2. Simplification of the conditions (G'), (A') and (B'). We have the following fundamental result which completely determines α up to multiplication by ± 1.

LEMMA 3.2.1. *For the a-series, we have that the square-free $\alpha | b_1$ and*

$$b_1 / |\alpha| = \widetilde{b}_1^2, \ \widetilde{b}_1 > 0,$$

is a square.

Respectively, for the b-series, we have that the square-free $\alpha | a_1$ and

$$a_1 / |\alpha| = \widetilde{a}_1^2, \ \widetilde{a}_1 > 0,$$

is a square.

PROOF. First let us prove that

$$\alpha | a_1 b_1. \tag{3.2.1}$$

Otherwise, $2a_1 b_1 / \alpha$ is odd. Let us consider the a-series. By (3.1.125), we have $\alpha(p - \mu\gamma q)^2 \equiv 0 \mod 4c^2$. Since α is square-free, it follows that $p - \mu\gamma q \equiv 0 \mod 2c$. It follows $p + \mu\gamma q \equiv 0 \mod 2$. Then $p^2 - \mu^2\gamma^2 \equiv 0 \mod 4c$. We have $\mu^2\gamma \equiv \delta \mod 4a_1 b_1 c^2/\gamma$. Then $\mu^2\gamma^2 \equiv \gamma\delta \mod 4c^2$. Thus, we obtain $p^2 - \gamma\delta q^2 \equiv 0 \mod 4c$. On the other hand, $p^2 - \gamma\delta q^2 = 4a_1 b_1 c/\alpha \equiv 2c \mod 4c$ if $2a_1 b_1 / \alpha$ is odd. We get a contradiction. It proves (3.2.1).

Now let us consider the a-series, and let us prove that

$$\alpha | b_1. \tag{3.2.2}$$

Otherwise, for a prime l one has $l | \alpha$, $l | a_1$, but l does not divide b_1.

By (3.1.125), we have $\alpha(p - \mu\gamma q)^2 \equiv 0 \mod 4a_1^2 c^2$. Since α is square-free, it follows that $p - \mu\gamma q \equiv 0 \mod 2a_1 c$. It follows $p + \mu\gamma q \equiv 0 \mod 2$. Then $p^2 - \mu^2\gamma^2 \equiv 0 \mod 4a_1 c$. We have $\mu^2\gamma \equiv \delta \mod 4a_1 b_1 c^2/\gamma$. Then $\mu^2\gamma^2 \equiv \gamma\delta \mod 4a_1 c^2$. Thus, we obtain $p^2 - \gamma\delta q^2 \equiv 0 \mod 4a_1 c$. Thus, we obtain $p^2 - \gamma\delta q^2 = 4a_1 b_1 c/\alpha \equiv 0 \mod 4a_1 c$. Equivalently, $a_1 b_1 / \alpha \equiv 0 \mod a_1$. This gives a contradiction if for a prime l one has $l | \alpha$, $l | a_1$, but l does not divide b_1. This proves (3.2.2).

Now let us prove that

$$b_1 / |\alpha| \tag{3.2.3}$$

is a square. Otherwise, for a prime l we have $l^{2t-1} | b_1/\alpha$, but l^{2t} does not divide b_1/α where $t \geq 1$. By (3.1.118), $(\gamma q)^2 \equiv 0 \mod b_1/\alpha$. It follows that

$$l^t | \gamma q. \tag{3.2.4}$$

By (3.1.113), we have

$$p - \gamma\mu q \equiv 0 \mod l^t. \tag{3.2.5}$$

By (3.2.4) and (3.1.5), we obtain $l^t | p$.

From $p^2 - \gamma\delta q^2 = 4a_1(b_1/\alpha)c$ we then get a contradiction if l does not divide $2c$.

We have

$$\gamma\delta \equiv \gamma^2\mu^2 \mod 4a_1 b_1 c^2. \tag{3.2.6}$$

From $p^2 - \gamma\delta q^2 = 4a_1(b_1/\alpha)c$ and (3.2.6) we then get

$$p^2 - \gamma^2\mu^2 q^2 \equiv (p - \gamma\mu q)(p + \gamma\mu q) \equiv 4a_1(b_1/\alpha)c \mod 4a_1 b_1 c^2.$$

Then, using (3.2.5), we get

$$\frac{p - \gamma\mu q}{2a_1 c l^t}(p + \gamma\mu q) \equiv 2(b_1/\alpha)/l^t \mod 2(b_1/l^t)c \tag{3.2.7}$$

where $(p - \gamma\mu q)/(2a_1 c l^t)$ is an integer.

If $l|c$ and l is odd, (3.2.5) and (3.2.7) give a contradiction because $l^t|p + \gamma\mu q$, $l^t|2(b_1/l^t)c$, but l^t does not divide $2(b_1/\alpha)/l^t$.

Now assume that $l = 2|c$. If $2^{t+1}|\gamma\mu q$, then $2^{t+1}|p$ by (3.2.5), and we get a contradiction in the same way. Assume that 2^{t+1} does not divide $\gamma\mu q$. Then $\gamma\mu q/2^t$ and $p/2^t$ are both odd, and $2^{t+1}|p + \gamma\mu q$. It also leads to a contradiction in the same way.

Now assume that $l = 2$ and c is odd. By (3.1.126), we obtain

$$\gamma\delta q^2 \equiv (\gamma\mu q)^2 \mod 2^{4t}. \tag{3.2.8}$$

Assume that $2^{t+1}|\mu\gamma q$. By (3.2.8), then $\gamma\delta q^2 \equiv 0 \mod 2^{2t+2}$. By (3.2.5), $2^{t+1}|p$ and $2^{2t+2}|p^2$. Then $2^{2t+1}|p^2 - \gamma\delta q^2 = 4a_1(b_1/\alpha)c$ which gives a contradiction because $4a_1(b_1/\alpha)c$ is divisible by 2^{2t+1} only.

Now assume that 2^{t+1} does not divide $\mu\gamma q$. By (3.2.8), we get $\gamma\delta q^2 \equiv 2^{2t}$ mod 2^{2t+2}. By (3.2.5) then $2^t|p$, but 2^{t+1} does not divide p. It follows that $p^2 \equiv 2^{2t}$ mod 2^{2t+2}. Then $2^{2t+2}|p^2 - \gamma\delta q^2 = 4a_1(b_1/\alpha)c$ which again leads to a contradiction. This finishes the proof of the theorem.

LEMMA 3.2.2. *Assume that*

$$g.c.d.(\delta - \gamma\mu^2, \ 4a_1 b_1 c^2) = (4a_1 b_1 c^2/\gamma)\gamma_0$$

where $\gamma_0|\gamma$.

For any $u|a_1 b_1 c^2/(\gamma_a\gamma_b)$ and $g.c.d.(u, \gamma/\gamma_0) = 1$ we can choose

$$\mu \in (\mathbb{Z}/((2a_1 b_1 c^2/\gamma)\gamma_0 u))^*$$

such that

$$\delta \equiv \gamma\mu^2 \mod (4a_1 b_1 c^2/\gamma)\gamma_0 u.$$

PROOF. Assume $\mu_0 \mod 2a_1 b_1 c^2/\gamma \in (\mathbb{Z}/(2a_1 b_1 c^2/\gamma))^*$ and

$$\delta \equiv \gamma\mu_0^2 \mod 4a_1 b_1 c^2/\gamma.$$

Taking $\mu = \mu_0 + (2a_1 b_1 c^2/\gamma)k$, we get

$$\delta - \gamma\mu^2 = \delta - \gamma\mu_0^2 - 4\mu_0 a_1 b_1 c^2 k - (4a_1^2 b_1^2 c^4/\gamma)k^2.$$

Then

$$(\delta - \gamma\mu^2)/(4a_1 b_1 c^2/\gamma) \equiv (\delta - \gamma\mu_0^2)/(4a_1 b_1 c^2/\gamma) + \gamma\mu_0 k \mod \gamma_0 u.$$

Since $(\gamma/\gamma_0)\mu_0$ are invertible   mod u, we can choose k such that

$$(\delta - \gamma\mu^2)/(4a_1 b_1 c^2/\gamma) \equiv 0 \mod \gamma_0 u.$$

It follows the statement.

Further we consider the a-series (similar results will be valid for b-series). The congruence (3.1.125) implies (3.1.113). The (3.1.125) is equivalent to

$$p - \mu\gamma q \equiv 0 \mod 2a_1 \widetilde{b_1} c. \tag{3.2.9}$$

Thus, (3.1.113) and (3.1.125) are equivalent to (3.2.9).

The congruence (3.1.118) is equivalent to $\gamma_b q \equiv 0 \mod \widetilde{b}_1$. Equivalently,

$$q = \frac{\widetilde{b}_1 q_1}{\gamma_b}, \quad \widetilde{b}_1 q_1 \equiv 0 \mod \gamma_b \tag{3.2.10}$$

where q_1 is an integer.

By (3.2.9), we have $p - \mu \gamma_2 \gamma_a \widetilde{b}_1 q_1 \equiv 0 \mod 2 a_1 \widetilde{b}_1 c$, and then $\gamma_2 \gamma_a \widetilde{b}_1 | p$. Then

$$p = \gamma_2 \gamma_a \widetilde{b}_1 p_1 \tag{3.2.11}$$

where p_1 is an integer. The congruence (3.2.9) is then equivalent to

$$p_1 - \mu q_1 \equiv 0 \mod (2/\gamma_2)(a_1/\gamma_a)c. \tag{3.2.12}$$

Denoting

$$\alpha = \pm b_1 / \widetilde{b}_1^2, \tag{3.2.13}$$

we can rewrite $p^2 - \gamma \delta q^2 = 4 a_1 b_1 c / \alpha$ as

$$\gamma p_1^2 - \delta q_1^2 = \pm 2(2/\gamma_2)(a_1/\gamma_a)\gamma_b c. \tag{3.2.14}$$

Now we can rewrite conditions (G'), (A') and (B') using the introduced (p_1, q_1).

As we have seen, the conditions (3.1.113), (3.1.118) and (3.1.125) are equivalent to (3.2.12) and

$$\widetilde{b}_1 q_1 \equiv 0 \mod \gamma_b. \tag{3.2.15}$$

The condition (3.1.114) gives

$$p_1 q_1 (\delta - \gamma \mu^2) \equiv 0 \mod (2/\gamma_2)(a_1/\gamma_a)c^2 \gamma_b.$$

Since g.c.d$(\gamma, c) = $ g.c.d$(\gamma_b, a_1 c^2) = 1$ and $\delta - \gamma \mu^2 \equiv 0 \mod 4 a_1 b_1 c^2 / \gamma$, it is equivalent to

$$p_1 q_1 (\delta - \gamma \mu^2) \equiv 0 \mod (2/\gamma_2)\gamma_b. \tag{3.2.16}$$

The condition (3.1.115) gives

$$q_1 (p_1 - \mu q_1) \equiv 0 \mod (2/\gamma_2)(a_1/\gamma_a)c\gamma_b.$$

Since g.c.d$(\gamma, c) = $ g.c.d$(\gamma_b, a_1 c^2) = 1$ and $p_1 - \mu q_1 \equiv 0 \mod (2/\gamma_2)(a_1/\gamma_a)c$ (by (3.2.12)), it is similarly equivalent to

$$q_1 (p_1 - \mu q_1) \equiv 0 \mod (2/\gamma_2)\gamma_b. \tag{3.2.17}$$

The condition (3.1.120) gives

$$(\delta - \gamma \mu^2) p_1 q_1 \equiv 0 \mod (4/\gamma_2^2) b_1 c^2. \tag{3.2.18}$$

Since $(\delta - \mu^2 \gamma) \equiv 0 \mod 4 b_1 c^2 / (\gamma_2 \gamma_b)$, the congruence (3.2.18) is actually a congruence $\mod \gamma_b$ on $p_1 q_1$. It also implies (3.2.16).

The condition (3.1.121) gives $b_1 p_1 (p_1 - \mu q_1) \equiv 0 \mod (2/\gamma_2)(a_1/\gamma_a)\gamma_b$. It satisfies because of (3.2.12) and $\gamma_b | b_1$.

The condition (3.1.122) gives

$$q_1 (\delta q_1 + \mu \gamma p_1) \equiv 0 \mod 2\gamma_b^2. \tag{3.2.19}$$

The condition (3.1.123) gives $(b_1/\gamma_b)q_1(-p_1 + \mu q_1) \equiv 0 \mod (2/\gamma_2)(a_1/\gamma_a)$. It satisfies by (3.2.12).

The condition (3.1.124) gives

$$\gamma_a q_1 (p_1 + \mu q_1) \equiv 0 \mod (2/\gamma_2)\gamma_b. \tag{3.2.20}$$

It is easy to see that (3.2.17), (3.2.20) together with $\delta \equiv \mu^2\gamma \mod 4a_1b_1c^2/\gamma$ imply (3.2.18).

Taking $\pm\gamma_a$ (3.2.17) plus (3.2.20), we obtain

$$\gamma_2 p_1 q_1 \equiv 0 \mod \gamma_b \tag{3.2.21}$$

and $\gamma_2\mu q_1^2 \equiv 0 \mod \gamma_b$. Since μ can be always taken coprime to γ_b, the last congruence is equivalent to

$$\gamma_2 q_1^2 \equiv 0 \mod \gamma_b. \tag{3.2.22}$$

Any of them together with (3.2.17) can be taken to replace (3.2.20).

The condition (3.1.126) is equivalent to

$$(\delta - \gamma\mu^2)q_1^2 \equiv 0 \mod (4/\gamma_2)b_1c^2\gamma_b. \tag{3.2.23}$$

By (3.2.22) and $(\delta - \mu^2\gamma) \equiv 0 \mod 4b_1c^2/(\gamma_2\gamma_b)$, the congruence (3.2.23) is actually a congruence $\mod \gamma_b$.

It is easy to see that (3.2.23) and (3.2.20) imply (3.2.19).

The condition (3.1.119) gives

$$\mu\gamma p_1^2 - (\delta + \mu^2\gamma)p_1 q_1 + \mu\delta q_1^2 \equiv 0 \mod (4/\gamma_2^2)(a_1^2/\gamma_a^2)\gamma_b c^2.$$

Since a_1 and b_1 are coprime, this is equivalent to two congruences

$$\mu\gamma p_1^2 - (\delta + \mu^2\gamma)p_1 q_1 + \mu\delta q_1^2 \equiv 0 \mod (4/\gamma_2^2)c^2\gamma_b$$

and

$$\mu\gamma p_1^2 - (\delta + \mu^2\gamma)p_1 q_1 + \mu\delta q_1^2 \equiv 0 \mod (4/\gamma_2^2)c^2(a_1^2/\gamma_a^2). \tag{3.2.24}$$

By (3.2.23), we have $\delta q_1^2 \equiv \gamma\mu^2 q_1^2 \mod (4/\gamma_2^2)c^2\gamma_b$, and the first congruence gives

$$\mu\gamma(p_1 - \mu q_1)^2 - (\delta - \mu^2\gamma)p_1 q_1 \equiv 0 \mod (4/\gamma_2^2)c^2\gamma_b$$

It satisfies because of (3.2.12) and (3.2.18). The congruence (3.2.24) can be written as

$$\mu\gamma(p_1 - \mu q_1)^2 + (\delta - \mu^2\gamma)(p_1 - \mu q_1)q_1 \equiv 0 \mod (4/\gamma_2^2)c^2(a_1^2/\gamma_a^2).$$

It satisfies because of (3.2.12) and since $\delta - \gamma\mu^2 \equiv 0 \mod (4/\gamma_2)(a_1/\gamma_a)c^2$.

The condition (3.1.116) is equivalent to

$$\begin{cases} (b_1/\gamma_b)(\gamma p_1^2 + \delta q_1^2) \equiv 0 \mod 2\gamma_b l \\ (b_1/\gamma_b)\gamma_2\gamma_a p_1 q_1 \equiv 0 \mod l \\ \gamma^2 p_1^2 + \gamma\delta q_1^2 \equiv 2\gamma^2\mu p_1 q_1 \mod 4a_1 c^2\gamma_b^2 l \end{cases} \tag{3.2.25}$$

is not satisfied for any prime l such that $l^2|a_1b_1$.

By its meaning, the congruences (3.2.25) satisfies if we formally put $l = 1$. Assume that for a prime l we have $l^2|a_1b_1$ and g.c.d$(l, \gamma) = 1$. Then (3.2.25) is equivalent to

$$\begin{cases} b_1(\gamma p_1^2 + \delta q_1^2) \equiv 0 \mod 2l \\ b_1 p_1 q_1 \equiv 0 \mod l \\ \gamma p_1^2 + \delta q_1^2 \equiv 2\gamma\mu p_1 q_1 \mod (4/\gamma_2)(a_1/\gamma_a)c^2 l \end{cases}$$

is not satisfied. By Lemma 3.2.2, we can assume that
$\delta \equiv \gamma\mu^2 \mod (4/\gamma_2)(a_1/\gamma_a)lc^2$, and the last condition is equivalent to

$$\begin{cases} b_1(\gamma p_1^2 + \delta q_1^2) \equiv 0 \mod 2l \\ b_1 p_1 q_1 \equiv 0 \mod l \\ \gamma(p_1 - \mu q_1)^2 \equiv 0 \mod (4/\gamma_2)(a_1/\gamma_a)lc^2 \end{cases}$$

is not satisfied. By (3.2.12), this is equivalent to

$$\begin{cases} b_1(\gamma p_1^2 + \delta q_1^2) \equiv 0 \mod 2l \\ b_1 p_1 q_1 \equiv 0 \mod l \\ \gamma_a[(p_1 - \mu q_1)/(2c/\gamma_2)]^2 \equiv 0 \mod (a_1/\gamma_a)l \end{cases} \qquad (3.2.26)$$

is not satisfied. Assume that $l|b_1$. By (3.2.14), then the first and second congruences
satisfy and (3.2.26) is equivalent to

$$p_1 - \mu q_1 \equiv 0 \mod (2/\gamma_2)(a_1/\gamma_a)cl \qquad (3.2.27)$$

does not satisfy. Assume that $l|a_1$. By (3.2.12) then third congruence in (3.2.26)
satisfies, and by (3.2.12) and (3.2.14) the condition (3.2.26) is equivalent to

$$l \nmid p_1. \qquad (3.2.28)$$

The condition (3.1.117) is equivalent to

$$\begin{cases} (b_1/\gamma_b)(\gamma p_1^2 + \delta q_1^2) \equiv 0 \mod 2\gamma_b l \\ (b_1/\gamma_b)p_1 q_1 \equiv 0 \mod \gamma_b l \\ \mu\gamma^2 p_1^2 + \mu\gamma\delta q_1^2 \equiv 2\delta\gamma p_1 q_1 \mod 4a_1 c^2 \gamma_b^2 l \end{cases} \qquad (3.2.29)$$

is not satisfied for any prime $l \,|\, (2a_1 b_1/\gamma)$.

By its meaning, (3.2.29) satisfies if we formally put $l = 1$. Assume that
g.c.d$(l, \gamma) = 1$ and $l|2a_1 b_1/\gamma$. Then (3.2.29) is equivalent to

$$\begin{cases} b_1(\gamma p_1^2 + \delta q_1^2) \equiv 0 \mod 2l \\ b_1 p_1 q_1 \equiv 0 \mod l \\ \mu\gamma p_1^2 + \mu\delta q_1^2 \equiv 2\delta p_1 q_1 \mod (4/\gamma_2)(a_1/\gamma_2)c^2 l \end{cases}$$

is not satisfied.

First assume that $l|a_1 b_1$. By Lemma 3.2.2, we can assume that
$\delta \equiv \gamma\mu^2 \mod (4/\gamma_2)(a_1/\gamma_a)lc^2$, and the last condition is equivalent to

$$\begin{cases} b_1(\gamma p_1^2 + \delta q_1^2) \equiv 0 \mod 2l \\ b_1 p_1 q_1 \equiv 0 \mod l \\ \mu\gamma(p_1 - \mu q_1)^2 \equiv 0 \mod (4/\gamma_2)(a_1/\gamma_a)lc^2 \end{cases}$$

is not satisfied. We obtain exactly the same conditions (3.2.27) and (3.2.28) as
above.

Now assume that $l \nmid a_1 b_1$. Then $l = 2$ and a_1, b_1, γ, δ are odd. We then obtain
that

$$\begin{cases} \gamma p_1^2 + \delta q_1^2 \equiv 0 \mod 2l \\ p_1 q_1 \equiv 0 \mod l \\ \mu\gamma p_1^2 + \mu\delta q_1^2 \equiv 2\delta p_1 q_1 \mod (4/\gamma_2)(a_1/\gamma_2)c^2 l \end{cases} \qquad (3.2.30)$$

is not satisfied. From first two congruences we get that $p_1 \equiv q_1 \equiv 0 \mod 2$. Then we get $q_1\delta \equiv q_1\gamma\mu^2 \mod (4/\gamma_2)(a_1/\gamma_a)lc^2$, and we can rewrite (3.2.30) again as

$$\begin{cases} \gamma p_1^2 + \delta q_1^2 \equiv 0 \mod 2l \\ b_1 p_1 q_1 \equiv 0 \mod l \\ \mu\gamma(p_1 - \mu q_1)^2 \equiv 0 \mod (4/\gamma_2)(a_1/\gamma_a)lc^2 \end{cases} \tag{3.2.31}$$

is not satisfied. From the last congruence in (3.2.31) we get $\gamma(p_1 - \mu q_1) \equiv 0 \mod 4c$ and $p_1 + \mu q_1 \equiv 0 \mod 2$. It follows that $\gamma p_1^2 - \gamma\mu^2 q_1^2 \equiv 0 \mod 8c$. Moreover, we have $q_1\delta \equiv q_1\gamma\mu^2 \mod 8c$. Thus, we obtain $\gamma p_1^2 - \delta q_1^2 \equiv 0 \mod 8c$. It leads to a contradiction since $\gamma p_1^2 - \delta q_1^2 = \pm 2(2/\gamma_2)(a_1/\gamma_a)\gamma_b c \equiv 4c \mod 8c$.

In fact, (3.2.29) always follows from (3.2.29) if $l|2a_1b_1$ and g.c.d$(l,\gamma) = 1$. Really, by our construction, (3.2.29) means that the corresponding element $\tilde{h} \in N(X)$ of Theorem 3.1.3 is not divisible by l. We have $\mu \in (\mathbb{Z}/(2a_1b_1c^2/\gamma))^*$, $\delta \equiv \gamma\mu^2 \mod 4a_1b_1c^2/\gamma$, and $\det N(X) = -\gamma\delta$ is not divisible by l. Then $N(X) \cdot \tilde{h}$ is not divisible by l, and (3.2.29) follows. In particular, we can assume that $l^2|a_1b_1$.

Thus, we can rewrite the conditions (G') and (A') respectively in the form
(A): The conditions of a-series:
(AG): The general conditions of a-series:

$$p_1 - \mu q_1 \equiv 0 \mod (2/\gamma_2)(a_1/\gamma_a)c, \tag{3.2.32}$$

$$p_1 - \mu q_1 \not\equiv 0 \mod (2/\gamma_2)(a_1/\gamma_a)c\,l$$
$$\text{for any prime } l \text{ such that } l^2|b_1 \text{ and g.c.d}(l,\gamma) = 1, \tag{3.2.33}$$

$$l \nmid p_1 \text{ for any prime } l \text{ such that } l^2|a_1 \text{ and g.c.d}(l,\gamma) = 1. \tag{3.2.34}$$

(AS) The singular conditions of a-series:

$$\tilde{b}_1 q_1 \equiv 0 \mod \gamma_b, \tag{3.2.35}$$

$$q_1(p_1 - \mu q_1) \equiv 0 \mod (2/\gamma_2)\gamma_b, \tag{3.2.36}$$

$$\gamma_2 p_1 q_1 \equiv \gamma_2 q_1^2 \equiv 0 \mod \gamma_b, \tag{3.2.37}$$

$$(\delta - \gamma\mu^2)q_1^2 \equiv 0 \mod (4/\gamma_2)b_1 c^2\gamma_b. \tag{3.2.38}$$

$$\begin{cases} (b_1/\gamma_b)(\gamma p_1^2 + \delta q_1^2) \equiv 0 \mod 2\gamma_b l \\ (b_1/\gamma_b)\gamma_2\gamma_a p_1 q_1 \equiv 0 \mod l \\ \gamma^2 p_1^2 + \gamma\delta q_1^2 \equiv 2\gamma^2\mu p_1 q_1 \mod 4a_1 c^2\gamma_b^2 l \end{cases} \tag{3.2.39}$$

is not satisfied for any prime l such that $l^2|a_1b_1$ and $l|\gamma$,

$$\begin{cases} (b_1/\gamma_b)(\gamma p_1^2 + \delta q_1^2) \equiv 0 \mod 2\gamma_b l \\ (b_1/\gamma_b)p_1 q_1 \equiv 0 \mod \gamma_b l \\ \mu\gamma^2 p_1^2 + \mu\gamma\delta q_1^2 \equiv 2\delta\gamma p_1 q_1 \mod 4a_1 c^2\gamma_b^2 l \end{cases} \tag{3.2.40}$$

is not satisfied for any prime $l \,|\, (2a_1b_1/\gamma)$ and $l|\gamma$.

Similarly we can rewrite the conditions (G') and (B') respectively in the form
(B): The conditions of b-series:
(BG): The general conditions of b-series:

$$p_1 - \mu q_1 \equiv 0 \mod (2/\gamma_2)(b_1/\gamma_b)c, \tag{3.2.41}$$

$$p_1 - \mu q_1 \not\equiv 0 \mod (2/\gamma_2)(b_1/\gamma_b)c\,l$$
for any prime l such that $l^2|a_1$ and g.c.d$(l, \gamma) = 1$, $\qquad$ (3.2.42)

$$l \nmid p_1 \text{ for any prime } l \text{ such that } l^2|b_1 \text{ and g.c.d}(l, \gamma) = 1. \qquad (3.2.43)$$

(BS) The singular conditions of b-series:

$$\widetilde{a}_1 q_1 \equiv 0 \mod \gamma_a, \qquad (3.2.44)$$

$$q_1(p_1 - \mu q_1) \equiv 0 \mod (2/\gamma_2)\gamma_a, \qquad (3.2.45)$$

$$\gamma_2 p_1 q_1 \equiv \gamma_2 q_1^2 \equiv 0 \mod \gamma_a, \qquad (3.2.46)$$

$$(\delta - \gamma\mu^2)q_1^2 \equiv 0 \mod (4/\gamma_2)a_1 c^2 \gamma_a. \qquad (3.2.47)$$

$$\begin{cases} (a_1/\gamma_a)(\gamma p_1^2 + \delta q_1^2) \equiv 0 \mod 2\gamma_a l \\ (a_1/\gamma_a)\gamma_2\gamma_b p_1 q_1 \equiv 0 \mod l \\ \gamma^2 p_1^2 + \gamma\delta q_1^2 \equiv 2\gamma^2 \mu p_1 q_1 \mod 4b_1 c^2 \gamma_a^2 l \end{cases} \qquad (3.2.48)$$

is not satisfied for any prime l such that $l^2|a_1 b_1$ and $l|\gamma$,

$$\begin{cases} (a_1/\gamma_a)(\gamma p_1^2 + \delta q_1^2) \equiv 0 \mod 2\gamma_a l \\ (a_1/\gamma_a)p_1 q_1 \equiv 0 \mod \gamma_a l \\ \mu\gamma^2 p_1^2 + \mu\gamma\delta q_1^2 \equiv 2\delta\gamma p_1 q_1 \mod 4b_1 c^2 \gamma_a^2 l \end{cases} \qquad (3.2.49)$$

is not satisfied for any prime $l \,|\, (2a_1 b_1/\gamma)$ and $l|\gamma$.

We remind that here

$$\mu \in (\mathbb{Z}/2a_1 b_1 c^2/\gamma)^*, \qquad (3.2.50)$$

$$\delta \equiv \gamma\mu^2 \mod 4a_1 b_1 c^2/\gamma, \qquad (3.2.51)$$

$$\gamma p_1^2 - \delta q_1^2 = \pm 2(2/\gamma_2)(a_1/\gamma_a)\gamma_b c \qquad (3.2.52)$$

for the a-series, and

$$\gamma p_1^2 - \delta q_1^2 = \pm 2(2/\gamma_2)(b_1/\gamma_b)\gamma_a c \qquad (3.2.53)$$

for the b-series.

3.3. The resolution of the singular conditions (AS) and (BS).

Here we completely resolve the singular conditions (AS) and (BS) (assuming the corresponding general conditions (3.2.50), (3.2.51), (3.2.52), (3.2.53), (3.2.32), (3.2.41) of these series). It makes all our results very effective.

Below we consider the singular condition (AS) of a-series. The singular condition (AS) consists of congruences and non-congruences $\mod \gamma$. We denote by $(AS)^{(p)}$ the corresponding conditions over the prime number $p|\gamma$. It is enough to satisfy all conditions $(AS)^{(p)}$ for all $p|\gamma$. Below we consider several cases which all together cover all possible ones. For a prime p and a natural number n, we denote as $n^{(p)} = p^{\nu_p}$ the p-component of n. That is $n^{(p)} = p^{\nu_p(n)}|n$, and g.c.d$(n^{(p)}, n/n^{(p)}) = 1$.

There are several cases which we consider below.

Case $\gamma_2 = 2$, $\gamma_a \equiv \gamma_b \equiv 1 \mod 2$. Then c, a_1, b_1 are odd and $(AS)^{(2)}$ obviously satisfies.

Case $\gamma_2 = 2$, $2|\gamma_a$. Then $\gamma^{(2)} = 2a_1^{(2)} \geq 4$, $\gamma_b^{(2)} = 1$ and $b_1 \equiv 1 \mod 2$. By (3.2.51), we obtain that

$$\delta \equiv 0 \mod 2. \tag{3.3.1}$$

All conditions (AS) trivially satisfy over 2 except (3.2.39) for $l = 2$ which gives $\delta q_1^2 \not\equiv 0 \mod 4$ if $4|a_1$. It is equivalent to $\delta \equiv 2 \mod 4$ and $q_1 \equiv 1 \mod 2$ if $4|a_1$. This follows from (3.2.52) since $4|\gamma$. Thus (AS) over 2 satisfies.

Case $\gamma_2 = 2$, $2|\gamma_b$. Then $\gamma_a^{(2)} = 1$ and $\gamma^{(2)} = 2\gamma_b^{(2)} \geq 4$, $b_1^{(2)} = \gamma_b^{(2)}$. Denote $\gamma^{(2)} = 2^t$, $t \geq 2$. For $l = 2$, the condition (3.2.40) satisfies, and (3.2.35)—(3.2.39) give over 2 respectively

$$q_1 \equiv 0 \mod 2^{[t/2]}, \tag{3.3.2}$$

$$q_1(p_1 - \mu q_1) \equiv 0 \mod 2^{t-1}, \tag{3.3.3}$$

$$p_1 q_1 \equiv q_1^2 \equiv 0 \mod 2^{t-2}, \tag{3.3.4}$$

$$(\delta - \gamma\mu^2)q_1^2 \equiv 0 \mod 2^{2t-1}, \tag{3.3.5}$$

$$\gamma p_1^2 + \delta q_1^2 \not\equiv 0 \mod 2^{t+1} \ if \ t \geq 3. \tag{3.3.6}$$

By (3.2.51), δ is even.

Assume $t = 2$. Then (3.3.2)—(3.3.6) are equivalent to $q_1 \equiv 0 \mod 2$. By (3.2.52), then $p_1 \equiv 1 \mod 2$.

Assume that t is even and $t \geq 4$. By (3.3.2), we have $q_1 \equiv 0 \mod 2^{t/2}$. Then (3.3.6) is equivalent to $p_1 \equiv 1 \mod 2$. By (3.3.3), then $q_1 \equiv 0 \mod 2^{t-1}$. It follows (3.3.5). Thus, (3.3.2)—(3.3.6) are equivalent to $p_1 \equiv 1 \mod 2$ and $q_1 \equiv 0 \mod 2^{t-1}$. We had the same for $t = 2$.

Assume that t is odd and $t \geq 3$. Let us suppose that $p_1 \equiv 0 \mod 2$. Then (3.3.6) gives $\delta q_1^2 \not\equiv 0 \mod 2^{t+1}$. By (3.3.2), we have $q_1 \equiv 0 \mod 2^{(t-1)/2}$. Moreover δ is even. Then (3.3.6) is equivalent to $\delta \equiv 2 \mod 4$ and and $q_1 \equiv 2^{(t-1)/2} \mod 2^{(t-1)/2+1}$. Then $(\delta - \gamma\mu^2)q_1^2 \equiv 2^t \mod 2^{t+1}$ and (3.3.5) is not valid. This shows that $p_1 \equiv 1 \mod 2$. By (3.3.2), q_1 is even. By (3.3.3), then $q_1 \equiv 0 \mod 2^{t-1}$. These imply all conditions (3.3.2)—(3.3.6). Thus, we obtain the same conditions p_1 is odd and $q_1 \equiv 0 \mod 2^{t-1}$. Thus, in this case, the condition (AS) over 2 is

If $\gamma_2 = 2$ *and* $2|b_1$, *then* $p_1 \equiv 1 \mod 2$ *and* $q_1 \equiv 0 \mod \gamma^{(2)}/2$. (3.3.7)

Case $\gamma_2 = 1$, $2|\gamma$.

Case $\gamma_2 = 1$, $2|\gamma$ *and* $2|a_1$. Then $2 \leq \gamma^{(2)} = \gamma_a^{(2)}|a_1$, $\gamma_b^{(2)} = 1$ and $b_1 \equiv 1 \mod 2$.

By (3.2.50), (3.2.51) and (3.2.32), we get respectively

$$\mu \equiv 1 \mod 2, \tag{3.3.8}$$

$$\delta \equiv \gamma\mu^2 \mod 4(a_1^{(2)}/\gamma_a^{(2)}), \tag{3.3.9}$$

and then $\delta \equiv \gamma \mod 4$,

$$p_1 - \mu q_1 \equiv 0 \mod 2(a_1^{(2)}/\gamma_a^{(2)}). \tag{3.3.10}$$

It follows that over 2 all conditions (AS) satisfy except (3.2.39) and (3.2.40) which give respectively

$$\gamma p_1^2 + \delta q_1^2 \equiv 2\gamma\mu p_1 q_1 \mod 8(a_1^{(2)}/\gamma_a^{(2)}) \tag{3.3.11}$$

is not satisfied if $4|a_1$, and

$$\begin{cases} p_1 q_1 \equiv 0 \mod 2 \\ \mu\gamma p_1^2 + \mu\delta q_1^2 \equiv 2\delta p_1 q_1 \mod 8(a_1^{(2)}/\gamma_a^{(2)}) \end{cases} \tag{3.3.12}$$

is not satisfied.

By (3.3.10), we have $\gamma p_1^2 + \gamma\mu q_1^2 \equiv 2\gamma\mu p_1 q_1 \mod 4a_1(a_1^{(2)}/\gamma_a^{(2)})$. Since a_1 is even, it follows

$$\gamma p_1^2 + \gamma\mu q_1^2 \equiv 2\gamma\mu p_1 q_1 \mod 8(a_1^{(2)}/\gamma_a^{(2)}). \tag{3.3.13}$$

Then (3.3.11) is equivalent to $(\delta - \gamma\mu^2)q_1^2 \not\equiv 0 \mod 8(a_1^{(2)}/\gamma_a^{(2)})$ if $4|a_1$. By (3.3.9), this is equivalent to

$$q_1 \equiv 1 \mod 2 \text{ and } \delta - \gamma\mu^2 \not\equiv 0 \mod 8(a_1^{(2)}/\gamma_a^{(2)}) \tag{3.3.14}$$

if $4|a_1$.

Similarly, one can see that (3.3.12) is equivalent to

$$\begin{cases} p_1 q_1 \equiv 0 \mod 2 \\ (\delta - \gamma\mu^2)q_1^2 \equiv 0 \mod 8(a_1^{(2)}/\gamma_a^{(2)}) \end{cases}$$

is not valid. By above relations, it is equivalent to $p_1 \equiv 1 \mod 2$. Thus, in this case, (AS) over 2 is equivalent to two conditions:

$$if \ 2|\gamma, \ \gamma_2 = 1 \ and \ 2|a_1, \ then \ p_1 \equiv 1 \mod 2 \tag{3.3.15}$$

and

$$if \ 2|\gamma, \ \gamma_2 = 1 \ and \ 4|a_1, \ then \ \delta - \gamma\mu^2 \not\equiv 0 \mod (8a_1 b_1 c^2/\gamma). \tag{3.3.16}$$

Case $\gamma_2 = 1$, $2|\gamma$ and $2|b_1$. Then $2 \leq \gamma^{(2)} = \gamma_b^{(2)}|b_1$, $\gamma_a^{(2)} = 1$ and $a_1 \equiv 1 \mod 2$.

By (3.2.50), (3.2.51) and (3.2.32), we get respectively

$$\mu \equiv 1 \mod 2, \tag{3.3.17}$$

$$\delta \equiv \gamma\mu^2 \mod 4(b_1^{(2)}/\gamma_b^{(2)}), \tag{3.3.18}$$

and then $\delta \equiv \gamma \mod 4$,

$$p_1 - \mu q_1 \equiv 0 \mod 2. \tag{3.3.19}$$

Over 2, the conditions (3.2.35)—(3.2.40) give respectively

$$\widetilde{b}_1 q_1 \equiv 0 \mod \gamma_b^{(2)}, \tag{3.3.20}$$

$$q_1(p_1 - \mu q_1) \equiv 0 \mod 2\gamma_b^{(2)}, \tag{3.3.21}$$

$$p_1 q_1 \equiv q_1^2 \equiv 0 \mod \gamma_b^{(2)}, \tag{3.3.22}$$

$$(\delta - \gamma\mu^2)q_1^2 \equiv 0 \mod 4(\gamma_b^{(2)})^2(b_1^{(2)}/\gamma_b^{(2)}), \tag{3.3.23}$$

$$\gamma p_1^2 + \delta q_1^2 \equiv 2\gamma\mu p_1 q_1 \mod 8\gamma_b^{(2)} \tag{3.3.24}$$

is not satisfied if $4|b_1$,

$$\begin{cases} (b_1/\gamma_b)p_1 q_1 \equiv 0 \mod 2\gamma_b^{(2)} \\ \mu\gamma p_1^2 + \mu\delta q_1^2 \equiv 2\delta p_1 q_1 \mod 8\gamma_b^{(2)} \end{cases} \tag{3.3.25}$$

is not satisfied.

By (3.3.23), we have $\delta q_1^2 \equiv \gamma \mu^2 q_1^2 \mod 8\gamma_b^{(2)}$. It follows that (3.3.24) is equivalent to

$$p_1 - \mu q_1 \not\equiv 0 \mod 4 \tag{3.3.26}$$

if $4|b_1$. Similarly (3.3.25) is equivalent to

$$\begin{cases} (b_1/\gamma_b)p_1q_1 \equiv 0 \mod 2\gamma_b^{(2)} \\ p_1 - \mu q_1 \equiv 0 \mod 4 \end{cases} \tag{3.3.27}$$

is not satisfied.

Assume that

$$(b_1/\gamma_b)p_1q_1 \not\equiv 0 \mod 2\gamma_b^{(2)}. \tag{3.3.28}$$

By (3.3.22), we have $p_1q_1 \equiv 0 \mod \gamma_b^{(2)}$. Then (3.3.28) is equivalent to (b_1/γ_b) is odd and $p_1q_1 \equiv \gamma_b^{(2)} \mod 2\gamma_b^{(2)}$. By (3.3.21), then $q_1^2 \equiv \gamma_b^{(2)} \mod 2\gamma_b^{(2)}$. Then $\gamma_b^{(2)}$ is a square and $4|b_1$. By (3.3.26), then $p_1 - \mu q_1 \not\equiv 0 \mod 4$. Thus, (3.3.26) and (3.3.27) are equivalent to

$$p_1 - \mu q_1 \equiv 2 \mod 4 \tag{3.3.29}$$

since $p_1 \equiv q_1 \equiv 0 \mod 2$ by (3.3.17), (3.3.19) and (3.3.22). By (3.3.21), then $q_1 \equiv 0 \mod \gamma_b^{(2)}$, and all other conditions of (AS) over 2 follow from here.

Thus, in this case, (AS) over 2 is equivalent to

if $2|\gamma$, $\gamma_2 = 1$, and $2|b_1$, then $p_1 - \mu q_1 \not\equiv 0 \mod 4$ and $q_1 \equiv 0 \mod \gamma_b^{(2)}$. (3.3.30)

Case a prime odd $l|\gamma$ and $l|a_1$. Then $l \le \gamma^{(l)} = \gamma_a^{(l)}|a_1$, $\gamma_b^{(l)} = 1$ and $b_1 \not\equiv 0 \mod l$.

By (3.2.50), (3.2.51), (3.2.32) we have respectively

$$\mu \in (\mathbb{Z}/(a_1^{(l)}/\gamma_a^{(l)}))^*, \tag{3.3.31}$$

$$\delta \equiv \gamma\mu^2 \mod (a_1^{(l)}/\gamma_a^{(l)}), \tag{3.3.32}$$

$$p_1 - \mu q_1 \equiv 0 \mod (a_1^{(l)}/\gamma_a^{(l)}). \tag{3.3.33}$$

Conditions (AS) satisfy over l except (3.2.39) and (3.2.40) which give respectively

$$\begin{cases} \delta q_1^2 \equiv 0 \mod l \\ \gamma p_1^2 + \delta q_1^2 \equiv 2\gamma\mu p_1q_1 \mod (a_1^{(l)}/\gamma_a^{(l)})l \end{cases} \tag{3.3.34}$$

is not satisfied if $l^2|a_1$,

$$\begin{cases} \delta q_1^2 \equiv 0 \mod l \\ p_1q_1 \equiv 0 \mod l \\ \mu\gamma p_1^2 + \mu\delta q_1^2 \equiv 2\delta p_1q_1 \mod (a_1^{(l)}/\gamma_a^{(l)})l \end{cases} \tag{3.3.35}$$

is not satisfied if $l \mid (a_1^{(l)}/\gamma_a^{(l)})$.

Taking square of (3.3.33), we obtain

$$\gamma p_1^2 + \gamma\mu^2 q_1^2 \equiv 2\gamma\mu p_1q_1 \mod (a_1^{(l)}/\gamma_a^{(l)})l. \tag{3.3.36}$$

This shows that (3.3.34) is equivalent to

$$\begin{cases} \delta q_1^2 \equiv 0 \mod l \\ (\delta - \gamma\mu^2)q_1^2 \equiv 0 \mod (a_1^{(l)}/\gamma_a^{(l)})l \end{cases} \tag{3.3.37}$$

is not satisfied if $l^2|a_1$. By (3.3.33), this is equivalent to

If odd prime $l|\gamma$ and $l^2|a_1$, then $q_1 \not\equiv 0 \mod l$ and

either $\delta \not\equiv 0 \mod l$ or $(\delta - \gamma\mu^2) \not\equiv 0 \mod (a_1^{(l)}/\gamma_a^{(l)})l$.

(3.3.38)

If $l \,|\, (a_1^{(l)}/\gamma_a^{(l)})$, then $l^2|a_1$ and (3.3.38) is valid. Then $q_1 \not\equiv 0 \mod l$. By (3.3.31) and (3.3.33), then $p_1 \not\equiv 0 \mod l$. Thus (3.3.35) follows from (3.3.38).

Finally we get that (AS) over l is equivalent to (3.3.38) in this case.

Case a prime odd $l|\gamma$ and $l|b_1$. Then $l \le \gamma^{(l)} = \gamma_b^{(l)}|b_1$, $\gamma_a^{(l)} = 1$ and $a_1 \not\equiv 0$ mod l.

By (3.2.50), (3.2.51), (3.2.32) we have respectively

$$\mu \in (\mathbb{Z}/(b_1^{(l)}/\gamma_b^{(l)}))^*,$$

(3.3.39)

$$\delta \equiv \gamma\mu^2 \mod (b_1^{(l)}/\gamma_b^{(l)}),$$

(3.3.40)

$$\gamma p_1^2 - \delta q_1^2 = \pm 2(2/\gamma_2)(a_1/\gamma_a)\gamma_b c$$

(3.3.41)

Over l, the conditions (3.2.35)—(3.2.40) give respectively

$$\widetilde{b}_1 q_1 \equiv 0 \mod \gamma_b^{(l)},$$

(3.3.42)

$$q_1(p_1 - \mu q_1) \equiv 0 \mod \gamma_b^{(l)},$$

(3.3.43)

$$p_1 q_1 \equiv q_1^2 \equiv 0 \mod \gamma_b^{(l)},$$

(3.3.44)

$$(\delta - \gamma\mu^2)q_1^2 \equiv 0 \mod (b_1^{(l)}/\gamma_b^{(l)})(\gamma_b^{(l)})^2,$$

(3.3.45)

$$\begin{cases} (b_1^{(l)}/\gamma_b^{(l)})(\gamma p_1^2 + \delta q_1^2) \equiv 0 \mod \gamma_b^{(l)}l \\ \gamma p_1^2 + \delta q_1^2 \equiv 2\gamma\mu p_1 q_1 \mod \gamma_b^{(l)}l \end{cases}$$

(3.3.46)

is not satisfied if $l^2|b_1$,

$$\begin{cases} (b_1^{(l)}/\gamma_b^{(l)})(\gamma p_1^2 + \delta q_1^2) \equiv 0 \mod \gamma_b^{(l)}l \\ \mu\gamma p_1^2 + \mu\delta q_1^2 \equiv 2\delta p_1 q_1 \mod \gamma_b^{(l)}l \end{cases}$$

(3.3.47)

is not satisfied if $l \,|\, (b_1^{(l)}/\gamma_b^{(l)})$.

By (3.3.45), we have $\delta q_1^2 \equiv \gamma\mu^2 q_1^2 \mod \gamma_b l$. Then (3.3.46) is equivalent to

$$\begin{cases} (b_1^{(l)}/\gamma_b^{(l)})(p_1^2 + \mu^2 q_1^2) \equiv 0 \mod l \\ p_1 - \mu q_1 \equiv 0 \mod l \end{cases}$$

(3.3.48)

is not satisfied if $l^2|b_1$. Similarly (using also (3.3.39)) one can see that (3.3.47) is equivalent to

$$\begin{cases} (b_1^{(l)}/\gamma_b^{(l)})(p_1^2 + \mu^2 q_1^2) \equiv 0 \mod l \\ p_1 - \mu q_1 \equiv 0 \mod l \end{cases}$$

(3.3.49)

is not satisfied if $l \,|\, (b_1^{(l)}/\gamma_b^{(l)})$. Thus, (3.3.48) implies (3.3.49), and it is enough to satisfy (3.3.48).

Assume that $p_1 - \mu q_1 \not\equiv 0 \mod l$. By (3.3.43) we get $q_1 \equiv 0 \mod \gamma_b^{(l)}$, and all other conditions follow.

Assume that $p_1 - \mu q_1 \equiv 0 \mod l$. Then $p_1^2 + \mu^2 q_1^2 \equiv 2p_1^2 \mod l$, and (3.3.48) implies that $p_1 \not\equiv 0 \mod l$. By (3.3.44), $q_1 \equiv 0 \mod l$, and we get a contradiction.

Since $q_1 \equiv 0 \mod l$, the condition $p_1 \not\equiv \mu q_1 \mod l$ can be replaced by $p_1 \not\equiv 0$ mod l.

Thus, in this case, the condition (AS) over l is equivalent to two conditions

$$\text{If odd prime } l|\gamma \text{ and } l|b_1, \text{ then } q_1 \equiv 0 \mod \gamma_b^{(l)}. \tag{3.3.50}$$

$$\text{If odd prime } l|\gamma \text{ and } l^2|b_1, \text{ then } p_1 \not\equiv 0 \mod l. \tag{3.3.51}$$

Thus, we obtain

THEOREM 3.3.1. *The singular condition (AS) is equivalent to*

if odd prime $l|\gamma$ and $l^2|a_1$, then $q_1 \not\equiv 0 \mod l$ and

either $\delta \not\equiv 0 \mod l$ or $(\delta - \gamma\mu^2) \not\equiv 0 \mod (a_1^{(l)}/\gamma_a^{(l)})l$;

if odd prime $l|\gamma$ and $l|b_1$, then $q_1 \equiv 0 \mod \gamma_b^{(l)}$;

if odd prime $l|\gamma$ and $l^2|b_1$, then $p_1 \not\equiv 0 \mod l$;

if $2|\gamma$, $\gamma_2 = 1$ and $2|a_1$, then $p_1 \equiv 1 \mod 2$;

if $2|\gamma$, $\gamma_2 = 1$ and $4|a_1$, then $\delta - \gamma\mu^2 \not\equiv 0 \mod (8a_1b_1c^2/\gamma)$;

if $2|\gamma$, $\gamma_2 = 1$, and $2|b_1$, then $p_1 - \mu q_1 \not\equiv 0 \mod 4$ and $q_1 \equiv 0 \mod \gamma_b^{(2)}$;

if $2|\gamma$, $\gamma_2 = 2$ and $2|b_1$, then $p_1 \equiv 1 \mod 2$ and $q_1 \equiv 0 \mod \gamma^{(2)}/2$.

$$\tag{3.3.52}$$

The singular condition (BS) is equivalent to

if odd prime $l|\gamma$ and $l|a_1$, then $q_1 \equiv 0 \mod \gamma_a^{(l)}$;

if odd prime $l|\gamma$ and $l^2|a_1$, then $p_1 \not\equiv 0 \mod l$;

if odd prime $l|\gamma$ and $l^2|b_1$, then $q_1 \not\equiv 0 \mod l$ and

either $\delta \not\equiv 0 \mod l$ or $(\delta - \gamma\mu^2) \not\equiv 0 \mod (b_1^{(l)}/\gamma_b^{(l)})l$;

if $2|\gamma$, $\gamma_2 = 1$, and $2|a_1$, then $p_1 - \mu q_1 \not\equiv 0 \mod 4$ and $q_1 \equiv 0 \mod \gamma_a^{(2)}$;

if $2|\gamma$, $\gamma_2 = 1$ and $2|b_1$, then $p_1 \equiv 1 \mod 2$;

if $2|\gamma$, $\gamma_2 = 1$ and $4|b_1$, then $\delta - \gamma\mu^2 \not\equiv 0 \mod (8a_1b_1c^2/\gamma)$;

if $2|\gamma$, $\gamma_2 = 2$ and $2|a_1$, then $p_1 \equiv 1 \mod 2$ and $q_1 \equiv 0 \mod \gamma^{(2)}/2$.

$$\tag{3.3.53}$$

Thus, we can finally rewrite the conditions (A) of a-series, and the conditions (B) of b-series in the very efficient form which makes all our results very effective.

(A): The conditions of a-series:
(AG): The general conditions of a-series:

$$p_1 - \mu q_1 \equiv 0 \mod (2/\gamma_2)(a_1/\gamma_a)c, \tag{3.3.54}$$

$$p_1 - \mu q_1 \not\equiv 0 \mod (2/\gamma_2)(a_1/\gamma_a)c\,l$$
$$\text{for any prime } l \text{ such that } l^2|b_1 \text{ and g.c.d}(l,\gamma) = 1, \tag{3.3.55}$$

$$l \nmid p_1 \text{ for any prime } l \text{ such that } l^2|a_1 \text{ and g.c.d}(l,\gamma) = 1. \tag{3.3.56}$$

(AS) The singular conditions of a-series:

if odd prime $l|\gamma$ and $l^2|a_1$, then $q_1 \not\equiv 0 \mod l$ and

either $\delta \not\equiv 0 \mod l$ or $(\delta - \gamma\mu^2) \not\equiv 0 \mod (a_1^{(l)}/\gamma_a^{(l)})l$;

if odd prime $l|\gamma$ and $l|b_1$, then $q_1 \equiv 0 \mod \gamma_b^{(l)}$;

if odd prime $l|\gamma$ and $l^2|b_1$, then $p_1 \not\equiv 0 \mod l$;

if $2|\gamma$, $\gamma_2 = 1$ and $2|a_1$, then $p_1 \equiv 1 \mod 2$;

if $2|\gamma$, $\gamma_2 = 1$ and $4|a_1$, then $\delta - \gamma\mu^2 \not\equiv 0 \mod (8a_1b_1c^2/\gamma)$;

if $2|\gamma$, $\gamma_2 = 1$, and $2|b_1$, then $p_1 - \mu q_1 \not\equiv 0 \mod 4$ and $q_1 \equiv 0 \mod \gamma_b^{(2)}$;

if $2|\gamma$, $\gamma_2 = 2$ and $2|b_1$, then $p_1 \equiv 1 \mod 2$ and $q_1 \equiv 0 \mod \gamma^{(2)}/2$.

$$(3.3.57)$$

(B): The conditions of b-series:
(BG): The general conditions of b-series:

$$p_1 - \mu q_1 \equiv 0 \mod (2/\gamma_2)(b_1/\gamma_b)c, \qquad (3.3.58)$$

$$p_1 - \mu q_1 \not\equiv 0 \mod (2/\gamma_2)(b_1/\gamma_b)c\, l$$
$$\text{for any prime } l \text{ such that } l^2|a_1 \text{ and g.c.d}(l, \gamma) = 1, \qquad (3.3.59)$$

$$l \nmid p_1 \text{ for any prime } l \text{ such that } l^2|b_1 \text{ and g.c.d}(l, \gamma) = 1. \qquad (3.3.60)$$

(BS) The singular conditions of b-series:

if odd prime $l|\gamma$ and $l|a_1$, then $q_1 \equiv 0 \mod \gamma_a^{(l)}$;

if odd prime $l|\gamma$ and $l^2|a_1$, then $p_1 \not\equiv 0 \mod l$;

if odd prime $l|\gamma$ and $l^2|b_1$, then $q_1 \not\equiv 0 \mod l$ and

either $\delta \not\equiv 0 \mod l$ or $(\delta - \gamma\mu^2) \not\equiv 0 \mod (b_1^{(l)}/\gamma_b^{(l)})l$;

if $2|\gamma$, $\gamma_2 = 1$, and $2|a_1$, then $p_1 - \mu q_1 \not\equiv 0 \mod 4$ and $q_1 \equiv 0 \mod \gamma_a^{(2)}$;

if $2|\gamma$, $\gamma_2 = 1$ and $2|b_1$, then $p_1 \equiv 1 \mod 2$;

if $2|\gamma$, $\gamma_2 = 1$ and $4|b_1$, then $\delta - \gamma\mu^2 \not\equiv 0 \mod (8a_1b_1c^2/\gamma)$;

if $2|\gamma$, $\gamma_2 = 2$ and $2|a_1$, then $p_1 \equiv 1 \mod 2$ and $q_1 \equiv 0 \mod \gamma^{(2)}/2$.

$$(3.3.61)$$

We remind that here $\gamma|2a_1b_1$ and

$$\mu \in (\mathbb{Z}/2a_1b_1c^2/\gamma)^*, \qquad (3.3.62)$$

$$\delta \equiv \gamma\mu^2 \mod 4a_1b_1c^2/\gamma, \qquad (3.3.63)$$

$$\gamma p_1^2 - \delta q_1^2 = \pm 2(2/\gamma_2)(a_1/\gamma_a)\gamma_b c \qquad (3.3.64)$$

for the a-series, and

$$\gamma p_1^2 - \delta q_1^2 = \pm 2(2/\gamma_2)(b_1/\gamma_b)\gamma_a c \qquad (3.3.65)$$

for the b-series.

4. Final results for $\rho = 2$

Now we can formulate the main results which follow from Theorem 3.1.3 and the calculations above.

THEOREM 4.1. *Let X be a K3 surface with $\rho(X) = 2$, and H a polarization of X of degree $H^2 = 2rs$ where $r, s \in \mathbb{N}$. Assume that the Mukai vector (r, H, s) is primitive. Let Y be the moduli space of sheaves on X with the isotropic Mukai vector $v = (r, H, s)$. Let $\widetilde{H} = H/d$ be the corresponding primitive polarization, and $\widetilde{H} \cdot N(X) = \gamma\mathbb{Z}$. We denote by μ the invariant of the pair $(N(X), \widetilde{H})$ and use notations of Proposition 3.1.1.*

We have $Y \cong X$ if

$$g.c.d(c, d\gamma) = 1$$

and X belongs either to a-series or to b-series.

Here X belongs to a-series if for one of $\epsilon = \pm 1$ the equation

$$\gamma p_1^2 - \delta q_1^2 = \epsilon 2(2/\gamma_2)(a_1/\gamma_a)\gamma_b c. \tag{4.1}$$

has an integral solution (p_1, q_1) satisfying the conditions (A) of a-series (3.3.54)—(3.3.57).

These solutions (p_1, q_1) of (4.1) give all solution (x, y) of Theorem 3.1.3 from a-series as associated solutions

$$(x, y) = \pm\left(\frac{-2a_1b_1c}{\gamma} + \frac{\epsilon b_1\gamma_2\gamma_a p_1^2}{\gamma_b}, \frac{\epsilon b_1\gamma_2\gamma_a p_1 q_1}{\gamma_b}\right). \tag{4.2}$$

Here X belongs to b-series if for one of signs $\epsilon = \pm 1$ the equation

$$\gamma p_1^2 - \delta q_1^2 = \epsilon 2(2/\gamma_2)(b_1/\gamma_b)\gamma_a c. \tag{4.3}$$

has an integral solution (p_1, q_1) satisfying the conditions (B) of b-series (3.3.58)—(3.3.61).

These solutions (p_1, q_1) of (4.3) give all solutions (x, y) of Theorem 3.1.3 of b-series as associated solutions

$$(x, y) = \pm\left(\frac{-2a_1b_1c}{\gamma} + \frac{\epsilon a_1\gamma_2\gamma_b p_1^2}{\gamma_a}, \frac{\epsilon a_1\gamma_2\gamma_b p_1 q_1}{\gamma_a}\right). \tag{4.4}$$

These conditions are necessary to have $Y \cong X$ if X is a general K3 surface with $\rho(X) = 2$, i. e. the automorphism group of the transcendental periods $(T(X), H^{2,0}(X))$ is ± 1.

Now we want to interpret solutions (p_1, q_1) of Theorem 4.1 as elements of the Picard lattice $N(X)$.

Let (p_1, q_1) be a solution of Theorem 4.1 from a-series. Then

$$\gamma p_1^2 - \delta q_1^2 = \epsilon 2(2/\gamma_2)(a_1/\gamma_a)\gamma_b c. \tag{4.5}$$

By (3.3.54), we have

$$p_1 - \mu q_1 \equiv 0 \mod (2/\gamma_2)(a_1/\gamma_a)c. \tag{4.6}$$

Let us put

$$t = (b_1/\gamma_b)c. \tag{4.7}$$

Then

$$tp_1 - \mu t q_1 \equiv 0 \mod 2a_1 b_1 c^2/\gamma. \tag{4.8}$$

and

$$\widetilde{h}_1 = \frac{t(p_1\widetilde{H} + q_1 f(\widetilde{H}))}{2a_1 b_1 c^2/\gamma} \in N(X). \tag{4.9}$$

We have

$$\widetilde{h}_1^2 = \frac{t^2(\gamma p_1^2 - \delta q_1^2)}{2a_1 b_1 c^2/\gamma} = \epsilon 2b_1 c \tag{4.10}$$

and

$$\widetilde{H} \cdot \widetilde{h}_1 = \gamma(b_1/\gamma_b)cp_1 \equiv 0 \mod \gamma(b_1/\gamma_b)c \text{ and } p_1 = \frac{\widetilde{H} \cdot \widetilde{h}_1}{\gamma(b_1/\gamma_b)c}. \tag{4.11}$$

Also

$$-f(\widetilde{H}) \cdot \widetilde{h}_1 = \frac{b_1 c\delta q_1}{\gamma_b} \text{ and } q_1 = -\frac{f(\widetilde{H}) \cdot \widetilde{h}_1}{\delta(b_1/\gamma_b)c}. \tag{4.12}$$

Here

$$-\gamma\delta = \det N(X). \tag{4.13}$$

Another calculation of p_1 and q_1 is as follows. We have

$$\widetilde{h}_1 = \frac{u\widetilde{H} + vf(\widetilde{H})}{2a_1 b_1 c^2/\gamma} = \frac{p_1\widetilde{H} + q_1 f(\widetilde{H})}{(2/\gamma_2)(a_1/\gamma_a)c} \tag{4.14}$$

where

$$u \equiv 0 \mod (b_1/\gamma_b)c, \text{ and } p_1 = \frac{u}{(b_1/\gamma_b)c}, \tag{4.15}$$

$$v \equiv 0 \mod (b_1/\gamma_b)c, \text{ and } q_1 = \frac{v}{(b_1/\gamma_b)c}. \tag{4.16}$$

We remind that here $\mathbb{Z}f(\widetilde{H})$ is the orthogonal complement to $\widetilde{H}$ in $N(X)$. Both these calculations of p_1 and q_1 are equivalent.

By construction, (3.3.54) is equivalent to $\widetilde{h}_1 \in N(X)$, (3.3.55) is equivalent to $\widetilde{h}_1/l \notin N(X)$, (3.3.56) is equivalent to $\widetilde{H} \cdot \widetilde{h}_1 \not\equiv 0 \mod \gamma(b_1/\gamma_b)cl$.

Changing the letters a and b places we get the same results for the b-series.

Thus, we get

THEOREM 4.2. *Let X be a K3 surface with $\rho(X) = 2$, and H a polarization of X of degree $H^2 = 2rs$ where $r, s \in \mathbb{N}$. Assume that the Mukai vector (r, H, s) is primitive. Let Y be the moduli space of sheaves on X with the isotropic Mukai vector $v = (r, H, s)$. Let $\widetilde{H} = H/d$ be the corresponding primitive polarization, and $\widetilde{H} \cdot N(X) = \gamma\mathbb{Z}$. We denote by μ the invariant of the pair $(N(X), \widetilde{H})$ and use notations of Proposition 3.1.1.*
We have $Y \cong X$ if

$$g.c.d(c, d\gamma) = 1,$$

and at least for one $\epsilon = \pm 1$ there exists $\widetilde{h}_1 \in N(X)$ which belongs to the a-series or to the b-series described below:
$\widetilde{h}_1$ belongs to the a-series if

$$\widetilde{h}_1^2 = \epsilon 2b_1 c \text{ and } \widetilde{H} \cdot \widetilde{h}_1 \equiv 0 \mod \gamma(b_1/\gamma_b)c, \tag{4.17}$$

$$\widetilde{H} \cdot \widetilde{h}_1 \not\equiv 0 \mod \gamma(b_1/\gamma_b)cl_1, \ \widetilde{h}_1/l_2 \notin N(X) \tag{4.18}$$

for any prime l_1 such that $l_1^2|a_1$ and $g.c.d(l_1, \gamma) = 1$, and for any prime l_2 such that $l_2^2|b_1$ and $g.c.d(l_2, \gamma) = 1$, and

$$p_1 = \frac{\widetilde{H} \cdot \widetilde{h}_1}{\gamma(b_1/\gamma_b)c}, \quad q_1 = -\frac{f(\widetilde{H}) \cdot \widetilde{h}_1}{\delta(b_1/\gamma_b)c} \tag{4.19}$$

satisfy the singular condition (AG) (conditions (3.3.57) mod γ) of a-series.
   $\widetilde{h}_1$ belongs to the b-series if

$$\widetilde{h}_1^2 = \epsilon 2a_1 c \text{ and } \widetilde{H} \cdot \widetilde{h}_1 \equiv 0 \mod \gamma(a_1/\gamma_a)c, \tag{4.20}$$

$$\widetilde{H} \cdot \widetilde{h}_1 \not\equiv 0 \mod \gamma(a_1/\gamma_a)cl_1, \ \widetilde{h}_1/l_2 \notin N(X) \tag{4.21}$$

for any prime l_1 such that $l_1^2|b_1$ and $g.c.d(l_1, \gamma) = 1$, and for any prime l_2 such that $l_2^2|a_1$ and $g.c.d(l_2, \gamma) = 1$, and

$$p_1 = \frac{\widetilde{H} \cdot \widetilde{h}_1}{\gamma(a_1/\gamma_a)c}, \quad q_1 = -\frac{f(\widetilde{H}) \cdot \widetilde{h}_1}{\delta(a_1/\gamma_a)c} \tag{4.22}$$

satisfy the singular condition (BG) (conditions (3.3.61) mod γ) of b-series.
   These conditions are necessary to have $Y \cong X$ if X is a general K3 surface with $\rho(X) = 2$, i. e. the automorphism group of the transcendental periods $(T(X), H^{2,0}(X))$ is ± 1.

IMPORTANT REMARK 4.3. Applying Theorem 3.1.3 and the formulae (4.2), (4.4) for the associated solution, we get the following formulae in terms of X for the canonical primitive *nef* element $\widetilde{h}$ of Y defined by $(-a, 0, b)$ mod $\mathbb{Z}v$:

$$\widetilde{h}' = \begin{cases} \frac{-\widetilde{H}}{c} + \frac{\epsilon(\widetilde{H} \cdot \widetilde{h}_1)\widetilde{h}_1}{b_1 c^2} & \text{if } \widetilde{h}_1 \text{ is from } a\text{-series,} \\ \frac{-\widetilde{H}}{c} + \frac{\epsilon(\widetilde{H} \cdot \widetilde{h}_1)\widetilde{h}_1}{a_1 c^2} & \text{if } \widetilde{h}_1 \text{ is from } b\text{-series} \end{cases} \tag{4.23}$$

belongs to $N(X)$ and

$$(Y, \widetilde{h}) \cong (X, \pm w(\widetilde{h}')) \text{ for some } w \in W^{(-2)}(N(X)). \tag{4.24}$$

   Specialising (by Lemma 2.2.1) the theorem 4.2, we get the following sufficient condition to have $Y \cong X$ which is valid for X with any $\rho(X)$. This is one of the main results of the paper.
   In Theorem 4.4 below, for $\widetilde{H} \in N$ we apply the same definitions and notations: $f(\widetilde{H})$, δ, μ, as for $\widetilde{H} \in N = N(X)$ of Proposition 3.1.1.

   THEOREM 4.4. *Let X be a K3 surface and H a polarization of X of degree $H^2 = 2rs$ where $r, s \in \mathbb{N}$. Assume that the Mukai vector (r, H, s) is primitive. Let Y be the moduli space of sheaves on X with the isotropic Mukai vector $v = (r, H, s)$. Let $\widetilde{H} = H/d$ be the corresponding primitive polarization.*
   We have $Y \cong X$ if there exists $\widetilde{h}_1 \in N(X)$ such that $\widetilde{H}, \widetilde{h}_1$ belong to a 2-dimensional primitive sublattice $N \subset N(X)$ such that $\widetilde{H} \cdot N = \gamma \mathbb{Z}, \gamma > 0$, and

$$g.c.d(c, d\gamma) = 1, \tag{4.25}$$

moreover, for one of $\epsilon = \pm 1$ the element $\widetilde{h}_1$ belongs to the a-series or to the b-series described below:

$\widetilde{h}_1$ *belongs to the a-series if*

$$\widetilde{h}_1^2 = \epsilon 2b_1 c \text{ and } \widetilde{H} \cdot \widetilde{h}_1 \equiv 0 \mod \gamma(b_1/\gamma_b)c, \tag{4.26}$$

$$\widetilde{H} \cdot \widetilde{h}_1 \not\equiv 0 \mod \gamma(b_1/\gamma_b)cl_1, \ \widetilde{h}_1/l_2 \notin N(X) \tag{4.27}$$

for any prime l_1 such that $l_1^2|a_1$ and $g.c.d(l_1, \gamma) = 1$, and any prime l_2 such that $l_2^2|b_1$ and $g.c.d(l_2, \gamma) = 1$, and

$$p_1 = \frac{\widetilde{H} \cdot \widetilde{h}_1}{\gamma(b_1/\gamma_b)c}, \quad q_1 = -\frac{f(\widetilde{H}) \cdot \widetilde{h}_1}{\delta(b_1/\gamma_b)c} \tag{4.28}$$

satisfy the singular condition (AS) of a-series:

if odd prime $l|\gamma$ and $l^2|a_1$, then $q_1 \not\equiv 0 \mod l$ and

either $\delta \not\equiv 0 \mod l$ or $(\delta - \gamma\mu^2) \not\equiv 0 \mod (a_1^{(l)}/\gamma_a^{(l)})l$;

if odd prime $l|\gamma$ and $l|b_1$, then $q_1 \equiv 0 \mod \gamma_b^{(l)}$;

if odd prime $l|\gamma$ and $l^2|b_1$, then $p_1 \not\equiv 0 \mod l$;

if $2|\gamma$, $\gamma_2 = 1$ and $2|a_1$, then $p_1 \equiv 1 \mod 2$;

if $2|\gamma$, $\gamma_2 = 1$ and $4|a_1$, then $\delta - \gamma\mu^2 \not\equiv 0 \mod (8a_1b_1c^2/\gamma)$;

if $2|\gamma$, $\gamma_2 = 1$, and $2|b_1$, then $p_1 - \mu q_1 \not\equiv 0 \mod 4$ and $q_1 \equiv 0 \mod \gamma_b^{(2)}$;

if $2|\gamma$, $\gamma_2 = 2$ and $2|b_1$, then $p_1 \equiv 1 \mod 2$ and $q_1 \equiv 0 \mod \gamma^{(2)}/2$.

$\widetilde{h}_1$ *belongs to the b-series if*

$$\widetilde{h}_1^2 = \epsilon 2a_1 c \text{ and } \widetilde{H} \cdot \widetilde{h}_1 \equiv 0 \mod \gamma(a_1/\gamma_a)c, \tag{4.29}$$

$$\widetilde{H} \cdot \widetilde{h}_1 \not\equiv 0 \mod \gamma(a_1/\gamma_a)cl_1, \ \widetilde{h}_1/l_2 \notin N(X) \tag{4.30}$$

for any prime l_1 such that $l_1^2|b_1$ and $g.c.d(l_1, \gamma) = 1$ and any prime l_2 such that $l_2^2|a_1$ and $g.c.d(l_2, \gamma) = 1$, and

$$p_1 = \frac{\widetilde{H} \cdot \widetilde{h}_1}{\gamma(a_1/\gamma_a)c}, \quad q_1 = -\frac{f(\widetilde{H}) \cdot \widetilde{h}_1}{\delta(a_1/\gamma_a)c} \tag{4.31}$$

satisfy the singular condition (BS) of b-series:

if odd prime $l|\gamma$ and $l|a_1$, then $q_1 \equiv 0 \mod \gamma_a^{(l)}$;

if odd prime $l|\gamma$ and $l^2|a_1$, then $p_1 \not\equiv 0 \mod l$;

if odd prime $l|\gamma$ and $l^2|b_1$, then $q_1 \not\equiv 0 \mod l$ and

either $\delta \not\equiv 0 \mod l$ or $(\delta - \gamma\mu^2) \not\equiv 0 \mod (b_1^{(l)}/\gamma_b^{(l)})l$;

if $2|\gamma$, $\gamma_2 = 1$, and $2|a_1$, then $p_1 - \mu q_1 \not\equiv 0 \mod 4$ and $q_1 \equiv 0 \mod \gamma_a^{(2)}$;

if $2|\gamma$, $\gamma_2 = 1$ and $2|b_1$, then $p_1 \equiv 1 \mod 2$;

if $2|\gamma$, $\gamma_2 = 1$ and $4|b_1$, then $\delta - \gamma\mu^2 \not\equiv 0 \mod (8a_1b_1c^2/\gamma)$;

if $2|\gamma$, $\gamma_2 = 2$ and $2|a_1$, then $p_1 \equiv 1 \mod 2$ and $q_1 \equiv 0 \mod \gamma^{(2)}/2$.

Moreover, one has formulae (4.23) and (4.24) in terms of X for the canonical primitive nef element $\widetilde{h}$ of Y defined by $(-a, 0, b) \mod \mathbb{Z}v$.

These conditions are necessary to have $Y \cong X$ if $\rho(X) \leq 2$ and X is a general K3 surface with its Picard lattice, i. e. the automorphism group of the transcendental periods $(T(X), H^{2,0}(X))$ is ± 1.

See Sect. 6 about the cases $\gamma = 1$ and $\gamma = 2$.

5. Divisorial conditions on moduli of (X, H) which imply $Y \cong X$ and $\gamma(\widetilde{H}) = \gamma$

Using notations above, assuming g.c.d$(c, d\gamma) = 1$ for

$$\overline{\mu} = \{\mu, -\mu\} \subset (\mathbb{Z}/(2a_1 b_1 c^2/\gamma))^*, \ \epsilon = \pm 1$$

we denote by

$$\mathcal{D}(r, s, d, \gamma; A)_\epsilon^{\overline{\mu}} \tag{5.1}$$

the set of all $\delta \in \mathbb{N}$ such that $\delta \equiv \gamma\mu^2 \mod 4a_1 b_1 c^2/\gamma$ and the equation $\gamma p_1^2 - \delta q_1^2 = \epsilon 2(2/\gamma_2)(a_1/\gamma_a)\gamma_b c$ has an integral solution (p_1, q_1) satisfying the condition (A) (3.3.54)—(3.3.57) of the a-series. Similarly, we define

$$\mathcal{D}(r, s, d, \gamma; B)_\epsilon^{\overline{\mu}} \tag{5.2}$$

the b-series changing a and b places (see the equation (3.3.65) and conditions (B) (3.3.58)—(3.3.61)).

We denote

$$\mathcal{D}(r, s, d, \gamma)^{\overline{\mu}} = \left(\bigcup_{\epsilon \in \{-1,1\}} \mathcal{D}(r, s, d, \gamma; A)_\epsilon^{\overline{\mu}} \right)$$

$$\bigcup \left(\bigcup_{\epsilon \in \{-1,1\}} \mathcal{D}(r, s, d, \gamma; B)_\epsilon^{\overline{\mu}} \right). \tag{5.3}$$

By Theorem 4.1, the set $\mathcal{D}(r, s, d, \gamma)^{\overline{\mu}}$ describes all possible pairs $H \in N(X)$ of general polarized K3 surfaces (X, H) with rk $N(X) = 2$, the primitive polarization $\widetilde{H} = H/d \in N(X)$, the invariant $\gamma(\widetilde{H}) = \gamma$ (i. e. $\widetilde{H} \cdot N(X) = \gamma\mathbb{Z}$) and the invariant $\pm\mu$ for $\widetilde{H} \in N(X)$), such that $Y \cong X$. By general results of [N1] and [N2] the pair $\widetilde{H} \in N(X)$ defines the irreducible 18-dimensional moduli of such pairs $(X, \widetilde{H})$, i. e. a (irreducible) divisorial condition on 19-dimensional moduli of polarized K3 surfaces (X, H) which implies that $\gamma(\widetilde{H}) = \gamma$ and $Y \cong X$. Thus, we can interpret our results as follows.

THEOREM 5.1. *The set*

$$\mathcal{D}(r, s, d, \gamma) = \{(\overline{\mu}, \delta) \mid \{\mu, -\mu\} \subset (\mathbb{Z}/(2a_1 b_1 c^2/\gamma))^*, \ \delta \in \mathcal{D}(r, s, d, \gamma)^{\overline{\mu}}\} \tag{5.4}$$

describes all irreducible divisorial conditions on moduli of polarized K3 surfaces (X, H) with $H^2 = 2rs$ and the primitive polarization $\widetilde{H} = H/d$ (here $d^2|rs$), which imply $Y \cong X$ for any X, and $\widetilde{H} \cdot N(X) = \gamma\mathbb{Z}$ for a general X.

We have (see (5.3))

$$\mathcal{D}(r, s, d, \gamma)^{\overline{\mu}} = \left(\bigcup_{\epsilon \in \{-1,1\}} \mathcal{D}(r, s, d, \gamma; A)_\epsilon^{\overline{\mu}} \right)$$

$$\bigcup\left(\bigcup_{\epsilon\in\{-1,1\}}\mathcal{D}(r,s,d,\gamma;B)_{\epsilon}^{\overline{\mu}}\right). \tag{5.5}$$

where each set $\mathcal{D}(r,s,d,\gamma;A)_{\epsilon}^{\overline{\mu}}$ and $\mathcal{D}(r,s,d,\gamma;B)_{\epsilon}^{\overline{\mu}}$ is infinite if it is not empty.

PROOF. We need to prove the last statement only. Assume that $\mathcal{D}(r,s,d,\gamma;A)_{\epsilon}^{\overline{\mu}}$ is not empty. Thus, there exist integral (p_0,q_0) such that

$$\begin{cases} \frac{\gamma p_0^2 - \epsilon 2(2/\gamma_2)(a_1/\gamma_a)\gamma_b c}{q_0^2} \equiv \gamma\mu^2 \mod \frac{4a_1 b_1 c^2}{\gamma} \\ \gamma p_0^2 - \epsilon 2(2/\gamma_2)(a_1/\gamma_a)\gamma_b c > 0 \\ (p_0,q_0) \text{ satisfies } (A) \end{cases}. \tag{5.6}$$

Then

$$\delta_0 = \frac{\gamma p_0^2 - \epsilon 2(2/\gamma_2)(a_1/\gamma_a)\gamma_b c}{q_0^2} \in \mathcal{D}(r,s,d,\gamma;A)_{\epsilon}^{\overline{\mu}}. \tag{5.7}$$

The (5.6) is equivalent to

$$\begin{cases} \gamma p_0^2 - \epsilon 2(2/\gamma_2)(a_1/\gamma_a)\gamma_b c \equiv \gamma\mu^2 q_0^2 \mod 4a_1 b_1 c^2 q_0^2/\gamma \\ \gamma p_0^2 - \epsilon 2(2/\gamma_2)(a_1/\gamma_a)\gamma_b c > 0 \\ (p_0,q_0) \text{ satisfies } (A) \end{cases}. \tag{5.8}$$

Clearly, (p,q_0) where

$$p \equiv p_0 \mod 8a_1 b_1 c^2 q_0^2, \text{ and } \gamma p_0^2 - \epsilon 2(2/\gamma_2)(a_1/\gamma_a)\gamma_b c > 0 \tag{5.9}$$

also satisfies (5.8) and defines

$$\delta = \frac{\gamma p_0^2 - \epsilon 2(2/\gamma_2)(a_1/\gamma_a)\gamma_b c}{q_0^2} \in \mathcal{D}(r,s,d,\gamma;A)_{\epsilon}^{\overline{\mu}}. \tag{5.10}$$

Obviously, their number is infinite. This proves the statement.

The key question is:

PROBLEM 5.2. When $\mathcal{D}(r,s,d,\gamma;A)_{\epsilon}^{\overline{\mu}}$ and $\mathcal{D}(r,s,d,\gamma;B)_{\epsilon}^{\overline{\mu}}$ are non-empty?

We hope to consider this question in further publications on the subject. It was shown in [**N4**] (see also [**MN1**], [**MN2**]) that at least one of these sets is not empty if $d = 1$ and $\gamma = 1$. Theorem 4.1 and exactly the same considerations as in [**N4**] show that it is also valid for $\gamma = 1$ and any d because singular conditions (AS) and (BS) satisfy if $\gamma = 1$. Thus we have

THEOREM 5.3. At least for one of $\overline{\mu}$, ϵ one of sets $\mathcal{D}(r,s,d,\gamma = 1;A)_{\epsilon}^{\overline{\mu}}$ or $\mathcal{D}(r,s,d,\gamma = 1;B)_{\epsilon}^{\overline{\mu}}$ is not empty.

In particular, for any primitive isotropic Mukai vector (r,H,s) the set of divisorial conditions on moduli of X which imply that $Y \cong X$ and $\gamma = 1$ is not empty and is then infinite.

We hope to consider Problem 5.2 for other γ in further publications on the subject.

6. Examples of $\gamma = 1$ and $\gamma = 2$

As concrete examples of results of Sect. 4, we consider cases of $\gamma = 1$ and $\gamma = 2$.

When $\gamma = 1$, then singular conditions (AS) and (BS) are obviously valid, and we obtain especially simple results. We formulate only the analogy of Theorem 4.4.

THEOREM 6.1. *Let X be a K3 surface and H a polarization of X of degree $H^2 = 2rs$ where $r, s \in \mathbb{N}$. Assume that the Mukai vector (r, H, s) is primitive. Let Y be the moduli space of sheaves on X with the isotropic Mukai vector $v = (r, H, s)$. Let $\widetilde{H} = H/d$ be the corresponding primitive polarization.*

We have $Y \cong X$ if there exists $\widetilde{h}_1 \in N(X)$ such that $\widetilde{H}$, $\widetilde{h}_1$ belong to a 2-dimensional primitive sublattice $N \subset N(X)$ such that

$$\widetilde{H} \cdot N = \mathbb{Z} \tag{6.1}$$

(i. e. $\gamma = 1$), moreover, for one of $\epsilon = \pm 1$ the element $\widetilde{h}_1$ belongs to the a-series or to the b-series described below:

$\widetilde{h}_1$ belongs to the a-series if

$$\widetilde{h}_1^2 = \epsilon 2 b_1 c \text{ and } \widetilde{H} \cdot \widetilde{h}_1 \equiv 0 \mod b_1 c, \tag{6.2}$$

$$\widetilde{H} \cdot \widetilde{h}_1 \not\equiv 0 \mod b_1 c l_1, \ \widetilde{h}_1/l_2 \notin N(X) \tag{6.3}$$

for any prime l_1 such that $l_1^2 | a_1$, and any prime l_2 such that $l_2^2 | b_1$.

$\widetilde{h}_1$ belongs to the b-series if

$$\widetilde{h}_1^2 = \epsilon 2 a_1 c \text{ and } \widetilde{H} \cdot \widetilde{h}_1 \equiv 0 \mod a_1 c, \tag{6.4}$$

$$\widetilde{H} \cdot \widetilde{h}_1 \not\equiv 0 \mod a_1 c l_1, \ \widetilde{h}_1/l_2 \notin N(X) \tag{6.5}$$

for any prime l_1 such that $l_1^2 | b_1$ and any prime l_2 such that $l_2^2 | a_1$.

Moreover, one has formulae (4.23) and (4.24) in terms of X for the canonical primitive nef element $\widetilde{h}$ of Y defined by $(-a, 0, b) \mod \mathbb{Z}v$.

These conditions are necessary to have $Y \cong X$ and $\widetilde{H} \cdot N(X) = \mathbb{Z}$ if $\rho(X) \le 2$ and X is a general K3 surface with its Picard lattice, i. e. the automorphism group of the transcendental periods $(T(X), H^{2,0}(X))$ is ± 1.

This generalises results of [**MN1**], [**MN2**] and [**N4**] where the condition $H \cdot N = \mathbb{Z}$ had been imposed (i. e. $d = \gamma = 1$).

Now let us assume that $\gamma = 2$. By Theorem 4.4, we obtain three cases which all together cover all possibilities for $\gamma = 2$.

When $\gamma = 2$ and $a_1 \equiv b_1 \equiv 1 \mod 2$, then $\gamma_2 = 2$, and singular conditions (AS) and (BS) satisfy. Theorem 4.4 gives then

THEOREM 6.2. *Let X be a K3 surface and H a polarization of X of degree $H^2 = 2rs$ where $r, s \in \mathbb{N}$. Assume that the Mukai vector (r, H, s) is primitive. Let Y be the moduli space of sheaves on X with the isotropic Mukai vector $v = (r, H, s)$. Let $\widetilde{H} = H/d$ be the corresponding primitive polarization. Assume that*

$$g.c.d(2, c) = 1 \tag{6.6}$$

and

$$a_1 \equiv b_1 \equiv 1 \mod 2. \tag{6.7}$$

We have $Y \cong X$ *if there exists* $\widetilde{h}_1 \in N(X)$ *such that* $\widetilde{H}$, $\widetilde{h}_1$ *belong to a 2-dimensional primitive sublattice* $N \subset N(X)$ *such that*

$$\widetilde{H} \cdot N = 2\mathbb{Z} \tag{6.8}$$

(then $\gamma = 2$, $\gamma_a = \gamma_b = 1$ *and* $\gamma_2 = 2$*), moreover, for one of* $\epsilon = \pm 1$ *the element* $\widetilde{h}_1$ *belongs to the a-series or to the b-series described below:*

$\widetilde{h}_1$ *belongs to the a-series if*

$$\widetilde{h}_1^2 = \epsilon 2b_1 c \text{ and } \widetilde{H} \cdot \widetilde{h}_1 \equiv 0 \mod 2b_1 c, \tag{6.9}$$

$$\widetilde{H} \cdot \widetilde{h}_1 \not\equiv 0 \mod 2b_1 c l_1, \ \widetilde{h}_1/l_2 \notin N(X) \tag{6.10}$$

for any prime l_1 *such that* $l_1^2 | a_1$, *and any prime* l_2 *such that* $l_2^2 | b_1$.

$\widetilde{h}_1$ *belongs to the b-series if*

$$\widetilde{h}_1^2 = \epsilon 2a_1 c \text{ and } \widetilde{H} \cdot \widetilde{h}_1 \equiv 0 \mod 2a_1 c, \tag{6.11}$$

$$\widetilde{H} \cdot \widetilde{h}_1 \not\equiv 0 \mod 2a_1 c l_1, \ \widetilde{h}_1/l_2 \notin N(X) \tag{6.12}$$

for any prime l_1 *such that* $l_1^2 | b_1$ *and any prime* l_2 *such that* $l_2^2 | a_1$.

Moreover, one has formulae (4.23) and (4.24) in terms of X *for the canonical primitive nef element* $\widetilde{h}$ *of* Y *defined by* $(-a, 0, b) \mod \mathbb{Z}v$.

These conditions are necessary (for odd c, a_1 *and* b_1*) to have* $Y \cong X$ *and* $\widetilde{H} \cdot N(X) = 2\mathbb{Z}$ *if* $\rho(X) \leq 2$ *and* X *is a general K3 surface with its Picard lattice, i. e. the automorphism group of the transcendental periods* $(T(X), H^{2,0}(X))$ *is* ± 1.

In [**MN2**] the primitive isotropic Mukai vector (c, H, c) where $H^2 = 2c^2$ had been considered. Then $a = b = 1$, $d = 1$, $a_1 = b_1 = 1$ and $\gamma | 2$. The case $\gamma = 1$ had been described in [**MN2**]. Theorem 6.2 describes the remaining case $\gamma = 2$ and then c is odd which was not considered in [**MN2**].

Now assume that $\gamma = 2$ and $2 | a_1$. Then $\gamma_2 = 1$, $\gamma_a = 2$ and $\gamma_b = 1$. The singular condition (AS) gives then (6.18) and (6.19) below. The singular condition (BS) gives (6.22) and (6.23) below. Thus, Theorem 4.4 implies the following.

THEOREM 6.3. *Let* X *be a K3 surface and* H *a polarization of* X *of degree* $H^2 = 2rs$ *where* $r, s \in \mathbb{N}$. *Assume that the Mukai vector* (r, H, s) *is primitive. Let* Y *be the moduli space of sheaves on* X *with the isotropic Mukai vector* $v = (r, H, s)$. *Let* $\widetilde{H} = H/d$ *be the corresponding primitive polarization. Assume that*

$$g.c.d.(2, c) = 1 \tag{6.13}$$

and

$$a_1 \equiv 0 \mod 2, \ b_1 \equiv 1 \mod 2. \tag{6.14}$$

We have $Y \cong X$ *if there exists* $\widetilde{h}_1 \in N(X)$ *such that* $\widetilde{H}$, $\widetilde{h}_1$ *belong to a 2-dimensional primitive sublattice* $N \subset N(X)$ *such that*

$$\widetilde{H} \cdot N = 2\mathbb{Z} \tag{6.15}$$

(then $\gamma = 2$, $\gamma_a = 2, \gamma_b = 1$ *and* $\gamma_2 = 1$*), moreover, for one of* $\epsilon = \pm 1$ *the element* $\widetilde{h}_1$ *belongs to the a-series or to the b-series described below:*

$\widetilde{h}_1$ *belongs to the a-series if*

$$\widetilde{h}_1^2 = \epsilon 2b_1 c \text{ and } \widetilde{H} \cdot \widetilde{h}_1 \equiv 0 \mod 2b_1 c, \tag{6.16}$$

$$\widetilde{H} \cdot \widetilde{h}_1 \not\equiv 0 \mod 2b_1 c l_1, \ \widetilde{h}_1/l_2 \notin N(X) \tag{6.17}$$

for any prime l_1 such that $l_1^2|a_1$ and $g.c.d(l_1,2) = 1$, and any prime l_2 such that $l_2^2|b_1$; moreover (singular conditions),

$$\widetilde{H} \cdot \widetilde{h}_1 \not\equiv 0 \quad \mathrm{mod}\ 4b_1 c \tag{6.18}$$

and

$$\delta \not\equiv 2\mu^2 \quad \mathrm{mod}\ 4a_1 \ \text{if}\ 4|a_1. \tag{6.19}$$

$\widetilde{h}_1$ *belongs to the b-series if*

$$\widetilde{h}_1^2 = \epsilon 2a_1 c \ \text{and}\ \widetilde{H} \cdot \widetilde{h}_1 \equiv 0 \quad \mathrm{mod}\ a_1 c, \tag{6.20}$$

$$\widetilde{H} \cdot \widetilde{h}_1 \not\equiv 0 \quad \mathrm{mod}\ a_1 c l_1, \ \widetilde{h}_1/l_2 \notin N(X) \tag{6.21}$$

for any prime l_1 such that $l_1^2|b_1$ and any prime l_2 such that $l_2^2|a_1$ and $g.c.d(l_2,2) = 1$; moreover (singular conditions),

$$\widetilde{H} \cdot \widetilde{h}_1 \equiv 0 \quad \mathrm{mod}\ 2a_1 c \tag{6.22}$$

and

$$\widetilde{h}_1/2 \notin N(X). \tag{6.23}$$

Moreover, one has formulae (4.23) and (4.24) in terms of X for the canonical primitive nef element $\widetilde{h}$ of Y defined by $(-a,0,b) \mod \mathbb{Z}v$.

These conditions are necessary (for odd c, even a_1 and odd b_1) to have $Y \cong X$ and $\widetilde{H} \cdot N(X) = 2\mathbb{Z}$ if $\rho(X) \le 2$ and X is a general K3 surface with its Picard lattice, i. e. the automorphism group of the transcendental periods $(T(X), H^{2,0}(X))$ is ± 1.

Changing a and b places, we get from Theorem 4.4 the remaining case.

THEOREM 6.4. *Let X be a K3 surface and H a polarization of X of degree $H^2 = 2rs$ where $r, s \in \mathbb{N}$. Assume that the Mukai vector (r, H, s) is primitive. Let Y be the moduli space of sheaves on X with the isotropic Mukai vector $v = (r, H, s)$. Let $\widetilde{H} = H/d$ be the corresponding primitive polarization. Assume that*

$$g.c.d(2, c) = 1 \tag{6.24}$$

and

$$a_1 \equiv 1 \quad \mathrm{mod}\ 2, \ b_1 \equiv 0 \quad \mathrm{mod}\ 2. \tag{6.25}$$

We have $Y \cong X$ if there exists $\widetilde{h}_1 \in N(X)$ such that $\widetilde{H}, \widetilde{h}_1$ belong to a 2-dimensional primitive sublattice $N \subset N(X)$ such that

$$\widetilde{H} \cdot N = 2\mathbb{Z} \tag{6.26}$$

(then $\gamma = 2$, $\gamma_a = 1, \gamma_b = 2$ and $\gamma_2 = 1$), moreover, for one of $\epsilon = \pm 1$ the element $\widetilde{h}_1$ belongs to the a-series or to the b-series described below:

$\widetilde{h}_1$ *belongs to the a-series if*

$$\widetilde{h}_1^2 = \epsilon 2b_1 c \ \text{and}\ \widetilde{H} \cdot \widetilde{h}_1 \equiv 0 \quad \mathrm{mod}\ b_1 c, \tag{6.27}$$

$$\widetilde{H} \cdot \widetilde{h}_1 \not\equiv 0 \quad \mathrm{mod}\ b_1 c l_1, \ \widetilde{h}_1/l_2 \notin N(X) \tag{6.28}$$

for any prime l_1 such that $l_1^2|a_1$ and any prime l_2 such that $l_2^2|b_1$ and $g.c.d(l_2,2) = 1$; moreover (singular conditions),

$$\widetilde{H} \cdot \widetilde{h}_1 \equiv 0 \quad \mathrm{mod}\ 2b_1 c \tag{6.29}$$

and

$$\widetilde{h}_1/2 \notin N(X). \tag{6.30}$$

$\widetilde{h}_1$ *belongs to the b-series if*

$$\widetilde{h}_1^2 = \epsilon 2a_1 c \ \text{and} \ \widetilde{H} \cdot \widetilde{h}_1 \equiv 0 \quad \text{mod } 2a_1 c, \tag{6.31}$$

$$\widetilde{H} \cdot \widetilde{h}_1 \not\equiv 0 \quad \text{mod } 2a_1 c l_1, \ \widetilde{h}_1/l_2 \notin N(X) \tag{6.32}$$

for any prime l_1 such that $l_1^2|b_1$ and $g.c.d(l_1, 2) = 1$, and any prime l_2 such that $l_2^2|a_1$; moreover (singular conditions),

$$\widetilde{H} \cdot \widetilde{h}_1 \not\equiv 0 \quad \text{mod } 4a_1 c \tag{6.33}$$

and

$$\delta \not\equiv 2\mu^2 \quad \text{mod } 4b_1 \ \text{if} \ 4|b_1. \tag{6.34}$$

Moreover, one has formulae (4.23) and (4.24) in terms of X for the canonical primitive nef element $\widetilde{h}$ of Y defined by $(-a, 0, b)$ mod $\mathbb{Z}v$.

These conditions are necessary (for odd c, odd a_1 and even b_1) to have $Y \cong X$ and $\widetilde{H} \cdot N(X) = 2\mathbb{Z}$ if $\rho(X) \leq 2$ and X is a general K3 surface with its Picard lattice, i. e. the automorphism group of the transcendental periods $(T(X), H^{2,0}(X))$ is ± 1.

Theorems 6.2 — 6.4 cover all types of a primitive isotropic Mukai vector when it is in principle possible to have $Y \cong X$ and $\gamma = 2$.

Using results of Sect. 4, one can write down similar very concrete and effective results for any γ.

References

[A]   Abe T., *A remark on the 2-dimensional moduli spaces of vector bundles on K3 surfaces*, Math. Res. Lett. **7** (2002), no. 4, 463–470.

[C]   Cossec F.R., *Reye Congruences*, Trans. Amer. Math. Soc. **280** (1983), no. 2, 737–751.

[J]   James D.G., *On Witt's theorem for unimodular quadratic forms*, Pacific J. Math. **26** (1968), 303–316.

[MN1]   Madonna C., Nikulin V.V., *On a classical correspondence between K3 surfaces*, Proc. Steklov Math. Inst. **241** (2003), 120 – 153; (see also math.AG/0206158).

[MN2]   Madonna C., Nikulin V.V., *On a classical correspondence between K3 surfaces II*, Clay Mathematics Proceedings, Vol. 3 (Strings and Geometry), Douglas M. , Gauntlett J., Gross M. editors, 2004, pp. 285—300; (see also math.AG/0304415).

[Mu1]   Mukai Sh., *Symplectic structure of the moduli space of sheaves on an Abelian or K3 surface*, Invent. math. **77** (1984), 101-116.

[Mu2]   Mukai Sh., *On the moduli space of bundles on K3 surfaces*, Vector bundles on algebraic varieties (Bombay, 1984), Tata Inst. Fund. Res. Studies in Math. no. 11, Bombay, 1987, pp. 341–413.

[Mu3]   Mukai Sh., *Duality of polarized K3 surfaces*, Hulek K. (ed.) New trends in algebraic geometry. Selected papers presented at the Euro conference, Warwick, UK, July 1996, Cambridge University Press. London Math. Soc. Lect. Notes Ser. 264, Cambridge, 1999, pp. 311–326.

[Mu4]   Mukai Sh., *Non abelian Brill–Noether theory and Fano 3-folds,*, alg-geom 9704015.

[Mu5]   Mukai Sh., *Vector bundles on a K3 surface*, Proc. ICM 2002 in Beijing, vol. 3, pp. 495–502.

[N1]   Nikulin V.V., *Finite automorphism groups of Kählerian surfaces of type K3*, Trudy Mosk. Matem. Ob-va, **38** (1979), 75–137 (Russian); English transl. in Trans. Moscow Math. Soc. **38** (1980), no. 2, 71–135.

[N2]   Nikulin V.V., *Integral symmetric bilinear forms and some of their geometric applications*, Izv. Akad. Nauk SSSR Ser. Mat. **43** (1979), no. 1, 111–177 (Russian); English transl. in Math. USSR Izv. **14** (1980).

[N3]   Nikulin V.V., *On correspondences between K3 surfaces*, Izv. Akad. Nauk SSSR Ser. Mat. **51** (1987), no. 2, 402–411 (Russian); English transl. in Math. USSR Izv. **30** (1988), no. 2 .

[N4]   Nikulin V.V., *On correspondences of a K3 surface with itself I*, Proc. Steklov Math. Institute **246** (2004), 217–239; (see also math.AG/0307355).

[PS]   I.I. Pjatetckiĭ-Šapiro and I.R. Šafarevich, *A Torelli theorem for algebraic surfaces of type K3*, Izv. Akad. Nauk SSSR Ser. Mat. **35** (1971), no. 3, 530–572 (Russian); English transl. in Math. USSR Izv. **5** (1971), no. 3, 547–588.

[S-D] Saint-Donat B., *Projective models of K-3 surfaces*, Amer. J. of Mathem. **96** (1974), no. 4, 602–639.

[Sh]   Shafarevich I.R. (ed.), *Algebraic surfaces*, Trudy Matem. Inst. Steklov, T. 75, 1965 (Russian); English transl. in Proc. Stekov Inst. Math. **75** (1965).

[T1]   Tyurin A.N., *Special 0-cycles on a polarized K3 surface*, Izv. Akad. Nauk SSSR Ser. Mat. **51** (1987), no. 1, 131 – 151 (Russian); English transl. in Math. USSR Izv. **30** (1988), no. 1, 123–143.

[T2]   Tyurin A.N., *Cycles, curves and vector bundles on algebraic surfaces*, Duke Math. J. **54** (1987), no. 1, 1 – 26.

[T3]   Tyurin A.N., *Symplectic structures on the varieties of moduli of vector bundles on algebraic surfaces with $p_g > 0$.*, Izv. Akad. Nauk SSSR Ser. Mat. **52** (1988), no. 4, 139 – 178; English transl. in Math. USSR Izv. **33** (1989), no. 1, 139–177.

[V]   Verra A., *The étale double covering of an Enriques' surface*, Rend. Sem. Mat. Univers. Politecn. Torino **41** (1983), no. 3, 131–167.

DEPTM. OF PURE MATHEM. THE UNIVERSITY OF LIVERPOOL, LIVERPOOL L69 3BX, UK;
STEKLOV MATHEMATICAL INSTITUTE, UL. GUBKINA 8, MOSCOW 117966, GSP-1, RUSSIA
E-mail address: vnikulin@liv.ac.uk   vvnikulin@list.ru

Contemporary Mathematics
Volume **422**, 2007

AUTOMORPHISMS OF HYPERKÄHLER MANIFOLDS IN THE VIEW OF TOPOLOGICAL ENTROPY

KEIJI OGUISO

Dedicated to Professor Igor Dolgachev on the occasion of his sixtieth birthday

ABSTRACT. First we show that any group of automorphisms of null-entropy of a projective hyperkähler manifold M is almost abelian of rank at most $\rho(M) - 2$. We then characterize automorphisms of a K3 surface with null-entropy and those with positive entropy in algebro-geometric terms. We also give an example of a group of automorphisms which is not almost abelian in each dimension.

1. INTRODUCTION - BACKGROUND AND MAIN RESULTS

The aim of this note is to study groups of automorphisms of a hyperkähler manifold from two points of view: topological entropy and how close to (or how far from) being abelian groups. Our main results are Theorems (1.3) and (2.1).

1. Let M be a compact Kähler manifold. We denote the biholomorphic automorphism group of M by $\mathrm{Aut}\,(M)$. By the fundamental work of Yomdin, Gromov and Friedland, the *topological entropy* $e(g)$ of an automorphism $g \in \mathrm{Aut}\,(M)$ can be computed in three different ways: topological, differential-geometrical, and cohomological (see [Yo], [Gr], [Fr], also [DS2]). In this note, we employ the cohomological one as its definition:

$$e(g) \; := \log \delta(g) \; .$$

Here $\delta(g)$ is the spectral radius of the action of g on the cohomology ring $H^*(M, \mathbf{C})$, i.e. the maximum of the absolute values of eigenvalues of the $\mathbf{C}$-linear extension of $g^*|H^*(M, \mathbf{Z})$. One has $e(g) \geq 0$, and $e(g) = 0$ iff the eigenvalues of g^* are on the unit circle $S^1 := \{\, z \in \mathbf{C}\,|\,|z| = 1\,\}$. Furthermore, by Dinh and Sibony [DS1], $e(g) > 0$ iff some eigenvalues of $g^*|H^{1,1}(M)$ are *outside* the unit circle S^1. A subgroup G of $\mathrm{Aut}\,(M)$ is said to be of *null-entropy* (resp. of *positive entropy*) if $e(g) = 0$ for $\forall g \in G$ (resp. $e(g) > 0$ for $\exists g \in G$).

2. Next we recall a few facts about hyperkähler manifolds. K3 surfaces are nothing but 2-dimensinal hyperkähler manifolds. All what we need is reviewed in [Og2, Section 2]:

Definition 1.1. *A hyperkähler manifold* is a compact complex simply-connected Kähler manifold M admitting an everywhere non-degenerate global holomorphic 2-form ω_M such that $H^0(M, \Omega_M^2) = \mathbf{C}\omega_M$.

2000 *Mathematics Subject Classification.* 14J50, 14J40, 14J28, 37B40.

Let M be a hyperkähler manifold. Then the second cohomology group $H^2(M, \mathbf{Z})$ admits a natural $\mathbf{Z}$-valued symmetric bilinear form of signature $(3, 0, b_2(M) - 3)$, called Beauville-Bogomolov-Fujiki's form (BF-form for short). BF-form is exactly the cup-product when M is a K3 surface. The signature of the Néron-Severi group $NS(M)$ w.r.t. BF-form is either $(1, 0, \rho(M) - 1)$, $(0, 1, \rho(M) - 1)$, or $(0, 0, \rho(M))$. Here $\rho(M)$ is the Picard number of M. We call these three cases *hyperbolic*, *parabolic* and *elliptic* respectively. Due to Huybrechts [Hu], M is projective iff $NS(M)$ is hyperbolic. Note also that, in dimension 2, $NS(M)$ is hyperbolic, parabolic, elliptic iff the algebraic dimension $a(M)$ is 2, 1, 0, respectively (see eg. [BPV]).

In our previous note, we have shown the following:

Theorem 1.2. [Og2] *The bimeromorphic automorphism group* $\mathrm{Bir}\,(M)$ *of a non-projective hyperkähler manifold M is an almost abelian group of finite rank. More precisely,* $\mathrm{Bir}\,(M)$ *is an almost abelian group of rank at most 1 (resp. $\rho(M) - 1$) when $NS(M)$ is elliptic (resp. parabolic).*

We call a group G *almost abelian* (resp. *almost abelian of finite rank r*) if there are a normal subgroup $G^{(0)}$ of G of finite index, a finite group K and an abelian group A (resp. $A = \mathbf{Z}^r$) which fit in the exact sequence

$$1 \longrightarrow K \longrightarrow G^{(0)} \longrightarrow A \longrightarrow 0 \ .$$

The rank r is well-defined, and invariant under replacing G by a subgroup H of finite index and by a quotient group Q of G by a finite normal subgroup (cf. [Og2, Section 9]). It is then clear that a subgroup of an almost abelian group of rank r is an almost abelian group of rank at most r.

3. Our main result is the following:

Theorem 1.3. *Let M be a projective hyperkähler manifold. Let $G < \mathrm{Aut}\,(M)$ (resp. $G < \mathrm{Bir}\,(M)$). Assume that G is of null-entropy (resp. of null-entropy at H^2-level, that is, the eigenvalues of $g^*|H^2(M, \mathbf{Z})$ are on the unit circle S^1). Then G is an almost abelian group of rank at most $\rho(M) - 2$. Moreover, this estimate is optimal in* $\dim M = 2$.

The key ingredient is Theorem (2.1), a result of linear algebra, whose source has been back to an important observation of Burnside [Bu1], and Lie's Theorem (cf. [Hm]). We shall prove Theorem (1.3) in Section 3.

4. Next, as an application of Theorem (1.3), we shall reproduce the following algebro-geometric characterization of positivity of entropy of automorphisms of a K3 surface. We should notice that this result is essentially known and can be read from works of Cantat [Ca1, 2]:

Theorem 1.4. *Let M be a (not necessarily projective) K3 surface, $G < \mathrm{Aut}\,(M)$, and $g \in \mathrm{Aut}\,(M)$. Then:*

(1) G is of null-entropy iff either G is finite or G makes an elliptic fibration $\varphi : M \longrightarrow \mathbf{P}^1$ stable. In particular, if G is of null-entropy, then G has no Zariski dense orbit; if G makes an elliptic fibration stable, then G is almost abelian of finite rank.

(2) $e(g) > 0$ iff g has a Zariski dense orbit, i.e. there is a point $x \in M$ such that the set $\{g^n(x) | n \in \mathbf{Z}\}$ is Zariski dense in M.

Theorem (1.4) is proved in Sections 3 and 4. This theorem is also motivated by earlier observations of [Sn] and by the following question posed by McMullen:

Question 1.5. [Mc] Does a K3 automorphism g have a dense orbit (in the Euclidean topology) when a K3 surface is projective and $e(g) > 0$?

In the same paper [Mc], he constructed a K3 surface M of $\rho(M) = 0$ having an automorphism g with Siegel disk. For this (M, g), one knows that $e(g) > 0$, g has a Zariski dense orbit but no dense orbit in the Euclidean topology, and that $\mathrm{Aut}\,(M) \simeq \mathbf{Z}$ (See [ibid] and also [Og2]).

5. So far, all groups in consideration are almost abelian. As a sort of counter parts, we shall show the following Theorem in Section 5:

Theorem 1.6. *(1) Let M be a K3 surface admitting two different Jacobian fibrations of positive Mordell-Weil rank. Then $\mathrm{Aut}\,(M)$ is not almost abelian (and hence is of positive entropy). This happens, for instance, for a K3 surface M of maximum Picard number $\rho(M) = 20$.*
(2) In each dimension $2m$, there is a projective hyperkähler manifold M whose $\mathrm{Aut}\,(M)$ is not almost abelian (and hence is of positive entropy).

Theorem (1.6)(1) is a part of refinement of a result [Ca2] (see also [CF] for a generalization in foliated case) and also a slight generalization of a result of Shioda and Inose [SI]: $\mathrm{Aut}\,(M)$ *is an infinite group for a K3 surface M with $\rho(M) = 20$.* In [ibid], this has been shown by finding a Jacobian fibration of positive Mordell-Weil rank.

Acknowledgement. I would like to express my thanks to Professors S. Cantat, D. Huybrechts, Y. Ishii, S. Kawaguchi, Y. Kawamata, J.H. Keum, T. Shioda and De-Qi Zhang for their warm encouragements and valuable suggestions, and to the referee for his very careful reading. The semi-final version has been completed during my stay at KIAS March 2005. I would like to express my thanks to Professors J.M. Hwang and B. Kim for invitation.

2. Group of isometries of null-entropy of a hyperbolic lattice

The main result of this section is Theorem (2.1).

A lattice is a pair $L := (L, b)$ of a free abelian group $L \simeq \mathbf{Z}^r$ and a symmetric bilinear form

$$b : L \times L \longrightarrow \mathbf{Z} \ .$$

A submodule M (resp. an element $v \neq 0$) of L is *primitive* if and only if L/M (resp. $L/\mathbf{Z}v$) is free.

A scalar extension $(L \otimes K, b \otimes id_K)$ of (L, b) by K is written as (L_K, b_K). We often write $b_K(x, y)$, $b_K(x, x)$ $(x, y \in L_K)$ simply as (x, y), (x^2).

The non-negative integer r is called the rank of L. The signature of L is the signature, i.e. the numbers of positive-, zero-, negative-eigenvalues, of a symmetric matrix associated to $b_{\mathbf{R}}$. It is denoted by $\mathrm{sgn}\,L$. The lattice L is called *hyperbolic* (resp. *parabolic, elliptic*) if $\mathrm{sgn}\,L$ is $(1, 0, r-1)$ (resp. $(0, 1, r-1)$, $(0, 0, r)$).

In what follows, L is assumed to be a hyperbolic lattice.

The positive cone $\mathcal{P}(L)$ of L is one of the two connected components of:

$$\mathcal{P}'(L_{\mathbf{R}}) := \{x \in L_{\mathbf{R}} \mid (x,x) > 0\} \ .$$

The boundary (resp. the closure) of $\mathcal{P}(L)$ is denoted by $\partial\mathcal{P}(L)$ (resp. $\overline{\mathcal{P}}(L)$). Obviously, $(x^2) = 0$ if $x \in \partial\mathcal{P}(L)$. Let $x, x' \in \overline{\mathcal{P}}(L) \setminus \{0\}$. Then, by the Schwartz inequality, $(x, x') \geq 0$ and the equality holds exactly when x and x' are proportional boundary points.

We denote the group of isometries of L by:

$$\mathrm{O}(L) := \{g \in \mathrm{Isom}_{\mathrm{group}}(L) \mid (gx, gy) = (x, y)\ \forall x, y \in L\} \ .$$

We have an index 2 subgroup:

$$\mathrm{O}(L)' := \{g \in \mathrm{O}(L) \mid g(\mathcal{P}(L)) = \mathcal{P}(L)\} \ .$$

Let $g \in \mathrm{O}(L)'$. The spectral value $\delta(g)$ of g is the maximum of the absolute values of eigenvalues of $g_{\mathbf{C}}$. By abuse of notation, we call

$$e(g) := \log \delta(g)$$

the *entropy* of g. As we shall see in Proposition (2.2), $e(g) \geq 0$, and $e(g) = 0$ if and only if the eigenvalues of $g_{\mathbf{C}}$ lie on the unit circle $S^1 := \{z \in \mathbf{C} \mid |z| = 1\}$. An element g of $\mathrm{GL}\,(r, \mathbf{C})$ is called *unipotent* if all the eigenvalues are 1.

The aim of this section is to prove the following:

Theorem 2.1. *Let L be a hyperbolic lattice of rank r and G be a subgroup of $\mathrm{O}(L)'$. Assume that G is an infinite group of null-entropy, i.e. $|G| = \infty$ and that $e(g) = 0$ for all $g \in G$. Then:*

(1) There is $v \in \partial\mathcal{P}(L) \setminus \{0\}$ such that $g(v) = v$ for all $g \in G$. Moreover, the ray $\mathbf{R}_{>0}v$ in $\partial\mathcal{P}(L)$ is unique and defined over $\mathbf{Z}$. In other words, one can take unique such v which is primitive in L.

(2) There is a normal subgroup $G^{(0)}$ of G such that $[G : G^{(0)}] < \infty$ and $G^{(0)}$ is isomorphic to a free abelian group of rank at most $r - 2$. In particular, G is almost abelian of rank at most $r - 2$.

We shall prove Theorem (2.1) dividing into several steps. In what follows G is as in Theorem (2.1).

Proposition 2.2. *(1) The eigenvalues of $g \in G$ lie on the unit circle S^1.*
(2) There is a positive integer n such that g^n is unipotent for all $g \in G$.
(3) There is $g \in G$ such that $\mathrm{ord}\,g = \infty$.

Proof. Since $g \in \mathrm{O}(L)$ and L is non-degenerate, we have $\det(g) = \pm 1$. Let α_i $(1 \leq i \leq r)$ be the eigenvalues of $g_{\mathbf{C}}$ (counted with multiplicities). Then

$$1 = |\det g| = \Pi_{i=1}^r |\alpha_i| \ .$$

Thus, $e(g) \geq 0$, and $e(g) = 0$ if and only if the eigenvalues of $g_{\mathbf{C}}$ lie on S^1. This proves (1).

Since g is defined over $\mathbf{Z}$, the eigenvalues of $g_{\mathbf{C}}$ are all algebraic integers. Thus they are all roots of unity by (1) and by the Kronecker Theorem [Ta]. Since the characteristic polynomial of g is of degree r and it is now the product of cyclotomic polynomials, the eigenvalues of g lie on:

$$\cup_{\varphi(d) \leq r}\ \mu_d\ < S^1 \ .$$

Here $\mu_d = \langle \zeta_d \rangle := \{z \in \mathbf{C} \mid z^d = 1\}$ and $\varphi(d) := |\mathrm{Gal}(\mathbf{Q}(\zeta_d)/\mathbf{Q})|$ is the Euler function. There are finitely many d with $\varphi(d) \leq r$. Let n be their product. Then g^n is unipotent for all $g \in G$. This proves (2).

Let us show (3). We have $G < \mathrm{GL}(L_{\mathbf{C}}) \simeq \mathrm{GL}(r, \mathbf{C})$. Suppose to the contrary that G consists of elements of finite order. Then each $g \in G$ is diagonalizable over $L_{\mathbf{C}}$. Let n be as in (2). Then, $g^n = id$ for all $g \in G$, i.e. G has a finite exponent n. However, then $|G| < \infty$ by the following theorem due to Burnside [Bu2] (based on [Bu1]), a contradiction:

Theorem 2.3. *Any subgroup of* $\mathrm{GL}(r, \mathbf{C})$ *of finite exponent is finite, i.e.* $|G| < \infty$.

$\square$

Remark 2.4. After Theorem (2.3), Burnside asked if a group of finite exponent is finite or not (the Burnside problem). It is now known to be false in general even for finitely generated groups. The original Burnside problem has been now properly modified and completely solved by Zelmanov. (See for instance [Za] about Burnside problems.)

Set
$$H := \{g \subset G \mid g \text{ is unipotent}\} \ .$$

Lemma 2.5. H *is a normal subgroup of* G. *Moreover,* H *has an element of infinite order.*

Proof. Let g be as in Proposition (2.2)(3) and n be as in Proposition (2.2)(1). Then g^n is an element of H of infinite order. The only non-trivial part is the closedness of H under the product, that is, if $a, b \in H$ then $ab \in H$. Since $a^l b^m \in G$, this follows from the next slightly more general result together with Proposition (2.2)(1). $\square$

Proposition 2.6. *(1) Let* $z_i \in S^1$ *($1 \leq i \leq r$). Then*
$$\left| \sum_{i=1}^{r} z_i \right| \leq r \ ,$$

and
$$\sum_{i=1}^{r} z_i = r \iff z_i = 1 \ \forall i \ .$$

(2) Let $A, B \in \mathrm{GL}(r, \mathbf{C})$ *such that* A *and* B *are unipotent and such that the eigenvalues of* $A^l B^m$ *lie on* S^1 *for all* $l, m \in \mathbf{Z}_{\geq 0}$. *Then* $A^l B^m$ *is unipotent for all* $l, m \in \mathbf{Z}_{\geq 0}$.

Proof. The statement (1) follows from the triangle inequality. Let us show (2). We may assume that A is of the Jordan form
$$A := J(r_1, 1) \oplus \cdots \oplus J(r_k, 1) \ .$$

We fix m and decompose B^m into blocks as:
$$B^m = (B_{ij}) \ , \ B_{ij} \in \mathrm{M}(r_i, r_j, \mathbf{C}) \ .$$

We have
$$\mathrm{tr}\, A^l B^m = \sum_{i=1}^{k} \mathrm{tr} J(r_i, 1)^l B_{ii} \ .$$

Using an explicit form of $J(r_i, 1)^l$ and calculating its product with B_{ii}, we obtain:

$$\mathrm{tr}\, A^l B^m = b_s l^s + b_{s-1} l^{s-1} + \cdots + b_1 l + \mathrm{tr}\, B^m \ .$$

Here $s = \max \{r_i\} - 1$ and b_t are constant being independent of l.

On the other hand, by the assumption and (1), we have

$$|\mathrm{tr}\, A^l B^m| \leq r \ ,$$

for all l. Thus, by varying l larger and larger, we have

$$b_t = 0 \text{ for } t = s, \, s-1, \, \cdots, 1$$

inductively. This implies

$$\mathrm{tr}\, A^l B^m = \mathrm{tr}\, B^m \ ,$$

for all l. Since B is unipotent, so is B^m. Thus, $\mathrm{tr}\, B^m = r$ and therefore

$$\mathrm{tr}\, A^l B^m = r \ .$$

Since the eigenvalues of $A^l B^m$ lie on S^1, this implies the result. $\qquad\square$

Corollary 2.7. $H < T(r, \mathbf{Q})$ *under suitable basis* $\langle u_i \rangle_{i=1}^r$ *of* $L_{\mathbf{Q}}$. *Here* $T(r, \mathbf{Q})$ *is the subgroup of* $\mathrm{GL}(r, \mathbf{Q})$ *consisting of the upper triangle unipotent matrices.*

Proof. By Lemma (2.5), H is a unipotent subgroup of $\mathrm{GL}(r, \mathbf{Q})$. Thus, the result follows from Lie's Theorem (see for instance [Hm, 17.5 Theorem]). Here is a small remark. Lie's Theorem in [Hm] is formulated and proved over algebraically closed field of characteristic 0. So, precisely speaking, we have that $H < T(r, \mathbf{C})$ under suitable basis $\langle u_i \rangle_{i=1}^r$ of $L_{\mathbf{C}}$. On the other hand, $L_{\mathbf{Q}}$ and all the elements of H are defined over $\mathbf{Q}$. Thus, such basis $\langle u_i \rangle_{i=1}^r$ of $L_{\mathbf{C}}$ are nothing but solutions of a system of linear equations with rational coefficients in the range $\det (u_i) \neq 0$. Thus, exisitence of such basis over $\mathbf{C}$ implies the exisitence of the desired basis over $\mathbf{Q}$ as claimed. $\qquad\square$

Let G and H be as above. So far, we did not yet use the fact that G and H are subgroups of $\mathrm{O}(L)$. From now, we shall use this fact together with Corollary (2.7).

Lemma 2.8. *(1) Let* N *be a subgroup of* $\mathrm{O}(L)$. *Assume that there is an element* x *of* L *such that* $(x^2) > 0$ *and* $a(x) = x$ *for all* $a \in N$. *Then* $|N| < \infty$. *In particular, there is no such* x *for* H.

(2) There is a unique ray $\mathbf{R}_{>0} v (\neq 0)$ *in* $\partial \mathcal{P}(L)$ *such that* $a(v) = v$ *for all* $a \in H$. *Moreover, the ray* $\mathbf{R}_{>0} v$ *is defined over* $\mathbf{Z}$, *i.e. one can take unique such* v *which is primitive in* L.

(3) Let v *be as in (2). Then* $b(v) = v$ *for all* $b \in G$.

Proof. Let us first show (1). By assumption, N can be naturally embedded into $\mathrm{O}(x_L^\perp)$. Here $x_L^\perp$ is the orthogonal complement of x in L. Since L is hyperbolic and $(x^2) > 0$, the lattice $x_L^\perp$ is of negative definite. Thus $\mathrm{O}(x_L^\perp)$ is finite. This shows (1).

Let us show (2). By (1) and by $|H| = \infty$, we have $r \geq 2$. First, we find $v \in \partial \mathcal{P}(L) \cap L \setminus \{0\}$ such that $a(v) = v$ for all $a \in H$ by the induction on $r \geq 2$.

By Corollary (2.7), there is $u \in L \setminus \{0\}$ such that $a(u) = u$ for all $a \in H$. H is then naturally embedded into $\mathrm{O}(u_L^\perp)$ if $(u^2) \neq 0$.

By (1), we have $(u^2) \leq 0$. If $(u^2) = 0$, then we are done. If $(u^2) < 0$, then $u_L^\perp$ is of signature $(1, 0, r - 2)$. If in addition $r = 2$, then $u_L^\perp$ is of positive definite and $\mathrm{O}(u_L^\perp)$ is finite, a contradiction. Hence $(u^2) = 0$ when $r = 2$ and we are done when $r = 2$. If $r > 2$, then by the induction, we can find a desired v in $u_L^\perp$ and we are done.

If necessarily, by replacing v by $\pm v/m$, we have that $\mathbf{R}_{>0}v \subset \partial \mathcal{P}(L)$ and $v \in L$ is primitive as well.

Next, we shall show the uniqueness of $\mathbf{R}_{>0}v$. Suppose to the contrary that there is a ray $(0 \neq)\mathbf{R}_{>0}u \subset \partial \mathcal{P}(L)$ s.t. $a(u) = u$ for all $a \in H$ and $\mathbf{R}_{>0}u \neq \mathbf{R}_{>0}v$. Then the common eigenspace of eigenvalue 1

$$V := \cap_{h \in H} V(h, 1)$$

is a linear subspace of $L_{\mathbf{C}}$ of $\dim V \geq 2$. Since each $V(h, 1)$ is defined over $\mathbf{Q}$, so is V. Thus, we may find such u in L. So, from the first, we may assume that $u \in L$. We have $((v + u)^2) > 0$ by the Schwartz inequality. We have also $a(v + u) = v + u$ for all $a \in H$. However, we would then have $|H| < \infty$ by (1), a contradiction. Now the proof of (2) is completed.

Finally, we shall show (3). Let $b \in G$. Put $b(v) = u$. Let $a \in H$. Since H is a normal subgroup of G, there is $a' \in H$ such that $ab = ba'$. Then

$$a(u) = a(b(v)) = b(a'(v)) = b(v) = u .$$

Verying a in H and using (2), we find that $u \in \mathbf{R}_{>0}v$, i.e. $b(v) = \alpha v$ for some $\alpha > 0$. Since the eigenvalues of b are on S^1, we have $\alpha = 1$.

$\square$

Proposition 2.9. *Let L be a hyperbolic lattice of rank r and N be a subgroup of $\mathrm{O}(L)$. Assume that there is a primitive element $v \in \partial \mathcal{P}(L) \cap L \setminus \{0\}$ such that $h(v) = v$ for all $h \in N$. Let*

$$\overline{L} := v_L^\perp / \mathbf{Z}v .$$

Then $\overline{L}$ is elliptic, of rank $r - 2$, and the isometry N on L naturally descends to the isometry of $\overline{L}$, say $h \mapsto \overline{h}$. Set

$$N^0 := \mathrm{Ker}\,(N \longrightarrow \mathrm{O}(\overline{L}) \times \{\pm 1\} \; ; \; h \mapsto (\overline{h}, \det h)) .$$

Then N^0 is of finite index in N and N^0 is a free abelian group of rank at most $r - 2$. Moreover N is of null-entropy.

Proof. The first part of the proposition is clear. We shall show the last two assertions. Since $\overline{L}$ is elliptic, the group $\mathrm{O}(\overline{L})$ is finite. Hence $[N : N^0] \leq 2 \cdot |\mathrm{O}(\overline{L})| < \infty$. Choose an integral basis $\langle \overline{u}_i \rangle_{i=1}^{r-2}$ of $\overline{L}$. Let $u_i \in v_L^\perp$ be a lift of $\overline{u}_i$. Then

$$\langle v, u_i \rangle_{i=1}^{r-1}$$

forms an integral basis of $v_L^\perp$. Since $v_L^\perp$ is primitive in L, there is an element $w \in L$ such that

$$\langle v \, , \, u_i \; (1 \leq i \leq r - 2) \, , \, w \rangle$$

forms an integral basis of L.

Let $h \in N^0$. Using $\overline{h}(\overline{u}_i) = \overline{u}_i$, we calculate

$$h(v) = v \, , \, h(u_i) = u_i + \alpha_i(h)v \, ,$$

where $\alpha_i(h)$ $(1 \le i \le r-2)$ are integers uniquely determined by h . Since $\det h = 1$, it follows that $h(w)$ is of the form:

$$h(w) = w + \beta(h)v + \sum_{i=1}^{r-2} \gamma_i(h)u_i \ ,$$

where $\beta(h)$ and $\gamma_i(h)$ are also integers uniquely determined by h.

This already shows that N^0 is unipotent. Then N is of null-entropy, because $[N : N^0] < \infty$, as well.

Let us show that N^0 is an abelian group of rank at most $r - 2$. Varying h in N^0, we can define the map φ by:

$$\varphi : N^0 \longrightarrow \mathbf{Z}^{r-2} \ ; \ h \mapsto (\alpha_i(h))_{i=1}^{r-2} \ .$$

Now the next claim completes the proof of Proposition (2.9). $\square$

Claim 2.10. φ is an injective group homomorphism.

Proof. Let $h, h' \in N^0$. Then by the formula above, we calculate:

$$h'h(u_i) = h'(u_i + \alpha_i(h)v) = h'(u_i) + \alpha_i(h)v = u_i + (\alpha_i(h') + \alpha_i(h))v \ .$$

Thus $\alpha_i(h'h) = \alpha_i(h') + \alpha_i(h)$ and φ is a group homomorphism.

Let us show that φ is injective. Let $h \in \operatorname{Ker}\varphi$. Then,

$$h(v) = v \ , \ h(u_i) = u_i \ , \ h(w) = w + \beta(h)v + \sum_{i=1}^{r-2} \gamma_i(h)u_i \ .$$

It suffices to show that $h(w) = w$. Using $(v, u_i) = 0$ and $(h(x), h(y)) = (x, y)$, we calculate

$$(w, u_i) = (h(w), h(u_i)) = (w, u_i) + \sum_{j=1}^{r-2} \gamma_j(h)(u_j, u_i) \ ,$$

that is,

$$A(\gamma_j(h))_{j=1}^{r-2} = (0)_{j=1}^{r-2} \ .$$

Here $A := ((u_i, u_j)) \in \mathrm{M}(r-2, \mathbf{Z})$. Since $(u_i, u_j) = (\overline{u}_i, \overline{u}_j)$ and $\overline{L}$ is elliptic, we have $\det A \neq 0$. Thus, $(\gamma_j(h))_{j=1}^{r-2} = (0)_{j=1}^{r-2}$, and therefore

$$h(w) = w + \beta(h)v \ .$$

By using $(v^2) = 0$, we calculate

$$(w^2) = (h(w)^2) = (w^2) + 2\beta(h)(w, v) \ .$$

Thus $\beta(h)(w, v) = 0$. Since $(v, v) = (u_i, v) = 0$ and L is non-degenerate, we have $(w, v) \neq 0$. Hence $\beta(h) = 0$, and therefore $h(w) = w$. $\square$

Now Theorem (2.1) follows from Proposition (2.9) applied for $N = G$ and $v \in \partial\mathcal{P}(L) \cap L \setminus \{0\}$ in Lemma (2.8).

3. GROUPS OF AUTMORPHISMS OF NULL-ENTROPY

In this section, we prove Theorem (1.3) and Theorem (1.4)(1).

Let us show Theorem (1.3). Let M be a projective hyperkähler manifold and $G < \mathrm{Aut}\,(M)$ (resp. $G < \mathrm{Bir}\,(M)$) be a subgroup of null-entropy (resp. of null-entropy at H^2-level). Set $H := \mathrm{Im}(r_{NS} : G \longrightarrow \mathrm{O}(NS(M)))$. Since M is projective, $|\mathrm{Ker}\,r_{NS}| < \infty$ by [Og2, Corollary 2.7]. Thus G is almost abelian of rank, say s, iff so is H. However, H is almost abelian of rank $\leq \rho(M) - 2$ by Theorem (2.1).

Next we shall show the optimality of the estimate. There is a Jacobian K3 surface $\varphi : M \longrightarrow \mathbf{P}^1$ s.t. $\rho(M) = 20$ and the Mordell-Weil rank of φ is 18 (See e.g. [Co], [Ny], [Og1]). As we shall show in the next Section, the action of a Mordell-Weil group is of null-entropy. This completes the proof of Theorem (1.3).

Remark 3.1. Let M be a non-projective hyperkähler manifold and $G < \mathrm{Bir}\,(M)$. Then, by [Og2] (the proof of Theorem (1.2) there), we know:

(1) *If $NS(M)$ is elliptic, then G is of null-entropy at H^2-level iff $|G| < \infty$.* Indeed, if $g \in G$ is of null-entropy at H^2-level, then $g^*|T(M) = id$ by [Og2, Theorem 2.4]. Moreover, $|\mathrm{Im}(r_{NS} : G \longrightarrow \mathrm{O}(NS(M))| < \infty$. Thus, $\mathrm{Im}(r : G \longrightarrow \mathrm{O}(H^2(M, \mathbf{Z})))$ is finite. Then $|G| < \infty$ by [Og2, Theorem 2.3]. The other direction is clear.

(2) *If $NS(M)$ is parabolic, then G is always of null-entropy at H^2-level.* Indeed, by [Og2, Corollary 2.7], it suffices to show that $N := \mathrm{Im}\,(r_{NS} : G \longrightarrow \mathrm{O}(NS(M)))$ is of null-entropy (on $NS(M)$). However, in the proof of [Og2, Proposition (5.1)], we find a finite index normal unipotent subgroup $N^{(0)}$ of N. Thus, N is of null-entropy.

Let us show Theorem (1.4)(1). Let $a(M)$ be the algebraic dimension of a K3 surface M. We argue by dividing into three cases where $NS(M)$ is elliptic, parabolic, and hyperbolic.

If $NS(M)$ is elliptic, then the result follows from Remark (3.1)(1) above.

If $NS(M)$ is parabolic, then a subgroup G of $\mathrm{Aut}\,(M)$ is always of null-entropy by Remark (3.1)(2) above. On the other hand, when $NS(M)$ is parabolic, the algebraic dimension of M is 1 (by the classification theory of surfaces). The algebraic reduction map gives rise to an elliptic fibration $a : M \longrightarrow \mathbf{P}^1$ of M. This fibration is stable under $\mathrm{Aut}\,(M)$, and therefore so is under G. This completes the proof when $NS(M)$ is parabolic.

Let us consider the case where $NS(M)$ is hyperbolic. In this case M is projective.

First we show "only if" part. As before, set $H := \mathrm{Im}\,(r_{NS} : G \longrightarrow \mathrm{O}(NS(M)))$. We may assume that $|G| = \infty$. Then so is H by [Og2, Corollary 2.7]. Applying Theorem (2.1) for H, we find a primitive element $v \in \partial \mathcal{P}(M) \cap NS(M) \setminus \{0\}$ s.t. $h^*(v) = v$ for all $h \in H$. Here the positive cone $\mathcal{P}(M) := \mathcal{P}(NS(M))$ is taken to be the component which contains an ample class of M.

By the Riemann-Roch theorem, v is represented by a non-zero effective divisor, say D, with $h^0(\mathcal{O}_M(D)) \geq 2$. Decompose $|D|$ into the movable part and fixed part:

$$|D| = |E| + B \ .$$

Then $(E^2) \geq 0$ and the class $[E]$ is H-stable. Thus $(E^2) \leq 0$ by Lemma (2.8)(1). Therfore $(E^2) = 0$. Hence $|E|$ is free and defines an elliptic fibration

$$\varphi : M \longrightarrow \mathbf{P}^1$$

on M. This is G-stable. Moreover, φ is the unique elliptic fibration stable under G. Otherwise, there is another class $[C] \in \partial \mathcal{P}(M) \cap NS(M) \setminus \{0\}$ such that $[C] \notin \mathbf{R}_{>0}[E]$ and H-stable, a contradiction to Lemma (2.8)(2).

Next, we shall show "if part". The result is clear if G is finite. So, we may assume that $|G| = \infty$ and the existence of a G-stable elliptic fibration $\varphi : M \longrightarrow \mathbf{P}^1$.

Let C be a fiber of φ. Then $[C] \in NS(M) \cap \partial \mathcal{P}(M) \setminus \{0\}$. $[C]$ is G-invariant as well. Thus, the result follows from Proposition (2.9) and [Og2, Corollary 2.7].

Let us finally show the last two statements in Theorem (1.4)(1). The second one follows from the contra-position of Theorem (1.3). Let us show the first one. The result is clear if $|G| < \infty$. So, we may assume that $|G| = \infty$. Then G makes an elliptic fibration $\varphi : M \longrightarrow \mathbf{P}^1$ stable. Recall that any elliptic fibration on a K3 surface admits at least three singular fibers. (see eg. [Ca2]; This follows from $\chi_{\text{top}}(M) = 24$ and $\rho(M) \geq 20$. See also [VZ] for a more general account.) Thus, $\text{Im}\,(G \longrightarrow \text{Aut}\,(\mathbf{P}^1))$ is finite. Therefore, each orbit lies in finitely many fibers. This completes the proof of Theorem (1.4)(1).

4. Autmorphism of a projective K3 surface of positive entropy

In this section, we shall show Theorem (1.4)(2).

Let M be a K3 surface and $g \in \text{Aut}\,(M)$. If $e(g) = 0$, then $\langle g \rangle$ is of null-entropy. Thus $\langle g \rangle$ has no Zariski dense orbit by Theorem (1.4)(1). This shows "if part" of Theorem (1.4)(2). Let us show "only if part" of Theorem 1.6(2). Assuming $e(g) > 0$, we want to find a point $x \in M$ such that $\{g^n(x)|n \in \mathbf{Z}\}$ is Zariski dense in M. Note that $\text{ord}\,g = \infty$ by $e(g) > 0$.

Let $\mathcal{F} = \cup_{n \in \mathbf{Z} \setminus \{0\}} \mathcal{F}_n$, where

$$\mathcal{F}_n := \{\, y \in M \,|\, g^n(y) = y \,\}\,,$$

and

$$\mathcal{C} := \cup_{C \subset M, C \simeq \mathbf{P}^1} C\,.$$

Since $\text{ord}\,g = \infty$, each $\mathcal{F}_n$ is a proper closed analytic subset of M. Since M is not uniruled, $\mathcal{C}$ is also at most countable union of $\mathbf{P}^1$. Thus, $\mathcal{F} \cup \mathcal{C}$ is a countable union of proper closed analytic subsets of M. Hence

$$M \neq \mathcal{F} \cup \mathcal{C}\,.$$

Choose $x \in M \setminus (\mathcal{F} \cup \mathcal{C})$ and set $\mathcal{O}(x) := \{g^n(x)|n \in \mathbf{Z}\}$.

The next claim will complete the proof.

Claim 4.1. $\mathcal{O}(x)$ is Zariski dense in M.

Proof. Let S be the Zariski closure of $\mathcal{O}(x)$ in M. Suppose to the contrary that S is a proper subset of M. Since $\mathcal{O}(x)$ is an infinite set by $x \notin \mathcal{F}$, the set S is decomposed into non-empty finitely many complete irreducible curves and (possibly empty) finite set of closed points. Let C be a 1-dimensional irreducible component of S. Then, there is N such that $g^{Nk}(C) = C$ for all $k \in \mathbf{Z}$. Note that $g^n(x) \in C$ for some n. (Indeed, otherwise, $\overline{S \setminus C}$ would be a smaller closed subset containing $\mathcal{O}(x)$.) By the choice of x, we have $x \in g^{-n}(C)$ and therefore $C \simeq g^{-n}(C) \not\simeq \mathbf{P}^1$. Hence $(C^2) \geq 0$. If $(C^2) > 0$, then M is projective and g^N, whence g, is of finite order on $NS(M)$ by Proposition (2.8)(1). Then $\text{ord}\,g < \infty$ by [Og2, Corollary 2.7], a contradiction. If $(C^2) = 0$, then C defines an elliptic fibration which is stable under g^N. Then, g^N, whence g, is of null-entropy by Theorem (1.4)(1), a contradiction. Therefore $S = M$. $\square$

5. Non-abelian subgroup of automorphisms of a K3 surface

In this section, we shall prove Theorem (1.6).

Theorem $(1.6)(2)$ follows from Theorem $1.6(1)$ by considering the Hilbert scheme $\mathrm{Hilb}^m(S)$ of 0-dimensional closed subschemes of length m of a K3 surface S. Indeed, by [Be], $\mathrm{Hilb}^m(S)$ is a hyperkähler manifold of dimension $2m$ having natural inclusions $H^2(S, \mathbf{Z}) \subset H^2(\mathrm{Hilb}^m(S), \mathbf{Z})$ and $\mathrm{Aut}(S) \subset \mathrm{Aut}(\mathrm{Hilb}^m(S))$. Thus, $\mathrm{Hilb}^m(S)$ for a K3 surface S in Theorem $(1.6)(1)$ satisfies the requirement.

We shall show Theorem $(1.6)(1)$. Let M be a K3 surface and $\varphi_i : M \longrightarrow \mathbf{P}^1$ $(i = 1, 2)$ be two different Jacobian fibrartions of positive Mordell-Weil rank on M. Note that M is projective under this assumption.

Let $\mathrm{MW}(\varphi_i)$ be the Mordell-Weil group of φ_i. Choose $f_i \in \mathrm{MW}(\varphi_i)$ s.t. $\mathrm{ord}\, f_i = \infty$. We may regard f_i as an element of $\mathrm{Aut}(M)$. Let $G := \langle f_1, f_2 \rangle < \mathrm{Aut}(M)$.

First we shall show that $\mathrm{Aut}(M)$ is not almost abelian. Note that any subgroup of an almost abelian group is again almost abelian (by definition). So, it suffices to show that G is not almost abelian. Suppose to the contrary that G is almost abelian. Then, by definition, there are a normal subgroup H of G of finite index, a finite subgroup N of H and an abelian group A which fit in with the exact sequence:

$$1 \longrightarrow N \longrightarrow H \longrightarrow A \longrightarrow 0 .$$

Since $|G/H| < \infty$, there is a positive integer m s.t. $f_1^m, f_2^m \in H$. Set $g_i := f_i^m$. Let $n := |N|$. Since A is abelian, one has $g_1^{-j} g_2 g_1^j g_2^{-1} \in N$ for each $j \in \mathbf{Z}$. Then, we have $n + 1$-elements in N:

$$g_1^{-1} g_2 g_1 g_2^{-1} , \ g_1^{-2} g_2 g_1^2 g_2^{-1} , \ \cdots , \ g_1^{-n} g_2 g_1^n g_2^{-1} , \ g_1^{-n-1} g_2 g_1^{n+1} g_2^{-1} \ \in N .$$

Thus, at least two of them have to coinside. Hence $g_1^k g_2 = g_2 g_1^k$ for some positive integer k. Set $g := g_1^k$ and $h := g_2$. Then $gh = hg$ and $g^*(e_1) = e_1$, $h^*(e_2) = e_2$. Here e_i is the class of a general fiber of φ_i $(i = 1, 2)$. Set $e_3 := h^*(e_1)$. Then e_3 is a (primitive) class of elliptic pencils on M. If $e_3 = e_1$, then e_1 and e_2 are both h-stable. Therefore, $e_1 + e_2$ is also h-stable. However, since $(e_1 + e_2)^2 > 0$, we have then $\mathrm{ord}\, h < \infty$ on $NS(M)$, whence $\mathrm{ord}\, g_2 < \infty$ by [Og2, Corollary 2.7], a contradiction to $\mathrm{ord}\, g_2 = \infty$. Consider next the case $e_3 \neq e_1$, i.e. the case where e_1 and e_3 correspond different elliptic pencils. Using $g^* h^* = h^* g^*$ and $g^*(e_1) = e_1$, one has $g^*(e_3) = e_3$. Then $e_1 + e_3$ is g-stable and $(e_1 + e_3)^2 > 0$, a contradiction for the same reason as above. Thus, G is not almost abelian.

Since G is not almost abelian, G is of positive entropy by the contra-position of Theorem (1.3).

Next we shall show that a K3 surface M of Picard number 20 has at least two different Jacobian fibrations of positive Mordell-Weil rank. Note that M is necessarily projective. In what follows, U stands for an even unimodular hyperbolic lattice of rank 2, and A_*, D_*, E_* stands for the negative definite root lattice whose basis is given by the corresponding Dynkin diagram. Since the transcendental lattice $T(M)$ is positive definite and of rank 2, $T(M)$ can be primitively embedded into $U^{\oplus 2}$, and $T(M)$ can be also uniquely primitively embedded into the K3 lattice $\Lambda_{\mathrm{K3}} := U^3 \oplus E_8(-1)^2$ (See e.g. [Ni]). Note that $H^2(M, \mathbf{Z}) \simeq \Lambda_{\mathrm{K3}}$. Thus the Néron-Severi lattice $NS(M)$ of M is of the form:

$$NS(M) = U \oplus E_8(-1)^{\oplus 2} \oplus N ,$$

where N is a negative definite lattice of rank 2. Put $d := \det N$. If $d = 3, 4$, then $N = A_2$, A_1^2, and if $d \neq 3, 4$, then N is not a root lattice. Recall that the $\mathbf{Q}$-rational hull of the ample cone of a projective K3 surface is the fundamental domain of the group of (-2)-reflections on the rational hull of the positive cone (See e.g. [BPV]). Using this fact and the Riemann-Roch formula, one finds a Jacobian fibration $\varphi_1 : M \longrightarrow \mathbf{P}^1$ whose reducible singular fibers are $II^* + II^*$ if $d \neq 3, 4$, $II^* + II^* + I_3$ or $II^* + II^* + IV$ if $d = 3$ and $II^* + II^* + A + B$ ($A, B \in \{I_2, III\}$) if $d = 4$. Let MW (φ_1) be the Mordell-Weil group of φ_1. Then rank MW $(\varphi_1) > 0$ unless $d = 3$, 4 by Shioda's rank formula (See for instance [Sh]). Note also that this φ_1 is essentially the same Jacobian fibration studied in [SI].

Let us find another Jacobian fibrations of positive Mordell-Weil rank. Join two II^* singular fibers of φ_1 by the 0-section and throw out the components of multiplicity 2 at the edge of two II^*, say C_1 in one II^* and C_2 in the other II^*. In this way, one obtains a divisor of Kodaira's type I_{12}^*, say, D, on M. The pencil $|D|$ gives rise to a Jacobian fibration $\varphi_2 : M \longrightarrow \mathbf{P}^1$ with reducible singular fiber of type I_{12}^* and with sections C_1, C_2. Take C_1 as the 0 and consider C_2 as an element P of MW (φ_2). We have that rank MW $(\varphi_2) > 0$ (regardless of d).

Indeed, otherwise, by using Shioda's rank formula, we see that the remaining reducible singular fibers of φ_2 are either I_3, VI, $I_2 + I_2$, $I_2 + III$, or $III + III$. In each case, the value $\langle P, P \rangle$ of Shioda's height pairing of P is positive, as it is checked by his explicit height pairing formula. This, however, implies ord $P = \infty$, a contradiction to the assumption. (See [Sh] for the definition and basic properties on Shioda's height pairing on the Mordell-Weil group.)

Thus we obtain two Jacobian fibrations φ_1, φ_2 of positive Mordell-Weil rank unless $d = 3$, 4. When $d = 3$, 4, Vinberg [Vi] calculated the full automorphism group Aut (M). It contains a subgroup H isomorphic to the free product $\mathbf{Z}_2 * \mathbf{Z}_2 * \mathbf{Z}_2$ of three cyclic groups of order 2. This H is far from being almost abelian. Thus, by Theorem (1.4)(1), H does not make the Jacobian fibration φ_2 stable. Now, by transforming φ_2 by H, one can find another Jacobian fibration of positive rank also when $d = 3$, 4. This completes the proof of Theorem (1.6)(1).

Remark 5.1. Under the same asumption in Theorem (1.6)(1), a K3 surface M has a dense Aut (M) orbit in the Euclidean topology. This remarkable result has been found by Cantat [Ca2] in a slightly more general situation.

REFERENCES

[BPV] W. Barth, C. Peters, A. Van de Ven, *Compact complex surfaces*, Springer-Verlag (1984).

[Be] A. Beauville, *Variété Kähleriennes dont la premiére classe de chern est nulle*, J. Diff. Deom. **18** (1983) 755–782.

[Bu1] W. Burnside, *On the condition of reducibility of any group of linear substitutions*, Proc.London Math. Soc. **3** (1905) 430–434.

[Bu2] W. Burnside, *On criteria for the finiteness of the order of a group of linear substitutions*, Proc.London Math. Soc. **3** (1905) 435–440.

[Ca1] S. Cantat, *Dynamique des automorphismes des surfaces K3*, Acta Math. **187** (2001) 1–57.

[Ca2] S. Cantat, *Sur la dynamique du groupe d'automorphismes des surfaces K3*, Transform. Groups **6** (2001) 201–214.

[CF] S. Cantat, C. Favre, *Symétries birationnelles des surfaces feuilletées*, J. Reine Angew. Math. **561** (2003) 199–235.

[Co] D. A. Cox, *Mordell-Weil groups of elliptic curves over $\mathbf{C}(t)$ with $p_g = 0$ or 1*, Duke Math. J. **49** (1982) 677 - 689.

[DS1] T.C. Dinh and N. Sibony, *Groupes commutatifs d'automorphismes d'une variete Kahlerienne compacte*, Duke Math. J. **123** (2004) 311–328.

[DS2] T.C. Dinh and N. Sibony, *Une borne supérieure pour l'entropie topologique d'une application rationnelle*, to appear in Ann. of Math.

[Fr] S. Friedland, *Entropy of algebraic maps*, in: Proceedings of the Conference in Honor of Jean-Pierre Kahane (Orsay 1993), J. Fourier Anal. Appl. (1995) 215–228.

[Gr] M. Gromov, *Entropy, homology and semialgebraic geometry*, Astérique, **145-146** (1987) 225–240.

[Hm] J. E. Humphreys, *Linear algebraic groups*, Graduate Texts in Mathematics, **21** Springer-Verlag, New York-Heidelberg, 1975.

[Hu] D. Huybrechts, *Compact hyperkähler manifolds: basic results*, Invent. Math. **135** (1999) 63–113. Erratum: "Compact hyperkähler manifolds: basic results" Invent. Math. **152** (2003) 209–212.

[Mc] C. T. McMullen, *Dynamics on K3 surfaces: Salem numbers and Siegel disks*, J. Reine Angew. Math. **545** (2002) 201–233.

[Ni] V. V. Nikulin, *Integral symmetric bilinear forms and some of their geometric applications*, Math. USSR Izv. **14** (1980) 103-167.

[Ny] K. Nishiyama, *Examples of Jacobian fibrations on some K3 surfaces whose Mordell-Weil lattices have the maximal rank 18*, Comment. Math. Univ. St. Paul. **44** (1995) 219–223.

[Og1] K. Oguiso, *Local families of K3 surfaces and applications*, J. Algebraic Geom. **12** (2003) 405–433.

[Og2] K. Oguiso, *Bimeromorphic automorphism groups of non-projective hyperkähler manifolds - a note inspired by C. T. McMullen.*

[Sn] J. H. Silverman, *Rational points on K3 surfaces: A new canonical height*, Invent. math **105** (1991) 347–373.

[Sh] T. Shioda, *On the Mordell-Weil lattices*, Comment. Math. Univ. St. Paul. **39** (1990) 211–240.

[SI] T. Shioda and H. Inose, *On singular K3 surfaces*, In: Complex analysis and algebraic geometry, Iwanami Shoten (1977) 119–136.

[Ta] T. Takagi, *Algebraic number theory* 2nd edition (1971) Iwamani Shoten (in Japanese).

[Vi] E.B. Vinberg, *The two most algebraic K3 surfaces*, Math. Ann. **265** (1983) 1–21.

[VZ] E. Viehweg and K. Zuo, *On the isotriviality of families of projective manifolds over curves*, J. Algebraic Geom. **10** (2001) 781–799.

[Yo] Y. Yomdin, *Volume growth and entropy*, Israel J. Math. **57** (1987) 285–300.

[Ze] Zelmanov, Efim I. *On the restricted Burnside problem.* Fields Medallists' lectures, 623–632, World Sci. Ser. 20th Century Math., 5, World Sci. Publishing, River Edge, NJ, 1997.

GRADUATE SCHOOL OF MATHEMATICAL SCIENCES, UNIVERSITY OF TOKYO, KOMABA, MEGURO-KU, TOKYO 153-8914, JAPAN, AND KOREA INSTITUTE FOR ADVANCED STUDY, 207-43 CHEONRYANGNI-2DONG, DONGDAEMUN-GU, SEOUL 130-722, KOREA

E-mail address: oguiso@ms.u-tokyo.ac.jp

Contemporary Mathematics
Volume **422**, 2007

$K3$ surfaces with ten cusps

Ichiro Shimada and De-Qi Zhang

Dedicated to Professor Igor V. Dolgachev for his 60th birthday

ABSTRACT. We show that normal $K3$ surfaces with ten cusps exist in and only in characteristic 3. We determine these $K3$ surfaces according to the degrees of the polarizations. Explicit examples are given.

1. Introduction

We work over an algebraically closed field k. A $K3$ surface always means an algebraic $K3$ surface.

An isolated singular point of an algebraic surface is called a *cusp* if it is a rational double point of type A_2 (Artin [**1, 2, 4**]).

In characteristic 0, the number of cusps on a normal $K3$ surface is at most nine. Barth showed in [**5**] that a complex normal $K3$ surface Y has nine cusps as its only singularities if and only if Y is the quotient of an abelian surface by a cyclic group of order 3. This is a generalization of the result of [**14**], in which Nikulin showed that a complex normal $K3$ surface Y has sixteen nodes as its only singularities if and only if Y is the quotient of an abelian surface by the involution. In [**6**], Barth classified normal $K3$ surfaces with nine cusps according to the degrees of the polarizations.

In positive characteristics, however, there exist normal $K3$ surfaces Y such that the singular locus $\operatorname{Sing} Y$ of Y consists of *ten* cusps. The purpose of this paper is to investigate such $K3$ surfaces.

A smooth $K3$ surface X is called *supersingular* (in the sense of Shioda [**25**]) if the Néron-Severi lattice $NS(X)$ of X is of rank 22. Supersingular $K3$ surfaces exist only in positive characteristics. Let X be a supersingular $K3$ surface in characteristic $p > 0$. Artin [**3**] showed that there exists a positive integer $\sigma(X) \leq 10$ such that $\operatorname{disc} NS(X) = -p^{2\sigma(X)}$ holds. This integer $\sigma(X)$ is called the *Artin invariant of* X.

1991 *Mathematics Subject Classification.* Primary 14J28; Secondary 14G50.
Key words and phrases. $K3$ surface, supersingular.

We denote by $U(m)$ the lattice of rank 2 whose intersection matrix is equal to

$$\begin{pmatrix} 0 & m \\ m & 0 \end{pmatrix}.$$

Our main results are Theorems 1.1 and 1.4 - 1.6.

THEOREM 1.1. *Let Y be a normal $K3$ surface such that $\operatorname{Sing} Y$ consists of ten cusps, and $\rho : X \to Y$ the minimal resolution of Y. Let R_ρ be the sublattice of $NS(X)$ generated by the classes of the (-2)-curves that are contracted by ρ. Then the following hold:*

(1) *The characteristic of the ground field k is 3.*

(2) *The orthogonal complement $R_\rho^\perp$ of R_ρ in $NS(X)$ is isomorphic to either $U(1)$ or $U(3)$.*

(3) *If $R_\rho^\perp \cong U(1)$, then $\sigma(X) \leq 5$, while if $R_\rho^\perp \cong U(3)$, then $\sigma(X) \leq 6$.*

Before we state the other main results, we fix the terminology below.

DEFINITION 1.2. Let L be a line bundle on a smooth $K3$ surface X. We say that L is *very ample modulo (-2)-curves* if the following hold:

(i) The complete linear system $|L|$ has no fixed components, and hence has no base points by [18, Corollary 3.2]. In particular, $|L|$ defines a morphism $\Phi_{|L|} : X \to \mathbb{P}^N$, where $N = L^2/2 + 1$.

(ii) The morphism $\Phi_{|L|}$ is birational onto the image $Y_{(X,L)} := \Phi_{|L|}(X)$.

A *polarized $K3$ surface* is a pair (X, L) of a $K3$ surface X and a line bundle L on X that is very ample modulo (-2)-curves. The *degree* of a polarized $K3$ surface (X, L) is defined to be L^2.

DEFINITION 1.3. Let (X, L) be a polarized $K3$ surface. We denote by

$$\rho_L : X \to Y_{(X,L)}$$

the birational morphism induced by $|L|$. By [18, Theorem 6.1], ρ_L is a contraction of an ADE-configuration of (-2)-curves on X. We denote by $R_{(X,L)}$ the sublattice of $NS(X)$ generated by the classes of the (-2)-curves that are contracted by ρ_L. We also denote by $\mathcal{R}_{(X,L)}$ the ADE-type of the configuration of these (-2)-curves.

Note that $\mathcal{R}_{(X,L)} = 10A_2$ is equivalent to saying that $\operatorname{Sing} Y_{(X,L)}$ consists of ten cusps. The degree of (X, L) can be completely determined:

THEOREM 1.4. *The following conditions on a positive integer d are equivalent:*

(i) $d = 2ab$, *where a and b are integers ≥ 3 such that $a \neq b$.*

(ii) *There exists a polarized supersingular $K3$ surface (X, L) of degree d such that $\mathcal{R}_{(X,L)} = 10A_2$ and $R_{(X,L)}^\perp \cong U(1)$.*

(iii) *Every supersingular $K3$ surface X in characteristic 3 with $\sigma(X) \leq 5$ admits a line bundle L such that (X, L) is a polarized $K3$ surface of degree d satisfying $\mathcal{R}_{(X,L)} = 10A_2$ and $R_{(X,L)}^\perp \cong U(1)$.*

THEOREM 1.5. *The following conditions on a positive integer d are equivalent:*

(i) $d \equiv 0 \bmod 6$.

(ii) *There exists a polarized supersingular $K3$ surface (X, L) of degree d such that $\mathcal{R}_{(X,L)} = 10A_2$ and $R_{(X,L)}^\perp \cong U(3)$.*

(iii) *Every supersingular $K3$ surface X in characteristic 3 with $\sigma(X) \leq 6$ admits a line bundle L such that (X, L) is a polarized $K3$ surface of degree d satisfying $\mathcal{R}_{(X,L)} = 10A_2$ and $R_{(X,L)}^{\perp} \cong U(3)$.*

Supersingular $K3$ surfaces with ten cusps can be obtained as purely inseparable triple covers of the smooth quadric surface $Q = \mathbb{P}^1 \times \mathbb{P}^1$. From now on to the end of this paragraph, we assume that k is of characteristic 3. For integers a and b, we denote by $\mathcal{O}_Q(a, b)$ the invertible sheaf $\mathrm{pr}_1^* \mathcal{O}_{\mathbb{P}^1}(a) \otimes \mathrm{pr}_2^* \mathcal{O}_{\mathbb{P}^1}(b)$ of $Q = \mathbb{P}^1 \times \mathbb{P}^1$, and by $L_Q(a, b) \to Q = \mathbb{P}^1 \times \mathbb{P}^1$ the corresponding line bundle. Because we are in characteristic 3, the differential map

$$d : H^0(Q, \mathcal{O}_Q(3, 3)) \to H^0(Q, \Omega_Q^1(3, 3))$$

is well-defined by the isomorphism $L_Q(3, 3) \cong L_Q(1, 1)^{\otimes 3}$. For $G \in H^0(Q, \mathcal{O}_Q(3, 3))$, we denote by $Z(dG)$ the subscheme of Q defined by $dG = 0$. If $\dim Z(dG) = 0$, then

$$\mathrm{length}\, \mathcal{O}_{Z(dG)} = c_2(\Omega_Q^1(3, 3)) = 10$$

holds, where c_2 is the second Chern class. We put

$$\mathcal{U}_{3,3} := \{\, G \in H^0(Q, \mathcal{O}_Q(3, 3)) \mid Z(dG) \text{ is reduced and of dimension } 0 \,\},$$

which is a Zariski open dense subset of $H^0(Q, \mathcal{O}_Q(3, 3))$. For a non-zero $G \in H^0(Q, \mathcal{O}_Q(3, 3))$, we denote by

$$\pi_G : Y_G \to Q = \mathbb{P}^1 \times \mathbb{P}^1$$

the purely inseparable triple cover of Q defined by

$$W^3 = G,$$

where W is a fiber coordinate of the line bundle $L_Q(1, 1)$. It is easy to see that G is contained in $\mathcal{U}_{3,3}$ if and only if Y_G is a normal $K3$ surface such that $\mathrm{Sing}\, Y_G = \pi_G^{-1}(Z(dG))$ consists of ten cusps. In particular, if $G \in \mathcal{U}_{3,3}$, then the minimal resolution X_G of Y_G is a supersingular $K3$ surface with Artin invariant ≤ 6 by Theorem 1.1. Conversely, we have the following:

THEOREM 1.6. *Let X be a supersingular $K3$ surface in characteristic 3 with Artin invariant ≤ 6. Then there exists $G \in \mathcal{U}_{3,3}$ such that X is isomorphic to X_G.*

We put

$$\mathcal{V}_{1,1} := \{\, H^3 \in H^0(Q, \mathcal{O}_Q(3, 3)) \mid H \in H^0(Q, \mathcal{O}_Q(1, 1)) \,\},$$

which is an additive group acting on $\mathcal{U}_{3,3}$ by $G \mapsto G + H^3$ ($G \in \mathcal{U}_{3,3}, H^3 \in \mathcal{V}_{1,1}$). For $G, G' \in \mathcal{U}_{3,3}$, the triple covers Y_G and $Y_{G'}$ are isomorphic over Q if and only if $G = cG' + H^3$ holds for some $c \in k^{\times}$ and $H^3 \in \mathcal{V}_{1,1}$. Hence the space

$$\mathfrak{M} := (PGL(2, k) \times PGL(2, k)) \backslash \mathbb{P}_*(\mathcal{U}_{3,3}/\mathcal{V}_{1,1})$$

is a moduli space of supersingular $K3$ surfaces in characteristic 3 with Artin invariant ≤ 6. We remark that, since $\dim \mathcal{U}_{3,3} = 16$ and $\dim \mathcal{V}_{1,1} = 4$, we have

$$\dim \mathfrak{M} = 16 - 4 - 1 - (3 + 3) = 5,$$

as is predicted from the result of Artin [3]. In particular, the *unique* supersingular $K3$ surface of Artin invariant 1 has the following precise model:

EXAMPLE 1.7. We put

$$G_0 := (x^3 - x)(y^3 - y),$$

where x and y are affine coordinates of the two factors of $Q = \mathbb{P}^1 \times \mathbb{P}^1$. Then $Z(dG_0)$ is equal to

$$\{ (\alpha, \beta) \mid \alpha, \beta \in \mathbb{F}_3 \} \cup \{(\infty, \infty)\}.$$

Therefore $G_0 \in \mathcal{U}_{3,3}$. It can be shown that the Artin invariant of the supersingular $K3$ surface X_{G_0} is 1. See Example 7.8.

Supersingular $K3$ surfaces in characteristic 2 with 21 nodes are investigated in [21, 22, 23]. In particular, it was shown there that every supersingular $K3$ surface in characteristic 2 is birational to a purely inseparable double cover of the projective plane with 21 nodes; that is, every supersingular $K3$ surface in characteristic 2 is obtained as a generic Zariski surface [7].

Quasi-elliptic $K3$ surfaces in characteristic 3 with a section and ten singular fibers of type A_2^* are constructed explicitly in [13]. The Artin invariants of these supersingular $K3$ surfaces are ≤ 5.

A family of smooth quartic surfaces in characteristic 3 containing an ADE-configuration of lines of type $10A_2$ is constructed in [20]. A general member of the family is of Artin invariant 6. See Example 4.3 for details.

This paper is organized as follows. In §2, we review the theory of discriminant forms of lattices due to Nikulin [15]. In §3, we quote from Artin [3], Rudakov-Shafarevich [17], Saint-Donat [18] and Nikulin [16] some known facts about Néron-Severi lattices and polarizations of supersingular $K3$ surfaces. In §4, we prove Theorem 1.1 using the theory of discriminant forms. In §5 and §6, we prove Theorems 1.5 and 1.4. We reduce the problem of existence of the polarizations on a supersingular $K3$ surface to a problem of existence of ternary codes with certain properties, and solve the latter by computer. In §7, we prove Theorem 1.6. The proof presented here seems to be quite lattice-intensive. We think there should be a more elementary proof. See Question 7.7.

In our preprint [24], we determine all possible Dynkin types R of rational double points of total Milnor number 20 on supersingular $K3$ surfaces in characteristic prime to $2 \operatorname{disc}(R)$.

Acknowledgment. This work was done during the second author's visit to Hokkaido University who likes to express his thanks for the very warm hospitality.

2. Discriminant forms of lattices

For a finite abelian group A and a prime integer p, we denote by

$$A = A_{(p)} \times A_{(p')}$$

the decomposition of A into the p-part $A_{(p)}$ and the p-prime-part $A_{(p')}$ of A.

A lattice is, by definition, a free $\mathbb{Z}$-module of finite rank with a non-degenerate symmetric $\mathbb{Z}$-valued bilinear form. A lattice Λ is said to be *even* if $v^2 \in 2\mathbb{Z}$ holds for every $v \in \Lambda$. Let Λ be an even lattice. We denote by $\Lambda^\vee$ the *dual lattice* $\operatorname{Hom}(\Lambda, \mathbb{Z})$. We have a natural embedding $\Lambda \hookrightarrow \Lambda^\vee$ of finite cokernel, and a symmetric bilinear form $\Lambda^\vee \times \Lambda^\vee \to \mathbb{Q}$ that extends the $\mathbb{Z}$-valued symmetric bilinear form on Λ. We put

$$\operatorname{Disc}(\Lambda) := \Lambda^\vee / \Lambda,$$

and call it the *discriminant group of* Λ. We then define the *discriminant form*

$$q_\Lambda \quad : \quad \mathrm{Disc}(\Lambda) \to \mathbb{Q}/2\mathbb{Z} \quad \text{and}$$
$$b_\Lambda \quad : \quad \mathrm{Disc}(\Lambda) \times \mathrm{Disc}(\Lambda) \to \mathbb{Q}/\mathbb{Z}$$

by

$$q_\Lambda(\bar{v}) \quad := \quad v^2 \bmod 2\mathbb{Z} \quad \text{and}$$
$$b_\Lambda(\bar{v}, \bar{w}) \quad := \quad vw \bmod \mathbb{Z} = (q_\Lambda(\bar{v} + \bar{w}) - q_\Lambda(\bar{v}) - q_\Lambda(\bar{w}))/2,$$

where $v, w \in \Lambda^\vee$, and $\bar{v} := v \bmod \Lambda$, $\bar{w} := w \bmod \Lambda$. Let p be a prime integer dividing $|\mathrm{Disc}(\Lambda)| = |\mathrm{disc}\,\Lambda|$. Then $\mathrm{Disc}(\Lambda)_{(p)}$ and $\mathrm{Disc}(\Lambda)_{(p')}$ are orthogonal with respect to b_Λ. We put

$$q_{\Lambda(p)} := q_\Lambda | \mathrm{Disc}(\Lambda)_{(p)}, \qquad q_{\Lambda(p')} := q_\Lambda | \mathrm{Disc}(\Lambda)_{(p')},$$
$$b_{\Lambda(p)} := b_\Lambda | \mathrm{Disc}(\Lambda)_{(p)} \times \mathrm{Disc}(\Lambda)_{(p)}, \qquad b_{\Lambda(p')} := b_\Lambda | \mathrm{Disc}(\Lambda)_{(p')} \times \mathrm{Disc}(\Lambda)_{(p')}.$$

For a subgroup H of $\mathrm{Disc}(\Lambda)$, we denote by $H^\perp$ the orthogonal complement of H with respect to b_Λ. Note that $(H^\perp)_{(p)}$ is canonically isomorphic to

$$(H_{(p)})^\perp := \{\, x \in \mathrm{Disc}(\Lambda)_{(p)} \mid b_{\Lambda(p)}(x, y) = 0 \text{ for any } y \in H_{(p)} \,\}.$$

We will use the notation $H^\perp_{(p)}$ to denote $(H^\perp)_{(p)} = (H_{(p)})^\perp$. A subgroup $H \subset \mathrm{Disc}(\Lambda)$ is called *isotropic* if $q_\Lambda | H$ is constantly equal to 0. If H is isotropic, then H is contained in $H^\perp$. Note that we have

$$(H^\perp/H)_{(p)} = H^\perp_{(p)}/H_{(p)}.$$

An *overlattice* of Λ is, by definition, a submodule Λ' of $\Lambda^\vee$ containing Λ such that the $\mathbb{Q}$-valued symmetric bilinear form on $\Lambda^\vee$ takes values in $\mathbb{Z}$ on Λ'.

PROPOSITION 2.1 (Nikulin [**15**]). *Let* $\mathrm{pr}_\Lambda : \Lambda^\vee \to \mathrm{Disc}(\Lambda)$ *be the natural projection. The correspondence*

$$H \mapsto \Lambda_H := \mathrm{pr}_\Lambda^{-1}(H)$$

gives a bijection from the set of isotropic subgroups of $\mathrm{Disc}(\Lambda)$ *to the set of even overlattices of* Λ. *For an isotropic subgroup* H, *the discriminant group of* Λ_H *is isomorphic to* $H^\perp/H$.

REMARK 2.2. If Λ is of rank r, then $\mathrm{Disc}(\Lambda)$ is generated by at most r elements.

A vector v in an even *negative-definite* lattice Λ is called a *root* if $v^2 = -2$. We denote by $\mathrm{Roots}(\Lambda)$ the set of roots in Λ. It is known that $\mathrm{Roots}(\Lambda)$ forms a root system of type ADE ([**9, 12**]). An even negative-definite lattice Λ is called a *root lattice* if it is generated by $\mathrm{Roots}(\Lambda)$.

Let $\mathbb{Z}[10A_2]$ denote the root lattice of type $10A_2$. Then $\mathbb{Z}[10A_2]$ is generated by roots c_i, d_i $(i = 1, \ldots, 10)$ satisfying

$$c_i^2 = d_i^2 = -2, \quad c_i d_i = 1, \quad \text{and} \quad \langle c_i, d_i \rangle \perp \langle c_j, d_j \rangle \text{ if } i \neq j.$$

We have

$$\mathrm{Roots}(\mathbb{Z}[10A_2]) = \{\pm c_i, \pm d_i, \pm(c_i + d_i) \quad (i = 1, \ldots, 10)\},$$

and

$$\mathbb{Z}[10A_2]^\vee = \{\, \sum_{i=1}^{10} (s_i c_i + t_i d_i)/3 \mid s_i, t_i \in \mathbb{Z}, \ s_i + t_i \equiv 0 \bmod 3 \ (i = 1, \ldots, 10) \,\}.$$

We put
$$\gamma_i := (c_i + 2d_i)/3 \mod \mathbb{Z}[10A_2] \in \mathrm{Disc}(\mathbb{Z}[10A_2]).$$
Then we have
$$\mathrm{Disc}(\mathbb{Z}[10A_2]) = \mathbb{F}_3\gamma_1 \oplus \cdots \oplus \mathbb{F}_3\gamma_{10},$$
and
$$(2.1) \qquad q_{\mathbb{Z}[10A_2]}(x_1\gamma_1 + \cdots + x_{10}\gamma_{10}) = -2(x_1^2 + \cdots + x_{10}^2)/3 \in \mathbb{Q}/2\mathbb{Z}.$$
For a vector
$$\mathbf{x} = (x_1, \ldots, x_{10}) = x_1\gamma_1 + \cdots + x_{10}\gamma_{10} \in \mathrm{Disc}(\mathbb{Z}[10A_2]) \cong \mathbb{F}_3^{10},$$
we define the *Hamming weight* $\mathrm{wt}(\mathbf{x})$ of $\mathbf{x}$ by
$$\mathrm{wt}(\mathbf{x}) := |\{ i \mid x_i \neq 0 \}| \in \mathbb{Z}_{\geq 0}.$$
Then, for a vector $r \in \mathbb{Z}[10A_2]^\vee$, we have
$$(2.2) \qquad r^2 \leq -(2/3)\,\mathrm{wt}(\bar{r}), \quad \text{where } \bar{r} := r \mod \mathbb{Z}[10A_2] \in \mathrm{Disc}(\mathbb{Z}[10A_2]).$$
Moreover,

(2.3) for a vector $\mathbf{x} \in \mathrm{Disc}(\mathbb{Z}[10A_2])$, there exists a vector $r \in \mathbb{Z}[10A_2]^\vee$ such that $\bar{r} = \mathbf{x}$ and $r^2 = (-2/3)\,\mathrm{wt}(\mathbf{x})$ hold.

Let e and f be basis of the lattice $U(m)$ satisfying
$$e^2 = f^2 = 0, \qquad ef = m.$$
We put $e^\vee := f/m$ and $f^\vee := e/m$. Then $\mathrm{Disc}(U(m)) \cong (\mathbb{Z}/m\mathbb{Z})^2$ is generated by
$$\bar{e}^\vee := e^\vee \mod U(m) \quad \text{and} \quad \bar{f}^\vee := f^\vee \mod U(m),$$
and the discriminant form is given by
$$(2.4) \qquad q_{U(m)}(y_1\bar{e}^\vee + y_2\bar{f}^\vee) = 2y_1y_2/m \in \mathbb{Q}/2\mathbb{Z}.$$

3. Néron-Severi lattices of supersingular $K3$ surfaces

A lattice Λ is called *hyperbolic* if the signature of Λ is $(1, \mathrm{rank}\,\Lambda - 1)$. Let p be a prime integer. A lattice Λ is called *p-elementary* if $\mathrm{Disc}(\Lambda)$ is a p-elementary abelian group; that is, $p\Lambda^\vee \subseteq \Lambda$ holds. An overlattice of a hyperbolic p-elementary lattice is again hyperbolic and p-elementary.

The following is due to Artin [3] and Rudakov-Shafarevich [17].

THEOREM 3.1. *Let X be a supersingular $K3$ surface in characteristic $p > 0$. Then $NS(X)$ is an even hyperbolic p-elementary lattice.*

The following is due to Rudakov-Shafarevich [17, Section 1].

THEOREM 3.2. *Suppose that p is odd. Let σ be a positive integer ≤ 10. Then the lattice N with the following properties is unique up to isomorphisms:*
(i) *N is even, hyperbolic of rank 22, and*
(ii) *$\mathrm{Disc}(N) \cong \mathbb{F}_p^{2\sigma}$.*

From now on to the end of this section, we assume that p is *odd*. We denote the lattice N in Theorem 3.2 by $N_{p,\sigma}$. Let X be a supersingular $K3$ surface in characteristic p with $\sigma(X) = \sigma$. By Theorems 3.1 and 3.2, there exists an isometry

$$\phi : N_{p,\sigma} \xrightarrow{\sim} NS(X).$$

More precisely, we have the following:

PROPOSITION 3.3. *Let h be a vector of $N_{p,\sigma}$ such that $h^2 \geq 4$, and let X be a supersingular $K3$ surface in characteristic p with $\sigma(X) = \sigma$.*
 (1) *The following conditions are equivalent:*
  (i) *There exist no vectors $u \in N_{p,\sigma}$ satisfying $hu = 1$ or 2 and $u^2 = 0$, and there exist no vectors $b \in N_{p,\sigma}$ satisfying $h = 2b$ and $b^2 = 2$.*
  (ii) *There exists an isometry $\phi : N_{p,\sigma} \xrightarrow{\sim} NS(X)$ such that $\phi(h)$ is the class $[L]$ of a line bundle L that is very ample modulo (-2)-curves.*

(2) *Suppose that the conditions in (1) are fulfilled, and let L be a line bundle very ample modulo (-2)-curves such that $\phi(h) = [L]$ by some isometry ϕ. Then $Y_{(X,L)}$ has only rational double points as its singularities, and the ADE-type $\mathcal{R}_{(X,L)}$ of $\operatorname{Sing} Y_{(X,L)}$ is equal to that of the root system*

$$\operatorname{Roots}(h^\perp) := \{\, r \in N_{p,\sigma} \mid rh = 0,\ r^2 - -2 \,\}.$$

For the proof, we use the following results due to Nikulin [**16**, Proposition 0.1] and Saint-Donat [**18**, Section 5].

PROPOSITION 3.4 (Nikulin [**16**]). *Let L be a nef line bundle on a $K3$ surface X with $L^2 > 0$. If $|L|$ has a fixed component, then $|L|$ is equal to $m|U| + \Gamma$, where Γ is the fixed part of $|L|$, $|U|$ is a (quasi-)elliptic pencil, and $U^2 = 0$, $U\Gamma = 1$, $\Gamma^2 = -2$, $m = \dim|L| = L^2/2 + 1$ hold. If $|L|$ has no fixed components, then a general member of $|L|$ is irreducible and $\dim|L| = L^2/2 + 1$.*

PROPOSITION 3.5 (Saint-Donat [**18**]). *Let $|L|$ be a complete linear system without fixed components on a $K3$ surface X such that $L^2 \geq 4$. Then the morphism $\Phi_{|L|}$ fails to be birational onto its image if and only if one of the following holds:*
 (i) *There exists an irreducible curve U such that $U^2 = 0$ and $UL = 2$.*
 (ii) *There exists an irreducible curve B such that $B^2 = 2$ and $L = \mathcal{O}_X(2B)$.*

PROOF OF PROPOSITION 3.3. The assertion (2) follows from [**18**, Theorem 6.1] and [**21**, Lemma 2.4]. We now prove (1).

Suppose that the condition (i) in (1) holds. By [**17**, Section 3, Proposition 3], there exists an isometry $\phi : N_{p,\sigma} \xrightarrow{\sim} NS(X)$ such that $\phi(h)$ is the class of a *nef* line bundle L. By Proposition 3.4, $|L|$ is fixed component free. By Proposition 3.5, $\Phi_{|L|}$ is birational onto its image. So (ii) is true.

Conversely, suppose that (ii) holds. We assume that there exists a vector $u \in N_{p,\sigma}$ satisfying $hu = 1$ or 2 and $u^2 = 0$, and derive a contradiction by the argument in [**27**, Proof of Proposition 1.7]. By the Riemann-Roch theorem, $\phi(u)$ is the class $[U]$ of an effective divisor U such that $\dim|U| \geq 1$. Let $D + \Delta$ be a general member of $|U|$, where Δ is the fixed part of $|U|$. We have $D \neq 0$ and $D^2 \geq 0$. If $DL = 0$, then $D^2 < 0$ would follow by Hodge index theorem, a contradiction. Since L is nef, $\Delta L \geq 0$. Therefore, we have $DL = 1$ or 2. Then the image of D by $\Phi_{|L|}$ is either a line or a plane conic. In any case, we have $\dim|D| = 0$, which is a contradiction.

Next we assume that there exists a vector $b \in N_{p,\sigma}$ such that $h = 2b$ and $b^2 = 2$. Let B be an effective divisor such that $\phi(b) = [B]$. Since $[B] = [L]/2$, B is nef. If there exists an irreducible member in $|B|$, then Proposition 3.5 implies that $\Phi_{|L|}$ is not birational onto its image. If there exist no irreducible members in $|B|$, then Proposition 3.4 implies that $|B|$ has a fixed component, and $|B|$ is written as $2|U| + \Gamma$, where $UB = 1$ and $U^2 = 0$. Then $UL = 2$ follows. Hence $\Phi_{|L|}$ is not birational onto its image, and we get a contradiction. So (i) is true. Thus the assertion (1) is proved. $\square$

REMARK 3.6. If there exists a vector b such that $h = 2b$ and $b^2 = 2$, then h is of degree 8.

4. Proof of Theorem 1.1

Theorem 1.1 follows from the structure theorem of Néron-Severi lattices of supersingular $K3$ surfaces (Theorems 3.1 and 3.2), and a purely lattice-theoretic Lemma 4.1 below. A sublattice $\Lambda' \subset \Lambda$ is called *primitive in* Λ if $(\Lambda' \otimes \mathbb{Q}) \cap \Lambda = \Lambda'$ holds.

LEMMA 4.1. *Let N be an even hyperbolic p-elementary lattice of rank 22 such that $\mathrm{Disc}(N)$ is isomorphic to $\mathbb{F}_p^{2\sigma}$, where σ is a positive integer. Suppose that N contains a sublattice R isomorphic to $\mathbb{Z}[10A_2]$. Then $p = 3$, and the orthogonal complement $R^\perp$ of R in N is isomorphic to $U(1)$ or $U(3)$. If $R^\perp \cong U(1)$, then $\sigma \leq 5$, while if $R^\perp \cong U(3)$, then $\sigma \leq 6$.*

PROOF. We put $S := R^\perp$, which is an even hyperbolic lattice of rank 2 primitive in N. Then N is an overlattice of the orthogonal direct sum $R \oplus S$. We put

$$H := N/(R \oplus S).$$

Clearly, we may assume that $H \neq (0)$.

Note that H is an isotropic subgroup of $\mathrm{Disc}(R \oplus S) = \mathrm{Disc}(R) \oplus \mathrm{Disc}(S)$ with respect to $q_{R \oplus S} = q_R \oplus q_S$, and $\mathrm{Disc}(N) \cong H^\perp/H$ is a p-elementary abelian group. Since S is primitive in N, we have

(4.1) $H \cap (0 \oplus \mathrm{Disc}(S)) = 0.$

Let l be a prime integer different from 3 and p. Assume that $\mathrm{Disc}(S)_{(l)}$ is not 0. Since $\mathrm{Disc}(N)_{(l)} = 0$, we see that $H_{(l)}$ is not 0. Since $\mathrm{Disc}(\mathbb{Z}[10A_2])_{(l)} = 0$, we have $H_{(l)} \subset (0 \oplus \mathrm{Disc}(S)_{(l)})$, which contradicts (4.1). Hence we obtain

(4.2) $\mathrm{Disc}(S)_{(l)} = 0$ for any prime l distinct from 3 and p.

Let $m_3 : \mathrm{Disc}(S)_{(3)} \to \mathrm{Disc}(S)_{(3)}$ be the homomorphism given by $m_3(x) := 3x$. Since every element of $\mathrm{Disc}(R)$ is annihilated by multiplication by 3, the image $H_{(3)}^S \subset \mathrm{Disc}(S)_{(3)}$ of $H_{(3)} \subset \mathrm{Disc}(R)_{(3)} \oplus \mathrm{Disc}(S)_{(3)}$ by the projection to the factor $\mathrm{Disc}(S)_{(3)}$ is contained in $\mathrm{Ker}\, m_3$ by (4.1):

(4.3) $H_{(3)}^S \subseteq \mathrm{Ker}\, m_3.$

Therefore, $\mathrm{Im}\, m_3$ is contained in the orthogonal complement of $H_{(3)}^S$ with respect to $q_{S(3)}$. Hence we obtain

(4.4) $0 \oplus \mathrm{Im}\, m_3 \subset H_{(3)}^\perp.$

We assume $p \neq 3$, and derive a contradiction. By (4.2), we have

(4.5) $$\mathrm{Disc}(S) = \mathrm{Disc}(S)_{(3)} \times \mathrm{Disc}(S)_{(p)}.$$

Since $\mathrm{Disc}(R)_{(p)} = 0$, the property (4.1) implies $H_{(p)} = 0$. Therefore $\mathrm{Disc}(N) = \mathrm{Disc}(N)_{(p)}$ is isomorphic to $\mathrm{Disc}(S)_{(p)}$. Since $\dim_{\mathbb{F}_p} \mathrm{Disc}(N) = 2\sigma$ is positive and even, and S is of rank 2, we obtain

(4.6) $$\mathrm{Disc}(S)_{(p)} \cong \mathbb{F}_p^2.$$

On the other hand, from $\mathrm{Disc}(N)_{(3)} = 0$, we obtain

(4.7) $$H_{(3)} = H_{(3)}^\perp.$$

By (4.1), (4.4) and (4.7), we obtain $\mathrm{Im}\, m_3 = 0$; that is, $\mathrm{Disc}(S)_{(3)}$ is 3-elementary. From (4.7), we have $10 + \dim_{\mathbb{F}_3} \mathrm{Disc}(S)_{(3)} = 2\dim_{\mathbb{F}_3} H_{(3)}$, and hence $\dim_{\mathbb{F}_3} \mathrm{Disc}(S)_{(3)}$ is even. Since S is of rank 2, we obtain

(4.8) $$\mathrm{Disc}(S)_{(3)} \cong 0 \quad \text{or} \quad \mathbb{F}_3^2.$$

Suppose that $\mathrm{Disc}(S)_{(3)} \cong 0$. Then $H_{(3)}$ can be regarded as an isotropic subgroup of $\mathrm{Disc}(R)$ with respect to q_R. Because $H_{(3)} = H_{(3)}^\perp$, the corresponding overlattice of R would be an even unimodular negative-definite lattice of rank 20. This contradicts the classification of unimodular lattices ([**19**, Chapter V]).

Suppose that $\mathrm{Disc}(S)_{(3)} \cong \mathbb{F}_3^2$. By (4.5) and (4.6), S is an even indefinite lattice of rank 2 such that $\mathrm{Disc}(S) \cong (\mathbb{Z}/3p\mathbb{Z})^2$. By the classification of indefinite lattices of rank 2 ([**10**, Chapter 15, Section 3]), we see that the intersection matrix of S with respect to an appropriate basis is

$$\begin{pmatrix} 0 & 3p \\ 3p & 0 \end{pmatrix}, \quad \text{or} \quad p = 2 \text{ and } \begin{pmatrix} 6 & 6 \\ 6 & 0 \end{pmatrix}.$$

In any case, the quadratic form $(\mathrm{Disc}(S)_{(3)}, q_{S(3)})$ is isomorphic to

$$\left(\mathbb{F}_3^2, \begin{bmatrix} 0 & 1/3 \\ 1/3 & 0 \end{bmatrix} \right) \cong (\mathrm{Disc}(U(3)), q_{U(3)}).$$

Therefore the isotropic subgroup $H_{(3)}$ of $\mathrm{Disc}(R) \oplus \mathrm{Disc}(S)_{(3)}$ satisfying $H_{(3)} = H_{(3)}^\perp$ would yield an even hyperbolic unimodular lattice of rank 22 as an overlattice of $R \oplus U(3)$, which again contradicts the classification of unimodular lattices.

Therefore $p = 3$ is proved.

By (4.2), we have $\mathrm{Disc}(S) = \mathrm{Disc}(S)_{(3)}$, and hence $H = H_{(3)}$ holds. Suppose that $(\xi, \eta) \in H^\perp$, where $\xi \in \mathrm{Disc}(R)$ and $\eta \in \mathrm{Disc}(S)$. Since $H^\perp/H$ is 3-elementary, we have $(3\xi, 3\eta) = (0, 3\eta) \in H$. By (4.1), we have $3\eta = 0$. Therefore the image $(H^\perp)^S \subset \mathrm{Disc}(S)$ of $H^\perp \subset \mathrm{Disc}(R) \oplus \mathrm{Disc}(S)$ by the projection to the factor $\mathrm{Disc}(S)$ is contained in $\mathrm{Ker}\, m_3$:

(4.9) $$(H^\perp)^S \subset \mathrm{Ker}\, m_3.$$

Next we will show that S is isomorphic to $U(1)$ or $U(3)$. Since $H^\perp/H$ is 3-elementary, (4.1) and (4.4) implies that $m_3(\mathrm{Im}\, m_3) = 0$; that is, $9x = 0$ for any $x \in \mathrm{Disc}(S)$. Since

$$2\sigma = \dim_{\mathbb{F}_3}(H^\perp/H) = 10 + \log_3 |\mathrm{Disc}(S)| - 2\log_3 |H|$$

is even and S is of rank 2, $\mathrm{Disc}(S)$ is isomorphic to 0, $\mathbb{F}_3^2$, $\mathbb{Z}/9\mathbb{Z}$ or $(\mathbb{Z}/9\mathbb{Z})^2$.

We first assume that $\mathrm{Disc}(S)$ is a cyclic group of order 9, and derive a contradiction. Let γ be a generator of $\mathrm{Disc}(S)$. We have $\mathrm{Im}\,m_3 = \mathrm{Ker}\,m_3 = \langle 3\gamma \rangle$. Let $H^R \subset \mathrm{Disc}(R)$ and $H^S \subset \mathrm{Disc}(S)$ be the images of $H \subset \mathrm{Disc}(R) \oplus \mathrm{Disc}(S)$ by the projections to the factors $\mathrm{Disc}(R)$ and $\mathrm{Disc}(S)$, respectively.

CLAIM 4.2. We have

$$H^\perp = (H^R)^\perp \oplus (H^S)^\perp,$$

where $(H^R)^\perp \subset \mathrm{Disc}(R)$ and $(H^S)^\perp \subset \mathrm{Disc}(S)$ are the orthogonal complements of H^R and H^S with respect to q_R and q_S, respectively. In particular, we have $(H^S)^\perp = (H^\perp)^S$.

PROOF. It is obvious that $H^\perp$ contains $(H^R)^\perp \oplus (H^S)^\perp$. Suppose that $(\xi, \eta) \in H^\perp$, where $\xi \in \mathrm{Disc}(R)$ and $\eta \in \mathrm{Disc}(S)$. By (4.9), we have $\eta \in \mathrm{Ker}\,m_3 = \mathrm{Im}\,m_3$. By (4.4), we have $(0, \eta) \in H^\perp$ and hence $(\xi, 0) \in H^\perp$ hold. Because $(\xi, 0) \in (H^R)^\perp$ and $(0, \eta) \in (H^S)^\perp$, Claim 4.2 is proved. $\qquad\square$

Because $H^\perp / H$ is 3-elementary, we have $(0, \gamma) \notin H^\perp$ by (4.1). Hence we obtain

(4.10) $$(H^S)^\perp = (H^\perp)^S \neq \mathrm{Disc}(S).$$

By (4.3), H^S is either 0 or $\mathrm{Ker}\,m_3$. If $H^S = 0$, then $(H^S)^\perp = \mathrm{Disc}(S)$ and we get a contradiction to (4.10). Suppose that $H^S = \mathrm{Ker}\,m_3$. Then $(H^S)^\perp \supset \mathrm{Im}\,m_3$, and hence $(H^S)^\perp = \mathrm{Im}\,m_3$ by (4.10). In particular, we have

(4.11) $$\log_3 |(H^S)^\perp| = 1.$$

Since $|(H^R)^\perp| = 3^{10}/|H^R|$ and $H \cong H^R$ by (4.1), we see that

$$2\sigma = \log_3 |H^\perp/H| = \log_3 |(H^R)^\perp| + \log_3 |(H^S)^\perp| - \log_3 |H| = 10 - 2\log_3 |H| + 1$$

is odd by (4.11), which is absurd. Therefore $\mathrm{Disc}(S) \not\cong \mathbb{Z}/9\mathbb{Z}$.

Because S is an even lattice, the classification of indefinite lattices of rank 2 ([**10**, Chapter 15, Section 3]) implies the following:

$$\begin{aligned}
\mathrm{Disc}(S) = 0 &\implies S \cong U(1), \\
\mathrm{Disc}(S) \cong \mathbb{F}_3^2 &\implies S \cong U(3), \\
\mathrm{Disc}(S) \cong (\mathbb{Z}/9\mathbb{Z})^2 &\implies S \cong U(9).
\end{aligned}$$

Next we assume $S \cong U(9)$, and derive a contradiction. Note that $\mathrm{Ker}\,m_3$ is generated by

$$3\bar{e}^\vee = f/3 \ \bmod\ S \quad \text{and} \quad 3\bar{f}^\vee = e/3 \ \bmod\ S.$$

By (4.9), we have

(4.12) $$H^\perp \ \subset \ \mathrm{Disc}(R) \oplus \mathrm{Ker}\,m_3.$$

Then H is also contained in $\mathrm{Disc}(R) \oplus \mathrm{Ker}\,m_3$. Suppose that H is generated by

$$g^{(\nu)} = \xi_1^{(\nu)}\gamma_1 + \cdots + \xi_{10}^{(\nu)}\gamma_{10} + \eta_1^{(\nu)}(3\bar{e}^\vee) + \eta_2^{(\nu)}(3\bar{f}^\vee) \qquad (\nu = 1, \ldots, r)$$

where $\xi_i^{(\nu)}, \eta_j^{(\nu)} \in \mathbb{F}_3$. We put

$$M := \begin{bmatrix} \xi_1^{(1)} & \cdots & \cdots & \xi_{10}^{(1)} & \eta_2^{(1)} & \eta_1^{(1)} \\ & \cdots & \cdots & & & \\ & \cdots & \cdots & & & \\ \xi_1^{(r)} & \cdots & \cdots & \xi_{10}^{(r)} & \eta_2^{(r)} & \eta_1^{(r)} \end{bmatrix}.$$

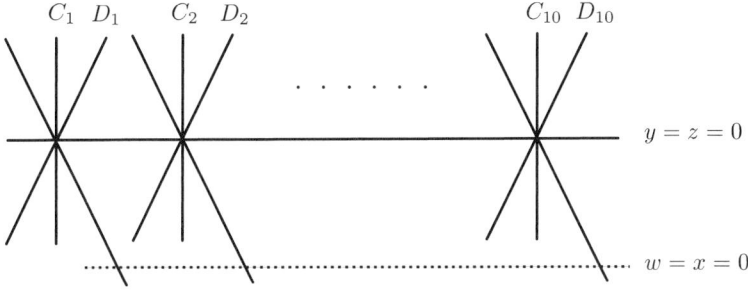

FIGURE 4.1. Lines on the quartic surface

From (2.1) and (2.4), an element

$$x_1\gamma_1 + \cdots + x_{10}\gamma_{10} + y_1\bar{e}^\vee + y_2\bar{f}^\vee \qquad (x_1, \ldots, x_{10} \in \mathbb{F}_3, \ \ y_1, y_2 \in \mathbb{Z}/9\mathbb{Z})$$

of $\mathrm{Disc}(R) \oplus \mathrm{Disc}(S)$ is contained in $H^\perp$ if and only if the vector $\mathbf{x} := [x_1, \ldots, x_{10}, y_1, y_2]$ satisfies the equation

(4.13) $$M \cdot {}^T\mathbf{x} \equiv \mathbf{0} \ \mathrm{mod} \ 3.$$

We consider (4.13) as a system of linear equations over $\mathbb{F}_3$. The property (4.12) of $H^\perp$ implies that every solution of (4.13) in $\mathbb{F}_3$ must satisfy

(4.14) $$y_1 = y_2 = 0.$$

Because of (4.1) and hence $H \cong H^R$, we can choose generators $g^{(1)}, \ldots, g^{(r)}$ of H in such a way that, after suitable permutations of 10 coordinates of $\mathrm{Disc}(R) = \mathbb{F}_3^{10}$ if necessary, the $r \times 12$ matrix M is of the form

$$M = \left[\begin{array}{c|c} I_r & * \end{array} \right],$$

where I_r $(r \leq 10)$ is a diagonal matrix whose diagonal entries are 1. Now non-zero elements of the subgroup of $H \cong H^R$ of $H^\perp$ should be solutions of (4.13) in $\mathbb{F}_3$, but do not satisfy (4.14). Thus we get a contradiction.

Hence S is isomorphic to $U(1)$ or $U(3)$. If $S \cong U(1)$, then $2\sigma = 10 - 2\dim_{\mathbb{F}_3} H \leq 10$, while if $S \cong U(3)$, then $2\sigma = 10 + 2 - 2\dim_{\mathbb{F}_3} H \leq 12$. $\square$

EXAMPLE 4.3. Let $[w : x : y : z]$ be homogeneous coordinates of $\mathbb{P}^3$. For homogeneous polynomials $f(y, z)$, $g(y, z)$ and $h(y, z)$ of degrees 3, 3 and 4, we consider the quartic surface X defined in $\mathbb{P}^3$ by

$$w^3 y + x^3 z + wf(y, z) + xg(y, z) + h(y, z) = 0.$$

When X is smooth, X is a supersingular $K3$ surface, because X contains a configuration of lines drawn by thick lines in Figure 4.1. It was shown in [**20**, Section 6] that, when f, g and h are general, the Artin invariant of X is 6, and hence the orthogonal complement $R^\perp$ of the sublattice $R \subset NS(X)$ generated by the classes of the lines $C_1, D_1, \ldots, C_{10}, D_{10}$ is isomorphic to $U(3)$. When f, g are general and $h = 0$, the Artin invariant of X is 5 by [**26**, Section 4]. In this case, the line ℓ defined by $w = x = 0$ is contained in X. Since $\ell^2 = -2$ and $[\ell] \in R^\perp$, $R^\perp$ is isomorphic to $U(1)$.

5. Proof of Theorem 1.5

The discriminant group D of $\mathbb{Z}[10A_2] \oplus U(3)$ is equal to

$$\mathbb{F}_3\gamma_1 \oplus \cdots \oplus \mathbb{F}_3\gamma_{10} \oplus \mathbb{F}_3\bar{e}^\vee \oplus \mathbb{F}_3\bar{f}^\vee,$$

and the discriminant form q of $\mathbb{Z}[10A_2] \oplus U(3)$ is given by

$$q(x_1, \ldots, x_{10}, y_1, y_2) = -2(x_1^2 + \cdots + x_{10}^2)/3 + 2y_1y_2/3 \quad \in \quad \mathbb{Q}/2\mathbb{Z}.$$

We consider subgroups of D as ternary codes. Recall from §2 that the Hamming weight of a word $\mathbf{x} = (x_1, \ldots, x_{10}) \in \mathrm{Disc}(\mathbb{Z}[10A_2])$ is defined by

$$\mathrm{wt}(\mathbf{x}) := |\{\, i \mid x_i \neq 0 \,\}|.$$

Then a ternary code $\mathcal{C} \subset D$ is isotropic with respect to q if and only if

(5.1) $\mathrm{wt}(\mathbf{x}) \equiv y_1y_2 \bmod 3 \quad \text{for any} \quad (\mathbf{x}, y_1, y_2) \in \mathcal{C}$

holds. A ternary code $\mathcal{C} \subset D$ satisfying (5.1) is therefore called an *isotropic code*. For an isotropic code $\mathcal{C}$, we denote by $N_\mathcal{C}$ the overlattice of $\mathbb{Z}[10A_2] \oplus U(3)$ corresponding to $\mathcal{C}$ by Proposition 2.1. By Theorem 3.2, $N_\mathcal{C}$ is isomorphic to the lattice $N_{3,\sigma}$, where $\sigma = 6 - \dim \mathcal{C}$.

It is easy to see that the following conditions for an isotropic code $\mathcal{C}$ are equivalent:

    (i) $\mathrm{wt}(\mathbf{x}) > 0$ for any non-zero word $(\mathbf{x}, y_1, y_2) \in \mathcal{C}$,
    (ii) $U(3)$ is primitive in $N_\mathcal{C}$, and
    (iii) $\mathbb{Z}[10A_2]^\perp = U(3)$ in $N_\mathcal{C}$.

We say that an isotropic code $\mathcal{C}$ is *admissible* if $\mathcal{C}$ satisfies the conditions above. Let $h = ae + bf$ be a vector of $U(3)$ with $a \geq 1$ and $b \geq 1$. We have $h^2 = 6ab$.

LEMMA 5.1. *Let $\mathcal{C}$ be an admissible isotropic code.*

(1) *There exists a vector $u \in N_\mathcal{C}$ satisfying $hu = 1$ or 2 and $u^2 = 0$ if and only if the following hold:*

    (α) $a = b = 1$, *and*
    (β) *there exists* $(\mathbf{x}, y_1, y_2) \in \mathcal{C}$ *such that* $\mathrm{wt}(\mathbf{x}) = 1$.

(2) *The set of roots* $\mathrm{Roots}(h^\perp) := \{r \in N_\mathcal{C} \mid rh = 0,\ r^2 = -2\}$ *in $h^\perp$ is strictly larger than* $\mathrm{Roots}(\mathbb{Z}[10A_2]) = \{\pm c_i, \pm d_i, \pm(c_i + d_i)\}$ *if and only if one of the following holds:*

    (a) *there exists* $(\mathbf{x}, y_1, y_2) \in \mathcal{C}$ *such that* $\mathrm{wt}(\mathbf{x}) = 3$ *and* $y_1 = y_2 = 0$, *or*
    (b) $a = b$, *and there exists* $(\mathbf{x}, y_1, y_2) \in \mathcal{C}$ *such that* $\mathrm{wt}(\mathbf{x}) = 2$, *or*
    (c) ($a = 2b$ *or* $b = 2a$) *and there exists* $(\mathbf{x}, y_1, y_2) \in \mathcal{C}$ *such that* $\mathrm{wt}(\mathbf{x}) = 1$.

PROOF. We prove (1) first. Suppose that a vector

(5.2) $u = r_u + \eta_1 e^\vee + \eta_2 f^\vee \quad (r_u \in \mathbb{Z}[10A_2]^\vee,\ \eta_1, \eta_2 \in \mathbb{Z})$

of $N_\mathcal{C}$ satisfies $hu = 1$ or 2 and $u^2 = 0$. Then we have

(5.3) $a\eta_1 + b\eta_2 = 1$ or 2,
(5.4) $r_u^2 + 2\eta_1\eta_2/3 = 0$.

Note that $(\eta_1, \eta_2) \not\equiv (0,0) \bmod 3$ by (5.3). Since $\mathcal{C}$ is admissible, we have $r_u \neq 0$, and hence $\eta_1\eta_2 > 0$ by (5.4). From (5.3), we obtain

$$a = b = 1, \qquad \eta_1 = \eta_2 = 1,$$

and hence, from (5.4), we have

$$r_u^2 = -2/3.$$

By (2.2), the word

$$\bar{u} = u \mod (\mathbb{Z}[10A_2] \oplus U(3)) = (\bar{r}_u, \bar{\eta}_1, \bar{\eta}_2) \qquad (\text{where } \bar{r}_u = r_u \mod \mathbb{Z}[10A_2])$$

of $\mathcal{C}$ has the property $\mathrm{wt}(\bar{r}_u) = 1$.

Conversely, suppose that $a = b = 1$ and that there exists a word $(\bar{r}, y_1, y_2) \in \mathcal{C}$ such that $\mathrm{wt}(\bar{r}) = 1$. Replacing $(\bar{r}, y_1, y_2)$ by $(-\bar{r}, -y_1, -y_2)$ if necessary, we can assume that $y_1 = y_2 = 1$ by (5.1). Then, by (2.3), there exists a vector

$$u = r + e^\vee + f^\vee \qquad (r \in \mathbb{Z}[10A_2]^\vee)$$

in $N_\mathcal{C}$ satisfying $r^2 = -2/3$. This vector u satisfies $hu = 2$ and $u^2 = 0$. Thus the assertion (1) is proved.

We now prove (2). Suppose that a vector $u \in N_\mathcal{C}$ given by (5.2) satisfies $hu = 0$, $u^2 = -2$ and $u \notin \mathrm{Roots}(\mathbb{Z}[10A_2])$. Then we have

$$(5.5) \qquad\qquad a\eta_1 + b\eta_2 = 0,$$

$$(5.6) \qquad\qquad r_u^2 + 2\eta_1\eta_2/3 = -2.$$

Suppose that $\eta_1 = 0$ or $\eta_2 = 0$. Then (5.5) implies $\eta_1 = \eta_2 = 0$ and hence $\mathrm{wt}(\bar{r}_u) \equiv 0 \mod 3$ holds because $\mathcal{C}$ is isotropic. By (2.2) and (5.6), we have $\mathrm{wt}(\bar{r}_u) \leq 3$. If $\mathrm{wt}(\bar{r}_u) = 0$, then $u = r_u$ is contained in $\mathrm{Roots}(\mathbb{Z}[10A_2])$. Hence we have $\mathrm{wt}(\bar{r}_u) = 3$, and therefore the condition (a) is satisfied. Suppose that $\eta_1 \neq 0$ and $\eta_2 \neq 0$. By (5.5), we have $\eta_1\eta_2 < 0$. By (2.2) and (5.6), we see that the pair $(\eta_1\eta_2, \mathrm{wt}(\bar{r}_u))$ is either $(-1, 2)$ or $(-2, 1)$. In the former case, we have $a = b$ by (5.5) and hence (b) is satisfied. In the latter case, we have $a = 2b$ or $b = 2a$ by (5.5) and hence (c) is satisfied.

Conversely, suppose that (a) is fulfilled. Using (2.3), we have a lift

$$u = r + 0 + 0 \in N_\mathcal{C} \qquad (r \in \mathbb{Z}[10A_2]^\vee)$$

of the word $(\bar{r}, 0, 0) \in \mathcal{C}$ with $\mathrm{wt}(\bar{r}) = 3$ such that $r^2 = -2$. Then $u \in \mathrm{Roots}(h^\perp) \setminus \mathrm{Roots}(\mathbb{Z}[10A_2])$. Suppose that (b) is satisfied. A vector

$$u = r + e^\vee - f^\vee \in N_\mathcal{C}$$

with $\mathrm{wt}(\bar{r}) = 2$ and $r^2 = -4/3$ satisfies $u \in \mathrm{Roots}(h^\perp) \setminus \mathrm{Roots}(\mathbb{Z}[10A_2])$. Suppose that (c) is satisfied and assume that $a = 2b$. A vector

$$u = r + e^\vee - 2f^\vee \in N_\mathcal{C}$$

with $\mathrm{wt}(\bar{r}) = 1$ and $r^2 = -2/3$ satisfies $u \in \mathrm{Roots}(h^\perp) \setminus \mathrm{Roots}(\mathbb{Z}[10A_2])$. Thus the assertion (2) is proved. $\qquad\square$

PROOF OF THEOREM 1.5. The implication (iii) $\Longrightarrow$ (ii) is obvious. Since every vector h of $U(3)$ satisfies $h^2 \equiv 0 \mod 6$, the implication (ii) $\Longrightarrow$ (i) is also obvious. Using computer, we can prove the following Claim 5.2. See Remark 5.3 and Table 5.1.

CLAIM 5.2. There exists an isotropic admissible code $\mathcal{C} \subset D$ of dimension 5 with the following property:

(5.7) $\qquad$ every non-zero word $(\mathbf{x}, y_1, y_2) \in \mathcal{C}$ satisfies the following:
$\qquad\qquad$ (i) $\mathrm{wt}(\mathbf{x}) \geq 3$, and (ii) if $\mathrm{wt}(\mathbf{x}) = 3$, then $(y_1, y_2) \neq (0, 0)$.

We now prove (i) $\implies$ (iii) Suppose that an integer $d = 6m$ $(m \in \mathbb{Z}_{>0})$ is given. Let X be a supersingular $K3$ surface in characteristic 3 with Artin invariant $\sigma \le 6$. For the basis e, f of $U(3)$ at the end of Section 2, we put

$$h := e + mf.$$

Then $h^2 = d$. Let $\mathcal{C}(\sigma)$ be a linear subspace of the code $\mathcal{C}$ in Claim 5.2 with $\dim \mathcal{C}(\sigma) = 6 - \sigma$. Since $\mathcal{C}(\sigma)$ is isotropic, the corresponding overlattice $N_{\mathcal{C}(\sigma)}$ of $\mathbb{Z}[10A_2] \oplus U(3)$ is isomorphic to $N_{3,\sigma}$ by Theorem 3.2. Hence there exists an isometry

$$\phi : N_{\mathcal{C}(\sigma)} \xrightarrow{\sim} NS(X)$$

by Theorem 3.1. Since every word of $\mathcal{C}(\sigma)$ satisfies the conditions (i) and (ii) in (5.7), Lemma 5.1 implies that there exist no vectors u in $N_{\mathcal{C}(\sigma)}$ satisfying $hu = 1$ or 2 and $u^2 = 0$, and that the set of roots in the orthogonal complement $h^\perp$ of h in $N_{\mathcal{C}(\sigma)}$ coincides with $\mathrm{Roots}(\mathbb{Z}[10A_2])$. By Proposition 3.3 and Remark 3.6, we can choose the isometry $\phi : N_{\mathcal{C}(\sigma)} \xrightarrow{\sim} NS(X)$ in such a way that $\phi(h)$ is the class $[L]$ of a line bundle L very ample modulo (-2)-curves such that $\Phi_{|L|}$ induces a contraction $\rho_L : X \to Y_{(X,L)}$ of an ADE-configuration of (-2)-curves of type $10A_2$. Since $\mathcal{C}(\sigma)$ is admissible, we see that $R_{(X,L)}^\perp \subset NS(X)$ is isomorphic to $U(3)$. Thus X admits a polarization L of degree d with the hoped-for properties. $\qquad\square$

REMARK 5.3. Let G denote the group of linear automorphisms of $D \cong \mathbb{F}_3^{10} \oplus \mathbb{F}_3^2$ generated by

$$(x_1, \ldots, x_{10}, y_1, y_2) \mapsto (x_{\sigma(1)}, \ldots, x_{\sigma(10)}, y_{\tau(1)}, y_{\tau(2)}) \qquad (\sigma \in \mathfrak{S}_{10}, \tau \in \mathfrak{S}_2), \quad \text{and}$$
$$(x_1, \ldots, x_{10}, y_1, y_2) \mapsto ((-1)^{\alpha_1} x_1, \ldots, (-1)^{\alpha_{10}} x_{10}, (-1)^\beta y_1, (-1)^\beta y_2)$$
$$(\alpha_1, \ldots, \alpha_{10} \in \mathbb{F}_2, \beta \in \mathbb{F}_2).$$

Note that, if $\mathcal{C} \subset D$ is an isotropic admissible code, then so is $g(\mathcal{C})$ for any $g \in G$. We define the weight enumerator of a ternary code $\mathcal{C}$ by

$$\mathrm{we}(\mathcal{C}) := \sum_{(\mathbf{x}, y_1, y_2) \in \mathcal{C}} z^{\mathrm{wt}(\mathbf{x})}.$$

Using computer, we have proved that there exist at least seven isomorphism classes of isotropic admissible codes of dimension 5 with the property (5.7). The representative codes $\mathcal{C}_1, \ldots, \mathcal{C}_7$ of these classes are given in Table 5.1. Their weight-enumerators are given in Table 5.2.

COROLLARY 5.4. *Let X be a supersingular $K3$ surface in characteristic 3 with Artin invariant 1. Then there exist at least seven line bundles $L_1, \ldots, L_7$ of degree 6 on X that are mutually non-isomorphic and that induce contractions of $10A_2$-configurations of (-2)-curves on X.*

See Example 7.8.

6. Proof of Theorem 1.4

The proof of Theorem 1.4 is similar to and simpler than that of Theorem 1.5.

The discriminant group D of $\mathbb{Z}[10A_2] \oplus U(1)$ is equal to

$$\mathbb{F}_3 \gamma_1 \oplus \cdots \oplus \mathbb{F}_3 \gamma_{10}.$$

$$
\mathcal{C}_1 \quad : \quad
\begin{bmatrix}
1 & 0 & 0 & 0 & 0 & 0 & 0 & 0 & 1 & 1 & 0 & 1 \\
0 & 1 & 0 & 0 & 0 & 0 & 0 & 1 & 0 & 1 & 2 & 0 \\
0 & 0 & 1 & 0 & 0 & 0 & 1 & 0 & 1 & 0 & 2 & 0 \\
0 & 0 & 0 & 1 & 0 & 0 & 1 & 1 & 0 & 0 & 0 & 1 \\
0 & 0 & 0 & 0 & 1 & 0 & 1 & 1 & 1 & 1 & 1 & 2
\end{bmatrix}
$$

$$
\mathcal{C}_2 \quad : \quad
\begin{bmatrix}
1 & 0 & 0 & 0 & 0 & 0 & 0 & 0 & 1 & 1 & 0 & 1 \\
0 & 1 & 0 & 0 & 0 & 0 & 0 & 1 & 0 & 1 & 2 & 0 \\
0 & 0 & 1 & 0 & 0 & 0 & 1 & 0 & 1 & 0 & 2 & 0 \\
0 & 0 & 0 & 1 & 0 & 0 & 1 & 1 & 0 & 0 & 0 & 1 \\
0 & 0 & 0 & 0 & 1 & 1 & 1 & 2 & 2 & 1 & 0 & 0
\end{bmatrix}
$$

$$
\mathcal{C}_3 \quad : \quad
\begin{bmatrix}
1 & 0 & 0 & 0 & 0 & 0 & 0 & 0 & 1 & 1 & 0 & 1 \\
0 & 1 & 0 & 0 & 0 & 0 & 1 & 1 & 0 & 0 & 0 & 1 \\
0 & 0 & 1 & 0 & 0 & 1 & 0 & 1 & 0 & 1 & 2 & 2 \\
0 & 0 & 0 & 1 & 0 & 1 & 1 & 0 & 1 & 0 & 2 & 2 \\
0 & 0 & 0 & 0 & 1 & 1 & 1 & 2 & 2 & 1 & 0 & 1
\end{bmatrix}
$$

$$
\mathcal{C}_4 \quad : \quad
\begin{bmatrix}
1 & 0 & 0 & 0 & 0 & 0 & 0 & 0 & 1 & 1 & 0 & 1 \\
0 & 1 & 0 & 0 & 0 & 0 & 1 & 1 & 0 & 0 & 0 & 1 \\
0 & 0 & 1 & 0 & 0 & 1 & 0 & 1 & 0 & 1 & 2 & 2 \\
0 & 0 & 0 & 1 & 0 & 1 & 1 & 0 & 1 & 0 & 2 & 2 \\
0 & 0 & 0 & 0 & 1 & 1 & 2 & 2 & 2 & 2 & 2 & 0
\end{bmatrix}
$$

$$
\mathcal{C}_5 \quad : \quad
\begin{bmatrix}
1 & 0 & 0 & 0 & 0 & 0 & 0 & 1 & 1 & 1 & 1 & 1 \\
0 & 1 & 0 & 0 & 0 & 0 & 1 & 0 & 1 & 1 & 2 & 2 \\
0 & 0 & 1 & 0 & 0 & 1 & 0 & 1 & 0 & 1 & 2 & 2 \\
0 & 0 & 0 & 1 & 0 & 1 & 1 & 0 & 0 & 1 & 1 & 1 \\
0 & 0 & 0 & 0 & 1 & 1 & 1 & 1 & 1 & 1 & 0 & 0
\end{bmatrix}
$$

$$
\mathcal{C}_6 \quad : \quad
\begin{bmatrix}
1 & 0 & 0 & 0 & 0 & 0 & 0 & 1 & 1 & 1 & 1 & 1 \\
0 & 1 & 0 & 0 & 0 & 0 & 1 & 0 & 1 & 1 & 2 & 2 \\
0 & 0 & 1 & 0 & 0 & 1 & 0 & 1 & 0 & 1 & 2 & 2 \\
0 & 0 & 0 & 1 & 0 & 1 & 1 & 0 & 0 & 1 & 1 & 1 \\
0 & 0 & 0 & 0 & 1 & 1 & 2 & 2 & 1 & 0 & 1 & 2
\end{bmatrix}
$$

$$
\mathcal{C}_7 \quad : \quad
\begin{bmatrix}
1 & 0 & 0 & 0 & 0 & 0 & 1 & 1 & 1 & 1 & 1 & 2 \\
0 & 1 & 0 & 0 & 0 & 1 & 0 & 1 & 1 & 2 & 2 & 1 \\
0 & 0 & 1 & 0 & 0 & 1 & 1 & 0 & 2 & 1 & 2 & 1 \\
0 & 0 & 0 & 1 & 0 & 1 & 1 & 2 & 0 & 2 & 1 & 2 \\
0 & 0 & 0 & 0 & 1 & 1 & 2 & 1 & 2 & 0 & 1 & 2
\end{bmatrix}
$$

TABLE 5.1. Bases of the codes $\mathcal{C}_1, \ldots, \mathcal{C}_7$

$$\mathrm{we}(\mathcal{C}_1) \;=\; 1 + 12\,z^3 + 18\,z^4 + 36\,z^5 + 108\,z^6 + 36\,z^7 + 18\,z^8 + 14\,z^9,$$
$$\mathrm{we}(\mathcal{C}_2) \;=\; 1 + 8\,z^3 + 10\,z^4 + 24\,z^5 + 86\,z^6 + 40\,z^7 + 30\,z^8 + 40\,z^9 + 4\,z^{10},$$
$$\mathrm{we}(\mathcal{C}_3) \;=\; 1 + 4\,z^3 + 8\,z^4 + 24\,z^5 + 94\,z^6 + 44\,z^7 + 30\,z^8 + 36\,z^9 + 2\,z^{10},$$
$$\mathrm{we}(\mathcal{C}_4) \;=\; 1 + 6\,z^3 + 6\,z^4 + 18\,z^5 + 102\,z^6 + 42\,z^7 + 36\,z^8 + 26\,z^9 + 6\,z^{10},$$
$$\mathrm{we}(\mathcal{C}_5) \;=\; 1 + 30\,z^4 + 60\,z^6 + 120\,z^7 + 20\,z^9 + 12\,z^{10},$$
$$\mathrm{we}(\mathcal{C}_6) \;=\; 1 + 18\,z^4 + 18\,z^5 + 96\,z^6 + 36\,z^7 + 36\,z^8 + 38\,z^9,$$
$$\mathrm{we}(\mathcal{C}_7) \;=\; 1 + 72\,z^5 + 60\,z^6 + 90\,z^8 + 20\,z^9.$$

TABLE 5.2. Weight-enumerators

A ternary code $\mathcal{C} \subset D$ is isotropic with respect to the discriminant form q of $\mathbb{Z}[10A_2] \oplus U(1)$ if and only if

$$(6.1) \qquad\qquad \mathrm{wt}(\mathbf{x}) \equiv 0 \mod 3 \quad \text{for any} \quad \mathbf{x} \in \mathcal{C}$$

holds. For an isotropic code $\mathcal{C}$, we denote by $N_{\mathcal{C}}$ the overlattice of $\mathbb{Z}[10A_2] \oplus U(1)$ corresponding to $\mathcal{C}$. By Theorem 3.2, $N_{\mathcal{C}}$ is isomorphic to the lattice $N_{3,\sigma}$, where $\sigma = 5 - \dim \mathcal{C}$.

Let $h = ae + bf$ be a vector of $U(1)$ with $a \geq 1$ and $b \geq 1$. We have $h^2 = 2ab$.

LEMMA 6.1. *Let $\mathcal{C}$ be an isotropic code in $D \cong \mathbb{F}_3^{10}$.*

(1) There exists a vector $u \in N_{\mathcal{C}}$ satisfying $hu = 1$ or 2 and $u^2 = 0$ if and only if $a \leq 2$ or $b \leq 2$.

(2) The set of roots $\mathrm{Roots}(h^\perp) := \{r \in N_{\mathcal{C}} \mid rh = 0,\ r^2 = -2\}$ in $h^\perp$ is strictly larger than $\mathrm{Roots}(\mathbb{Z}[10A_2])$ if and only if one of the following holds;

    (a) *there exists $\mathbf{x} \in \mathcal{C}$ such that $\mathrm{wt}(\mathbf{x}) = 3$, or*
    (b) *$a = b$.*

PROOF. We prove (1) first. Suppose that a vector

$$(6.2) \qquad\qquad u = r_u + \eta_1 f + \eta_2 e \quad (r_u \in \mathbb{Z}[10A_2]^\vee,\ \eta_1, \eta_2 \in \mathbb{Z})$$

of $\mathbb{Z}[10A_2]^\vee \oplus U(1)^\vee = \mathbb{Z}[10A_2]^\vee \oplus U(1)$ satisfies $hu = 1$ or 2 and $u^2 = 0$. Then we have

$$(6.3) \qquad\qquad a\eta_1 + b\eta_2 = 1 \text{ or } 2,$$
$$(6.4) \qquad\qquad r_u^2 + 2\eta_1\eta_2 = 0.$$

By (6.4), we have $\eta_1\eta_2 \geq 0$. Using (6.3), we have $a \leq 2$ or $b \leq 2$. Conversely, if $a \leq 2$, then $u = f$ satisfies $hu = a = 1$ or 2 and $u^2 = 0$. Thus (1) is proved.

Next we prove (2). Suppose that a vector u given by (6.2) satisfies $hu = 0$, $u^2 = -2$ and $u \notin \mathrm{Roots}(\mathbb{Z}[10A_2])$. Then we have

$$(6.5) \qquad\qquad a\eta_1 + b\eta_2 = 0,$$
$$(6.6) \qquad\qquad r_u^2 + 2\eta_1\eta_2 = -2.$$

Because $\mathcal{C}$ is isotropic, $\mathrm{wt}(\bar{r}_u) \equiv 0 \mod 3$ holds. If $\eta_1 = 0$ or $\eta_2 = 0$, then (6.5) implies $\eta_1 = \eta_2 = 0$. By (2.2) and (6.6), we have $\mathrm{wt}(\bar{r}_u) \leq 3$. If $\mathrm{wt}(\bar{r}_u) = 0$, then $u = r_u$ is contained in $\mathrm{Roots}(\mathbb{Z}[10A_2])$. Hence we have $\mathrm{wt}(\bar{r}_u) = 3$, and therefore the condition (a) is satisfied. Suppose that $\eta_1 \neq 0$ and $\eta_2 \neq 0$. By (6.5), we have

$\eta_1 \eta_2 < 0$. By (6.6), we have $r_u = 0$ and $\eta_1 \eta_2 = -1$, and hence $a = b$ follows from (6.5).

Conversely, suppose that (a) is fulfilled. Using (2.3), we have a lift

$$u = r + 0 + 0 \in N_C \qquad (r \in \mathbb{Z}[10A_2]^\vee)$$

of the word $\bar{r} \in C$ with $\mathrm{wt}(\bar{r}) = 3$ such that $r^2 = -2$. Then u is contained in $\mathrm{Roots}(h^\perp) \setminus \mathrm{Roots}(\mathbb{Z}[10A_2])$. Suppose that (b) is satisfied. The vector

$$u = e - f \ \in \ N_C$$

satisfies $u \in \mathrm{Roots}(h^\perp) \setminus \mathrm{Roots}(\mathbb{Z}[10A_2])$. $\qquad\qquad\qquad\qquad\qquad \square$

In order to prove Theorem 1.4, it is therefore enough to show the following:

CLAIM 6.2. There exists an isotropic code $C \subset D \cong \mathbb{F}_3^{10}$ of dimension 4 such that $\mathrm{wt}(\mathbf{x}) \geq 6$ holds for any $\mathbf{x} \in C$.

The code C generated by the row vectors of

$$\begin{bmatrix} 1 & 0 & 0 & 0 & 0 & 1 & 1 & 1 & 1 & 1 \\ 0 & 1 & 0 & 0 & 1 & 0 & 1 & 1 & 2 & 2 \\ 0 & 0 & 1 & 0 & 1 & 1 & 0 & 2 & 1 & 2 \\ 0 & 0 & 0 & 1 & 1 & 1 & 2 & 0 & 2 & 1 \end{bmatrix}$$

satisfies $\mathrm{wt}(\mathbf{x}) \geq 6$ for any $\mathbf{x} \in C$. The weight-enumerator $\sum_{\mathbf{x} \in C} z^{\mathrm{wt}(\mathbf{x})}$ of this code C is

$$1 + 60z^6 + 20z^9.$$

REMARK 6.3. The code C above is obtained as a subcode of the extended ternary Golay code in $\mathbb{F}_3^{12}$. See [12, Chapter 5, Section 2].

7. Proof of Theorem 1.6

Let (X, L) be a polarized $K3$ surface of degree 6. Then $Y_{(X,L)}$ is a complete intersection of multi-degree $(2, 3)$ in $\mathbb{P}^4$ by [18, Theorem 6.1]. Let $\widetilde{Q}_{(X,L)}$ denote the unique quadric hypersurface in $\mathbb{P}^4$ containing $Y_{(X,L)}$.

PROPOSITION 7.1. Suppose that $\mathcal{R}_{(X,L)} = 10A_2$ and $R^\perp_{(X,L)} \cong U(3)$. Then $\widetilde{Q}_{(X,L)}$ is a cone over a non-singular quadric surface $Q = \mathbb{P}^1 \times \mathbb{P}^1$, and $Y_{(X,L)}$ does not pass through the vertex P of the cone $\widetilde{Q}_{(X,L)}$.

PROOF. By the assumption, $R^\perp_{(X,L)}$ is generated by the numerical equivalence classes $[E]$ and $[F]$ of divisors E and F satisfying

(7.1) $\qquad\qquad E^2 = F^2 = 0, \quad EF = 3, \quad [L] = [E] + [F].$

By the Riemann-Roch theorem, we can assume that E and F are effective. Suppose that $|E|$ has a fixed component. Let $M + \Gamma$ be a general member of $|E|$, where Γ is the fixed part of $|E|$. Because $\rho_L : X \to Y_{(X,L)}$ is birational, ρ_L induces a birational map from M to $\rho_L(M)$. Note that $\rho_L(M + \Gamma)$ is a cubic curve. If $\rho_L(\Gamma)$ is of dimension 1, then $\rho_L(M)$ is a line or a plane conic, and hence contradicts $\dim |M| > 0$. Therefore, ρ_L contracts every irreducible component of Γ to a point, and hence $[\Gamma] \in R_{(X,L)}$. From $[E] \in R^\perp_{(X,L)}$, we obtain $E\Gamma = 0$ and hence $M^2 = E^2 + \Gamma^2 < 0$. Thus we get a contradiction again. Hence $|E|$ has no fixed components. In particular, E is nef. Since ρ_L is birational and E is primitive in $R^\perp_{(X,L)}$ (being part of its basis), a general member E of $|E|$ is mapped by ρ_L birationally to a plane cubic curve in $\mathbb{P}^4$. Therefore a general member of $|E|$ is irreducible, and hence $|E|$ is a (quasi-)elliptic pencil by [**15**, Proposition 0.1]. Therefore the quadric hypersurface $\widetilde{Q}_{(X,L)}$ contains a one-dimensional family $\{\Pi_t^E\}$ of planes such that

$$|E| = \{\rho_L^*(\Pi_t^E \cap Y_{(X,L)})\}.$$

Hence $\widetilde{Q}_{(X,L)}$ is singular. Since $\widetilde{Q}_{(X,L)}$ contains two irreducible families $\{\Pi_t^E\}$ and $\{\Pi_t^F\}$ of planes corresponding to $|E|$ and $|F|$, we have $\dim \operatorname{Sing} \widetilde{Q}_{(X,L)} = 0$, and $\widetilde{Q}_{(X,L)}$ is a cone over a non-singular quadric surface $Q = \mathbb{P}^1 \times \mathbb{P}^1$. If $Y_{(X,L)}$ passed through the vertex P of the cone $\widetilde{Q}_{(X,L)}$, then the linear system $|E|$ would have a fixed component that is contracted to the point P. Hence P is not contained in $Y_{(X,L)}$. $\qquad\square$

Note that a non-ordered pair of the numerical equivalence classes $[E]$ and $[F]$ in $R_{(X,L)}^\perp$ satisfying (7.1) is unique. The following has been shown in the proof above:

COROLLARY 7.2. *The divisors E and F are nef. The complete linear systems $|E|$ and $|F|$ are (quasi-)elliptic pencils.*

We denote by

$$\pi_P : Y_{(X,L)} \to Q = \mathbb{P}^1 \times \mathbb{P}^1$$

the projection from the vertex P of the cone $\widetilde{Q}_{(X,L)}$. Let x and y be affine coordinates of the two factors of $\mathbb{P}^1 \times \mathbb{P}^1$. The surface $Y_{(X,L)}$ is defined by an equation

(7.2) $$\Psi := W^3 + a(x,y)\,W^2 + b(x,y)\,W + c(x,y) = 0,$$

where W is a fiber coordinate of the affine line bundle $\widetilde{Q}_{(X,L)} \setminus \{P\} \cong L_Q(1,1)$ on $Q = \mathbb{P}^1 \times \mathbb{P}^1$, and a, b, c are polynomials of degrees 1, 2 and 3, respectively.

Let us consider the fibrations

$$\Phi_{|E|} = \mathrm{pr}_1 \circ \pi_P \circ \rho_L \ : \ X \to \mathbb{P}^1, \qquad \text{and}$$
$$\Phi_{|F|} = \mathrm{pr}_2 \circ \pi_P \circ \rho_L \ : \ X \to \mathbb{P}^1,$$

where $\mathrm{pr}_i : \mathbb{P}^1 \times \mathbb{P}^1 \to \mathbb{P}^1$ is the projection onto the i-th factor. Because $Y_{(X,L)}$ has ten cusps, the classification of fibers of (quasi-)elliptic fibrations and the criterion [**17**, Section 4] for quasi-ellipticity imply the following:

PROPOSITION 7.3. *The fibrations $\Phi_{|E|}$ and $\Phi_{|F|}$ are quasi-elliptic. Let Θ be a fiber of the quasi-elliptic fibration $\Phi_{|E|}$. Then Θ is either of type II, of type IV or of type IV*. (See Figure 7.1). Moreover, we have*

$$\Theta \text{ is of type II} \iff \rho_L(\Theta) \text{ does not pass through any cusps of } Y_{(X,L)},$$
$$\Theta \text{ is of type IV} \iff \rho_L(\Theta) \text{ passes through exactly one cusp of } Y_{(X,L)},$$
$$\Theta \text{ is of type IV*} \iff \rho_L(\Theta) \text{ is a line with multiplicity 3 passing through}$$
$$\text{exactly three cusps of } Y_{(X,L)}.$$

Same hold for fibers of $\Phi_{|F|}$.

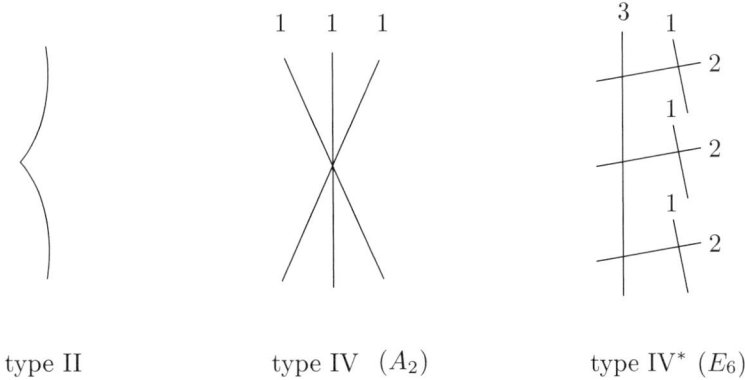

FIGURE 7.1. Fibers of quasi-elliptic fibrations

PROOF OF THEOREM 1.6. Let X be a supersingular $K3$ surface with $\sigma(X) \leq 6$. We choose a subcode $\mathcal{C}$ of the isotropic admissible code $\mathcal{C}_7$ in Table 5.1 with

$$\dim \mathcal{C} = 6 - \sigma(X),$$

and consider the corresponding overlattice $N_{\mathcal{C}}$ of $\mathbb{Z}[10A_2] \oplus U(3)$. There exists an isometry

$$\phi : N_{\mathcal{C}} \xrightarrow{\sim} NS(X)$$

such that $\phi(e + f)$ is the class $[L]$ of a line bundle L that is very ample modulo (-2)-curves, where e, f form the canonical basis of $U(3)$; see the proof of Theorem 1.5. Then $Y_{(X,L)}$ is a complete intersection in $\mathbb{P}^4$ with multi-degree $(2, 3)$ that has ten cusps as its only singularities. We will prove Theorem 1.6 by showing that, for this polarized supersingular $K3$ surface (X, L), the morphism π_P from $Y_{(X,L)}$ to $\mathbb{P}^1 \times \mathbb{P}^1$ is purely inseparable; that is, the polynomials a and b in (7.2) are zero.

We assume that π_P is separable, and derive a contradiction.

For $i = 1, \ldots, 10$, let C_i and D_i be the (-2)-curves contracted by ρ_L satisfying

$$C_i^2 = D_i^2 = -2, \qquad C_i D_i = 1, \qquad \langle [C_i], [D_i] \rangle \perp \langle [C_j], [D_j] \rangle \quad (i \neq j),$$

and let E, F be divisors such that $\phi(e) = [E]$ and $\phi(f) = [F]$. Then E and F satisfy $[E], [F] \in R_{(X,L)}^{\perp}$ and (7.1). We put

$$\gamma_i \quad := \quad ([C_i] + 2[D_i])/3 \bmod (R_{(X,L)} \oplus R_{(X,L)}^{\perp}),$$
$$\bar{f}^{\vee} \quad := \quad [E]/3 \bmod (R_{(X,L)} \oplus R_{(X,L)}^{\perp}),$$
$$\bar{e}^{\vee} \quad := \quad [F]/3 \bmod (R_{(X,L)} \oplus R_{(X,L)}^{\perp}).$$

The code $\mathcal{C}_{(X,L)}$ defined by

$$\mathcal{C}_{(X,L)} := NS(X)/(R_{(X,L)} \oplus R_{(X,L)}^{\perp}) \quad \subset \quad \mathrm{Disc}(R_{(X,L)} \oplus R_{(X,L)}^{\perp}) \cong \mathbb{F}_3^{10} \oplus \mathbb{F}_3^2$$

is isomorphic to the subcode $\mathcal{C}$ of $\mathcal{C}_7$ chosen above. Let G be a divisor on X. Then $[G] \in NS(X)$ is written as

$$\frac{1}{3} \sum_{i=1}^{10} (s_i[C_i] + t_i[D_i]) + \frac{\alpha}{3}[E] + \frac{\beta}{3}[F],$$

where s_i, t_i, α, β are integers satisfying $s_i + t_i \equiv 0 \bmod 3$. We denote by

$$\langle G \rangle := [G] \bmod (R_{(X,L)} \oplus R^{\perp}_{(X,L)})$$

the word of $\mathcal{C}_{(X,L)}$ corresponding to $[G]$, which is written as

$$(\mathbf{x}(G), \bar{\alpha}, \bar{\beta}) = \sum_{i=1}^{10} x_i \gamma_i + \bar{\alpha} \bar{f}^{\vee} + \bar{\beta} \bar{e}^{\vee},$$

where $\bar{\alpha} = \alpha \bmod 3$, $\bar{\beta} = \beta \bmod 3$, and

$$x_i = \begin{cases} 0 & \text{if } (s_i, t_i) \equiv (0,0) \bmod 3, \\ 1 & \text{if } (s_i, t_i) \equiv (1,2) \bmod 3, \\ 2 & \text{if } (s_i, t_i) \equiv (2,1) \bmod 3. \end{cases}$$

We put

$$\begin{aligned} s(G) &:= \{ i \mid (s_i, t_i) \neq (0,0) \} = \{ i \mid C_i G \neq 0 \text{ or } D_i G \neq 0 \}, \\ s_1(\mathbf{x}(G)) &:= \{ i \mid x_i \neq 0 \} = \{ i \mid (s_i, t_i) \not\equiv (0,0) \bmod 3 \}, \\ s_2(G) &:= \{ i \mid (s_i, t_i) \neq (0,0) \text{ and } (s_i, t_i) \equiv (0,0) \bmod 3 \}. \end{aligned}$$

By definition, we have

$$s(G) = s_1(\mathbf{x}(G)) \sqcup s_2(G).$$

LEMMA 7.4. *Suppose that G is a reduced irreducible curve on X. Then the following holds:*

(7.3) $$|s_2(G)| \leq \frac{1}{3}(\alpha\beta - |s_1(\mathbf{x}(G))|) + 1.$$

In particular, we have $\alpha\beta - |s_1(\mathbf{x}(G))| \geq -3$.

PROOF. Let s and t be integers such that $s + t \equiv 0 \bmod 3$. If $(s,t) \neq (0,0)$, then

$$((sC_i + tD_i)/3)^2 \leq -2/3$$

holds. If $(s,t) \neq (0,0)$ and $(s,t) \equiv (0,0) \bmod 3$, then

$$((sC_i + tD_i)/3)^2 \leq -2$$

holds. Therefore we have

$$G^2 \leq -\frac{2}{3}|s_1(\mathbf{x}(G))| - 2|s_2(G)| + \frac{2}{3}\alpha\beta.$$

On the other hand, we have $G^2 \geq -2$. Hence we get the inequality (7.3). $\square$

Let us denote by $\bar{T}$ the Cartier divisor on $Y_{(X,L)}$ cut out by the equation

(7.4) $$\frac{\partial \Psi}{\partial W} = -aW + b = 0,$$

and let T be the proper transform of $\bar{T}$ by ρ_L. By the assumption that π_P is separable, $\bar{T}$ is a divisor and π_P is étale outside $\bar{T}$. Hence the divisor $\bar{T}$ contains the ten cusps of $Y_{(X,L)}$. Therefore we have

(7.5) $$s(T) = \{1, 2, \ldots, 10\}.$$

From the defining equation (7.4) of $\bar{T}$ on $Y_{(X,L)}$, we have

(7.6) $$ET = FT = 6.$$

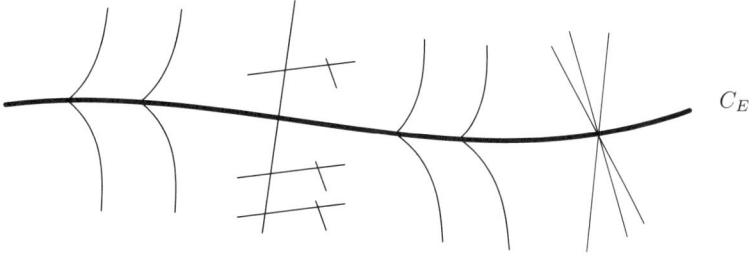

FIGURE 7.2. The curve C_E

We denote by C_E the closure of the locus

$$\{\, x \in X \mid \text{the fiber of } \Phi_{|E|} \text{ passing through } x \text{ is of type II and is singular at } x \,\},$$

and equip C_E with the reduced structure. A general member E of $|E|$ intersects C_E at one point with multiplicity 3 ([**8**]). See Figure 7.2. We define C_F in the same way. Both of C_E and C_F are irreducible, and we have

(7.7) $$C_E E = C_F F = 3.$$

Because $\mathrm{pr}_1 : \mathbb{P}^1 \times \mathbb{P}^1 \to \mathbb{P}^1$ is smooth, if $\mathrm{pr}_1 \circ \pi_P$ is not smooth at a non-singular point of $Y_{(X,L)}$, then π_P is not smooth at that point. Therefore the divisor T contains C_E as a reduced irreducible component. Same holds for C_F.

CLAIM 7.5. *The two curves C_E and C_F are distinct.*

PROOF. Suppose that $C_E = C_F$ holds. Let x be a general point of $C_E = C_F$. Since the fibers E_x of $\Phi_{|E|}$ and F_x of $\Phi_{|F|}$ passing through x are both singular at x, we have $E_x F_x \geq 4$, which contradicts $EF = 3$. □

Let

$$T = C_E + C_F + T_1 + \cdots + T_t$$

be the decomposition of T into reduced irreducible components. We put

$$
\begin{aligned}
[C_E] &= \sum (s_{E,i}[C_i] + t_{E,i}[D_i])/3 + (\alpha_E[E] + \beta_E[F])/3, \\
[C_F] &= \sum (s_{F,i}[C_i] + t_{F,i}[D_i])/3 + (\alpha_F[E] + \beta_F[F])/3, \\
[T_\nu] &= \sum (s_{\nu,i}[C_i] + t_{\nu,i}[D_i])/3 + (\alpha_\nu[E] + \beta_\nu[F])/3 \quad (\nu = 1, \ldots, t).
\end{aligned}
$$

Since E and F are nef, we have

(7.8) $$\alpha_E \geq 0, \ \ \beta_E \geq 0, \ \ \alpha_F \geq 0, \ \ \beta_F \geq 0, \ \ \alpha_\nu \geq 0, \ \ \beta_\nu \geq 0 \quad (\nu = 1, \ldots, t).$$

Since π_P is finite, $\pi_P \circ \rho_L$ maps each irreducible component of T to a curve on $\mathbb{P}^1 \times \mathbb{P}^1$. Therefore we have

(7.9) $$\alpha_\nu > 0 \ \text{ or } \ \beta_\nu > 0 \quad (\nu = 1, \ldots, t).$$

By (7.7), we have

(7.10) $$\beta_E = 3, \quad \alpha_F = 3.$$

Then, from (7.6), we have

(7.11) $$\alpha_E + \sum_{\nu=1}^{t} \alpha_\nu = 3, \quad \text{and} \quad \beta_F + \sum_{\nu=1}^{t} \beta_\nu = 3.$$

Consider the words

$$\langle C_E \rangle = (\mathbf{x}_E, \bar\alpha_E, 0), \quad \langle C_F \rangle = (\mathbf{x}_F, 0, \bar\beta_F), \quad \langle T_\nu \rangle = (\mathbf{x}_\nu, \bar\alpha_\nu, \bar\beta_\nu) \qquad (\nu = 1, \ldots, t)$$

in the code $\mathcal{C}_{(X,L)}$. From Lemma 7.4, we have

$$(7.12) \qquad\qquad -\operatorname{wt}(\mathbf{x}_\nu) + \alpha_\nu \beta_\nu \geq -3 \qquad (\nu = 1, \ldots, t).$$

CLAIM 7.6. $\mathbf{x}_E = \mathbf{x}_F = \mathbf{0}$.

PROOF. Let Θ be a fiber of $\Phi_{|E|}$ such that $\rho_L(\Theta)$ passes through a cusp $q_i :=$ $\rho_L(C_i) = \rho_L(D_i)$ of $Y_{(X,L)}$. Then Θ is of type IV or IV*. Suppose that Θ is of type IV. Then Θ consists of three irreducible components of multiplicity one, two of which are C_i and D_i, that intersect at one point. The curve C_E passes through the intersection point. Since $\Theta C_E = 3$, we have $C_E C_i = C_E D_i = 1$, and therefore $s_{i,E} = t_{i,E} = -3$ holds. Suppose that Θ is of type IV*. Then C_E passes through a point of the multiplicity 3 component of Θ, and does not intersect other irreducible components. This fact can be proved by considering the pull-back of the quasi-elliptic fibration $\Phi_{|E|}$ by the base change $\mathbb{P}^1 \to \mathbb{P}^1$ of degree 2 branching at the point $\Phi_{|E|}(\Theta)$, which makes the fiber Θ into type IV. Then it follows that $s_{i,E} = t_{i,E} = 0$. In any case, we have $i \notin s_1(\mathbf{x}_E)$. Since this holds for any cusp q_i of $Y_{(X,L)}$, we have $s_1(\mathbf{x}_E) = \emptyset$. □

Since $\bar T$ is a Cartier divisor of $Y_{(X,L)}$, the total transform $\rho_L^*(\bar T)$ is contained in $R_{(X,L)}^\perp = \mathbb{Z}[E] \oplus \mathbb{Z}[F]$, and hence $\langle T \rangle = 0$. Therefore we obtain

$$(7.13) \qquad\qquad \mathbf{x}_1 + \cdots + \mathbf{x}_t = \mathbf{0}.$$

By (7.5), we have

$$s(C_E) \cup s(C_F) \cup s(T_1) \cup \cdots \cup s(T_t) = \{1, 2, \ldots, 10\}.$$

Using Lemma 7.4, we obtain

$$(7.14) \quad t + 2 + \frac{1}{3}\left(\alpha_E \beta_E + \alpha_F \beta_F + \sum_{\nu=1}^{t} (\alpha_\nu \beta_\nu - |s_1(\mathbf{x}_\nu)|) \right) \geq$$
$$10 - |s_1(\mathbf{x}_1) \cup \cdots \cup s_1(\mathbf{x}_t)|.$$

Because $\mathcal{C}_{(X,L)}$ is isomorphic to a subcode of $\mathcal{C}_7$, we have shown that there exist integers α_E, β_F, α_ν, β_ν ($\nu = 1, \ldots, t$) and words

$$(\mathbf{0}, \bar\alpha_E, 0), \quad (\mathbf{0}, 0, \bar\beta_F), \quad (\mathbf{x}_\nu, \bar\alpha_\nu, \bar\beta_\nu) \qquad (\nu = 1, \ldots, t)$$

in the code $\mathcal{C}_7$ satisfying (7.8)-(7.14). Using computer, however, we can show that such integers and words do not exist. Thus we get a contradiction. □

Instead of the code $\mathcal{C}_7$, we can use the codes $\mathcal{C}_3, \ldots, \mathcal{C}_6$ in Table 5.1. However, we cannot use $\mathcal{C}_2$ or $\mathcal{C}_1$. Indeed, in $\mathcal{C}_1$, for example, we have the following integers and words:

$$[0,0,0,0,0,0,0,0,0,0,3,0],$$
$$[0,0,0,0,0,0,0,0,0,0,0,3],$$
$$[1,0,0,0,0,0,0,0,0,1,1,0,1],$$
$$[0,2,0,0,0,0,0,2,0,2,1,0],$$
$$[0,0,2,0,0,0,2,0,2,0,1,0],$$
$$[0,0,0,1,0,0,1,1,0,0,0,1],$$
$$[2,1,1,2,0,0,0,0,0,0,1,1].$$

Nevertheless, we can ask the following:

QUESTION 7.7. Is $\pi_P : Y_{(X,L)} \to \mathbb{P}^1 \times \mathbb{P}^1$ inseparable for any polarized supersingular $K3$ surface (X, L) of degree 6 with $\mathcal{R}_{(X,L)} = 10A_2$ and $R^{\perp}_{(X,L)} \cong U(3)$?

EXAMPLE 7.8. Consider the purely inseparable triple cover of $\mathbb{P}^1 \times \mathbb{P}^1$ defined by

$$W^3 = (x^3 - x)(y^3 - y),$$

and the corresponding polarized supersingular $K3$ surface (X, L). We will show that the Artin invariant of X is 1, and that the 5-dimensional ternary code $\mathcal{C}_{(X,L)}$ is isomorphic to $\mathcal{C}_1$. For $\alpha \in \mathbb{F}_3$, let l_α and m_α be the lines on $\mathbb{P}^1 \times \mathbb{P}^1$ defined by $x = \alpha$ and $y = \alpha$, respectively. The strict transforms of l_α and m_α by $\pi_P \circ \rho_L$ are written as $3\tilde{l}_\alpha$ and $3\tilde{m}_\alpha$, respectively. Numbering the twenty (-2)-curves $C_1, D_1, \ldots, C_{10}, D_{10}$ in an appropriate way, we can write the numerical equivalence classes $[\tilde{l}_\alpha]$, $[\tilde{m}_\alpha]$ as follows:

$$
\begin{aligned}
[\tilde{l}_0] &= A_1 + A_2 + A_3 + [E]/3, \\
[\tilde{m}_0] &= A'_1 + A'_4 + A'_7 + [F]/3, \\
[\tilde{l}_1] &= A_4 + A_5 + A_6 + [E]/3, \\
[\tilde{m}_1] &= A'_2 + A'_5 + A'_8 + [F]/3, \\
[\tilde{l}_2] &= A_7 + A_8 + A_9 + [E]/3, \\
[\tilde{m}_2] &= A'_3 + A'_6 + A'_9 + [F]/3,
\end{aligned}
$$

where

$$A_i = -([C_i] + 2[D_i])/3, \qquad A'_i = -(2[C_i] + [D_i])/3.$$

The discriminant of the sublattice of $NS(X)$ generated by the classes $[E], [F]$, the classes of the twenty exceptional curves, and the 6 classes above is equal to -9. Hence these classes span $NS(X)$, and the Artin invariant of X is 1. The 6 words $\langle \tilde{l}_\alpha \rangle$, $\langle \tilde{m}_\alpha \rangle$ generate a 5-dimensional ternary code isomorphic to $\mathcal{C}_1$.

QUESTION 7.9. Find the defining equations of purely inseparable triple covers of $Q = \mathbb{P}^1 \times \mathbb{P}^1$ corresponding to the other ternary codes $\mathcal{C}_2, \ldots, \mathcal{C}_7$ of dimension 5 in Table 5.1. (See Corollary 5.4.)

In [**11**], Dolgachev and Kondo gave various defining equations of the supersingular $K3$ surface in characteristic 2 with Artin invariant 1, and determined the full automorphism group of this $K3$ surface. We expect that various defining equations of the supersingular $K3$ surface in characteristic 3 with Artin invariant 1 would be also helpful in the study of the automorphism group of this surface.

References

[1] M. Artin, *Some numerical criteria for contractability of curves on algebraic surfaces*, Amer. J. Math. **84** (1962), 485–496.

[2] ———, *On isolated rational singularities of surfaces*, Amer. J. Math. **88** (1966), 129–136.

[3] ———, *Supersingular K3 surfaces*, Ann. Sci. École Norm. Sup. (4) **7** (1974), 543–567 (1975).

[4] ———, *Coverings of the rational double points in characteristic p*, Complex analysis and algebraic geometry, (a collection of papers dedicated to K. Kodaira. Edited by W. L. Baily, Jr., and T. Shioda), Iwanami Shoten, Tokyo, 1977, pp. 11–22.

[5] W. Barth, *K3 surfaces with nine cusps*, Geom. Dedicata, **72** (1998), 171–178.

[6] ———, *On the classification of K3 surfaces with nine cusps*, Complex analysis and algebraic geometry, de Gruyter, Berlin, 2000, pp. 41–59.

[7] P. Blass and J. Lang, *Zariski surfaces and differential equations in characteristic p > 0*, Marcel Dekker Inc., New York, 1987.

[8] E. Bombieri and D. Mumford, *Enriques' classification of surfaces in char. p. III*, Invent. Math. **35** (1979), 197–232.

[9] N. Bourbaki, *Éléments de mathématique. Fasc. XXXIV. Groupes et algèbres de Lie. Chapitre IV: Groupes de Coxeter et systèmes de Tits. Chapitre V: Groupes engendrés par des réflexions. Chapitre VI: Systèmes de racines*, Hermann, Paris, 1968.

[10] J. H. Conway and N. J. A. Sloane, *Sphere packings, lattices and groups*, third ed., Springer-Verlag, New York, 1999.

[11] I. Dolgachev and S. Kondo, *A supersingular K3 surface in characteristic 2 and the Leech lattice*, Int. Math. Res. Not. 2003, no. 1, 1–23 (2001).

[12] W. Ebeling, *Lattices and codes*, A course partially based on lectures by F. Hirzebruch. Second revised edition. Friedr. Vieweg & Sohn, Braunschweig, 2002.

[13] H. Ito, *The Mordell-Weil groups of unirational quasi-elliptic surfaces in characteristic 3*, Math. Z. **211** (1992), no. 1, 1–39.

[14] V. V. Nikulin, *Kummer surfaces*, Izv. Akad. Nauk SSSR Ser. Mat. **39** (1975), no. 2, 278–293, 471.

[15] ———, *Integer symmetric bilinear forms and some of their geometric applications*, Math. USSR-Izv. **14** (1979), no. 1, 103–167.

[16] ———, *Weil linear systems on singular K3 surfaces*, Algebraic geometry and analytic geometry (Tokyo, 1990), Springer, Tokyo, 1991, pp. 138–164.

[17] A. N. Rudakov and I. R. Šafarevič, *Surfaces of type K3 over fields of finite characteristic*, Current problems in mathematics, Vol. 18, Akad. Nauk SSSR, Vsesoyuz. Inst. Nauchn. i Tekhn. Informatsii, Moscow, 1981, pp. 115–207: Igor R. Shafarevich, Collected mathematical papers, Springer-Verlag, Berlin, 1989, pp. 657–714.

[18] B. Saint-Donat, *Projective models of K − 3 surfaces*, Amer. J. Math. **96** (1974), 602–639.

[19] J.-P. Serre, *A course in arithmetic*, Graduate Texts in Mathematics, No. 7, Springer-Verlag, New York, 1973.

[20] I. Shimada, *On supercuspidal families of curves on a surface in positive characteristic*, Math. Ann. **292** (1992), no. 4, 645–669.

[21] ———, *Rational double points on supersingular K3 surfaces*, Math. Comp. **73** (2004), no. 248, 1989–2017 (electronic).

[22] ———, *Supersingular K3 surfaces in characteristic 2 as double covers of a projective plane*, Asian J. Math. **8** (2004), no. 3, 531–586.

[23] ———, *Moduli curves of supersingular K3 surfaces in characteristic 2 with Artin invariant 2*, Proc. Edinburgh Math. Soc. **49** (2006), 435–503.

[24] I. Shimada and De-Qi Zhang, *Dynkin diagrams of rank 20 on supersingular K3 surfaces*, Preprint, http://www.math.sci.hokudai.ac.jp/~shimada/preprints.html.

[25] T. Shioda, *Supersingular K3 surfaces*, Algebraic geometry (Proc. Summer Meeting, Univ. Copenhagen, Copenhagen, 1978), Lecture Notes in Math., Vol. 732, Springer, Berlin, 1979, pp. 564–591.

[26] ———, *Some results on unirationality of algebraic surfaces*, Math. Ann. **230** (1977), no. 2, 153–168.

[27] T. Urabe, *Dynkin graphs and combinations of singularities on plane sextic curves*, Singularities (Iowa City, IA, 1986), Amer. Math. Soc., Providence, RI, 1989, pp. 295–316.

DEPARTMENT OF MATHEMATICS, FACULTY OF SCIENCE, HOKKAIDO UNIVERSITY, SAPPORO 060-0810, JAPAN

E-mail address: shimada@math.sci.hokudai.ac.jp

DEPARTMENT OF MATHEMATICS, NATIONAL UNIVERSITY OF SINGAPORE, 2 SCIENCE DRIVE 2, SINGAPORE 117543, SINGAPORE

E-mail address: matzdq@math.nus.edu.sg

Contemporary Mathematics
Volume **422**, 2007

Classical Kummer Surfaces and Mordell-Weil Lattices

Tetsuji Shioda

To Igor Dolgachev

ABSTRACT. Suggested by the classical theory, we study the Kummer surface of a genus two Jacobian variety as an elliptic surface with special type of singular fibres. We determine the Mordell-Weil lattice together with the explicit generators in the general case.

1. Introduction

Let C be a curve of genus two and let $J = J(C)$ be its Jacobian variety. The Kummer surface $S = \mathrm{Km}(J)$ is a smooth K3 surface obtained from the quotient surface J/ι_J of J by the inversion ι_J, by resolving the 16 singular points corresponding to the points of order 2 on J. We assume for simplicity that the base field k is algebraically closed and $\mathrm{char}(k) \neq 2, 3$.

It is known in the classical theory of Kummer surfaces ([**2**]) that there is a beautiful symmetry called the 16_6-configuration. There are two sets of 16 disjoint (-2)-curves on $S = \mathrm{Km}(J)$ which can be labeled as $\{A_{ij}\}$ and $\{Z_{ij}\}$ ($i, j \in I = \{1, 2, 3, 4\}$) in such a way that A_{ij} and Z_{kl} intersect if and only if $i = k$ or $j = l$ but not both; for instance, Z_{11} meets the 6 curves $A_{12}, A_{13}, A_{14}, A_{21}, A_{31}, A_{41}$ and only these curves. Geometrically the 16 curves $\{A_{ij}\}$ are the exceptional curves corresponding to the 2-torsion points on J, while $\{Z_{ij}\}$ arise from embedding C into J as symmetric theta divisors (cf. [**5**], [**7**]).

Using the curves in the 16_6-configuration, we can define various elliptic fibrations on S, since on a K3 surface it is equivalent to giving a divisor consisting of (-2)-curves which has the same type as in Kodaira's list of singular fibres ([**3**], [**7**]). In this note, we focus on an especially neat elliptic fibration $f : S \to \mathbf{P}^1$ which has two disjoint singular fibres of type I_0^*:

$$\Phi_1 = 2Z_{11} + A_{12} + A_{13} + A_{21} + A_{31},$$

$$\Phi_2 = 2Z_{44} + A_{24} + A_{34} + A_{42} + A_{43}.$$

For general C, we have six more singular fibres of type I_2. On the other hand, the curves $Z_{12}, Z_{13}, Z_{21}, Z_{24}, Z_{31}, Z_{34}, Z_{42}, Z_{43}$ are sections of this elliptic surface, since each of them intersects Φ_1 with intersection number 1. Choose one of them as the

2000 *Mathematics Subject Classification.* 14J27, 14J28, 14H40 .

©2007 American Mathematical Society

zero-section. Then, using the height formula ([9]), we can see that three of them form the 2-torsion sections and that the remaining four are sections of height 1.

In this paper, we study more closely the elliptic surface in question (called an *elliptic Kummer surface* for short) and some related ones, in terms of explicit equations. (The geometric theory of 16_6-configuration is used only for the motivation.) First we look at the twisted rational elliptic surface which has six I_2 fibres (or some confluent I_4 fibres) (§3). We determine the generators of the Mordell-Weil lattice (MWL) by using the height formula (§4). Then we turn to the study of the elliptic Kummer surface (§5). By using the correspondence on the curve C and its image in the Kummer surface $S = \mathrm{Km}(J)$, we obtain some "new" elements in the Néron-Severi group $\mathrm{NS}(S)$, which give rise to some nontrivial rational points in the Mordell-Weil lattice (§6). Also we clarify the relation of some automorphisms of C and the confluence of singular fibres (producing type I_4). In §7, we find a rational point of height 1 which gives an explicit generator of the MWL modulo torsion in the general case. As a byproduct, we obtain elliptic K3 surfaces with twelve I_2 fibres with positive rank, depending on 3 moduli.

2. The defining equation

Let us take the equation of a genus two curve C as follows:

$$(2.1) \qquad y^2 = f_5(x) = x(x^4 + c_1 x^3 + c_2 x^2 + c_3 x + c_4) = x \prod_{i=1}^{4} (x - d_i)$$

where we always assume that $\{d_1, d_2, d_3, d_4, 0, \infty\}$ are mutually distinct 6 points of the x-line, normalized so that

$$(2.2) \qquad c_4 = \prod_{i=1}^{4} d_i = 1.$$

As the Jacobian variety J of a genus g curve C is birationally equivalent to the g-th symmetric product of C in general ([11]), the function field of J, $k(J)$, is generated (in case $g = 2$) by the symmetric functions

$$x_1 + x_2, x_1 x_2, y_1 + y_2, y_1 y_2, x_1 y_2 + x_2 y_1$$

of two independent generic points (x_1, y_1) and (x_2, y_2) of C over k. As the inversion ι_J is induced by the hyperelliptic involution $(x, y) \mapsto (x, -y)$, we see that the function field $k(S) = k(J/\iota_J)$ is generated by

$$(2.3) \qquad\qquad X = x_1 + x_2, \ t = x_1 x_2, \ Y = y_1 y_2.$$

Then we have $k(S) = k(X, Y, t)$ with the relation

$$(2.4) \qquad Y^2 = f_5(x_1) f_5(x_2) = x_1 x_2 \prod_{i=1}^{4} (x_1 - d_i)(x_2 - d_i),$$

which can be rewritten, under (2.2), as

$$(2.5) \qquad\qquad E : Y^2 = t \prod_{i=1}^{4} (X - (d_i + \frac{t}{d_i})).$$

This equation defines an elliptic curve E over $k(t)$, which gives an elliptic fibration on the K3 surface $S = \mathrm{Km}(J)$:

(2.6) $$f : S \to \mathbf{P}^1$$

3. The twisted rational elliptic surface

First we consider the quadratic twist $\mathcal{E}$ of the elliptic curve E with respect to $k(\sqrt{t})$:

(3.1) $$\mathcal{E} : Y^2 = \prod_{i=1}^{4}(X - (d_i + \frac{t}{d_i})).$$

By an elementary algorithm as in [**1**, Ch.8], we can transform it into the Weierstrass form:

(3.2) $$\mathcal{E}_W : y^2 = x^3 + a_4(t)x + a_6(t)$$

where $a_4(t), a_6(t)$ are polynomials in t of degree 4 or 6 with coefficients in $k_0 = \mathbf{Q}_0(d_1, d_2, d_3, d_4)$. (Here $\mathbf{Q}_0$ denotes the prime field in k.) We do not write down $a_4(t), a_6(t)$ here, but they can be derived easily from (4.5). The discriminant $\Delta(\mathcal{E}_W)$ is equal, up to a constant, to

(3.3) $$\delta(t) = \prod_{i<j}(t - d_i d_j)^2.$$

Therefore the elliptic surface associated with (3.2) is a rational elliptic surface, and it has 6 singular fibres of type I_2 at $t = d_i d_j$ (since $a_4 \neq 0$ there) provided that the six values $d_i d_j (i < j)$ are distinct.

LEMMA 3.1. *Given d_1, d_2, d_3, d_4 with $d_i \neq d_j (i < j)$ and $\prod_{i=1}^{4} d_i = 1$, we have the following three cases:*
(i) the six values $d_i d_j (i < j)$ are distinct; (ii) there is exactly one pair such that $d_i d_j = d_k d_l$; (iii) there are two such pairs.
In terms of the coefficients c_i of (2.1), the above correspond to the three cases:
(i) $c_1 \neq \pm c_3$; (ii) $c_1 = \pm c_3 \neq 0$; (iii) $c_1 = c_3 = 0$.

LEMMA 3.2. *In case $c_1 = c_3$ or $c_1 = -c_3$, the curve C admits an involutive automorphism*

(3.4) $$\phi : (x, y) \mapsto (\frac{1}{x}, \frac{y}{x^3}) \text{ or } (\frac{-1}{x}, \frac{\sqrt{-1}y}{x^3}),$$

and the quotient curve C/ϕ is an elliptic curve. Hence the Jacobian variety J is isogenous to a product of elliptic curves.

The proof of these lemmas is immediate, and we omit it.

PROPOSITION 3.3. *The rational elliptic surface defined by (3.2) has the following singular fibres: (i) $I_2 \times 6$, (ii) $I_2 \times 4 + I_4$, or (iii) $I_2 \times 2 + I_4 \times 2$, according to the three cases in Lemma 3.1. The structure of the Mordell-Weil lattice on $\mathcal{E}_W(k(t))$ is accordingly isomorphic to the following:*
(i) $A_1^{*\oplus 2} \oplus (\mathbf{Z}/2\mathbf{Z})^{\oplus 2}$, (ii) $\langle \frac{1}{4} \rangle \oplus (\mathbf{Z}/2\mathbf{Z})^{\oplus 2}$, (iii) $\mathbf{Z}/2\mathbf{Z} \oplus \mathbf{Z}/4\mathbf{Z}$.

Proof The first part follows from Lemma 3.1. Then the second part follows from [**6**] where No.42, No.60 and No.74 are the relevant cases for (i), (ii) and (iii). *q.e.d.*

4. Explicit generators via the height formula

Can one write down the generators of the Mordell-Weil group explicitly? Yes, we can. To present a clear prescription for it, we proceed as follows. By (3.2) and (3.3), the RHS of (3.2) decomposes, at $t = d_1 d_2$, as $(x - \alpha)^2(x + 2\alpha)$, where the computation shows that α is given by

$$(4.1) \qquad N(d_1, d_2 | d_3, d_4) = \frac{(d_1 - d_3)(d_1 - d_4)(d_2 - d_3)(d_2 - d_4)}{12 \ d_3 d_4} \neq 0.$$

Now by the height formula (see [**9**] for what follows), we have

$$(4.2) \qquad \langle P, P \rangle = 2\chi + 2(PO) - \sum_v \mathrm{contr}_v(P) \ (\chi = 1)$$

for any $P \in \mathcal{E}_W(k(t))$, with $\mathrm{contr}_v(P) = 1/2$ or 0 for type I_2. Thus, for P 2-torsion, we have

$$(4.3) \qquad 0 = \langle P, P \rangle = 2 + 2 \cdot 0 - \frac{1}{2} \times 4 - 0 \times 2,$$

which means that the x-coordinate $x(P) = A + Bt + Ct^2$ is a degree 2 polynomial which takes the value $N(d_i, d_j | d_k, d_l)$ at $t = d_i d_j$ for 4 distinct pairs (ij).

By solving the linear equations in A, B, C:

$$(4.4) \qquad A + B(d_i d_j) + C(d_i d_j)^2 = N(d_i, d_j | d_k, d_l)$$

for $i = 1, 2$ and $j = 3, 4$, we find (under the condition (2.2)) that

$$A = C = \frac{1}{12}\{(d_1 + d_2)(d_3 + d_4) - 2(d_1 d_2 + d_3 d_4)\},$$

$$B = \frac{1}{12}\{(d_1 + d_2)(d_3 + d_4)(d_1 d_2 + d_3 d_4) - 2d_1 d_2(d_3^2 + d_4^2) - 2d_3 d_4(d_1^2 + d_2^2)\}.$$

We denote this 2-torsion point P by T_1, i.e. $T_1 = (x(T_1), 0)$ with $x(T_1) = A + Bt + Ct^2$ where A, B, C are determined above.

By permuting the indices $\{1, 2, 3, 4\}$, we obtain two more points T_2, T_3 and we have $\{O, T_1, T_2, T_3\} \cong (\mathbf{Z}/2\mathbf{Z})^{\oplus 2}$. It should be remarked that we can recover the Weierstrass equation (3.2) from these data, since it is equal to

$$(4.5) \qquad y^2 = (x - x(T_1))(x - x(T_2))(x - x(T_3)).$$

Next we determine the rational points of height 1/2. The height formula says this time that

$$(4.6) \qquad \frac{1}{2} = \langle P, P \rangle = 2 + 2 \cdot 0 - \frac{1}{2} \times 3 - 0 \times 3.$$

By solving the linear equations (4.4) for $i = 1, j = 2, 3, 4$, we obtain a rational point $P = Q_1$ with $(x(Q_1)$ omitted and)

$$(4.7) \qquad y(Q_1) = \frac{(d_1 - d_2)(d_1 - d_3)(d_1 - d_4)}{8d_1^2}(t - d_1 d_2)(t - d_1 d_3)(t - d_1 d_4).$$

By permutations of indices, we get four points Q_1, Q_2, Q_3, Q_4 of height 1/2. One can check that they are stable under translation by 2-torsions, i.e. 2-torsors.

Similarly, solving (4.4) for $(ij) = (23), (24), (34)$, we obtain $P = R_1$ with

$$(4.8) \qquad y(R_1) = \frac{(d_1 - d_2)(d_1 - d_3)(d_1 - d_4)}{8}(t - d_2 d_3)(t - d_2 d_4)(t - d_3 d_4).$$

By permutations, we get four points R_1, R_2, R_3, R_4 of height $1/2$. One can check again that they are 2-torsors.

Now we can state the following result on explicit generators.

THEOREM 4.1. *In the general case (i), the Mordell-Weil group $\mathcal{E}_W(k(t))$ is generated by Q_1, R_1, T_1, T_2, where Q_1, R_1 are the rational points of height $1/2$ such that $\langle Q_1, R_1 \rangle = 0$ and T_1, T_2 generate the torsion group. In the special case (ii), Q_1 (and R_1) becomes a rational point of height $1/4$, and Q_1, T_1, T_2 generate the Mordell-Weil group. In the very special case (iii), Q_1 (and R_1) reduces to a torsion point of order 4, and Q_1, T_1 generate the Mordell-Weil group isomorphic to $\mathbf{Z}/4\mathbf{Z} \oplus \mathbf{Z}/2\mathbf{Z}$.*

Proof The case (i) is proven above in view of Proposition 3.3. The verification of the case (ii) and (iii) will be left as an exercise to the reader.

5. The elliptic Kummer surface

Let us go back to the elliptic fibration on the Kummer surface $\mathrm{Km}(J)$ defined by (2.5).

First the Weierstrass normal form of (2.5) is given by the twist of (3.2):

$$(5.1) \qquad E_W : y^2 - x^3 + t^2 a_4(t)x + t^3 a_6(t),$$

whose discriminant $\Delta(E_W)$ is equal to $t^6 \delta(t)$ up to a constant (cf. (3.3)). Hence we have:

PROPOSITION 5.1. *The elliptic K3 surface defined by (5.1) has the two singular fibres of type I_0^* at $t = 0$ and ∞, in addition to the semi-stable fibres (i) $I_2 \times 6$, (ii) $I_2 \times 4 + I_4$, or (iii) $I_2 \times 2 + I_4 \times 2$, as in Proposition 3.3.*

Thus the trivial lattice $T \subset \mathrm{NS}(S)$ is given by

$$(5.2) \qquad T = U \oplus D_4^{\oplus 2} \oplus \begin{cases} A_1^{\oplus 6} & (i) \\ A_1^{\oplus 4} \oplus A_3 & (ii) \\ A_1^{\oplus 2} \oplus A_3^{\oplus 2} & (iii) \end{cases}$$

where U is a rank two unimodular lattice spanned by the fibre class and zero-section. In particular, we have

$$(5.3) \qquad \mathrm{rk}\, T = 16, 17 \text{ or } 18, \quad \det T = 2^{10}.$$

Now we consider the Mordell-Weil lattice $E_W(k(t))$. Its rank is given by the wellknown formula

$$(5.4) \qquad r := \mathrm{rk}\, E_W(k(t)) = \rho(S) - \mathrm{rk}\, T$$

where $\rho(S)$ is the Picard number of $S = \mathrm{Km}(J)$. Recall that $\rho(\mathrm{Km}(A)) = \rho(A) + 16$ for any abelian surface, and that $\rho(A)$ is equal to the rank of $\mathrm{End}(A)^{sym}$ which is the symmetric part of the endomorphism algebra of A (cf. [4]). In the case under consideration, we have by (5.3):

$$(5.5) \qquad r = \rho(J) - \begin{cases} 0 & (i) \\ 1 & (ii) \\ 2 & (iii) \end{cases}$$

It follows from Lemma 3.1 and 3.2 that J is isogenous to a product of two elliptic curves $C_1 \times C_2$ in case (ii) and to a self-product $C_1 \times C_1$ in case (iii). It implies that $\rho(J) \geq 2$ in case (ii) and $\rho(J) \geq 3$ in case (iii). Hence we have:

PROPOSITION 5.2. *The Mordell-Weil lattice $E_W(k(t))$ has always a positive rank r given by (5.5). In particular, the Kummer surface $\mathrm{Km}(J)$ of any genus two curve has an infinite group of automorphisms preserving the elliptic fibration (2.6).*

COROLLARY 5.3. *(to Proposition 5.1) The torsion subgroup of $E_W(k(t))$ is $(\mathbf{Z}/2\mathbf{Z})^{\oplus 2}$ in all three cases (i), (ii), (iii), at least if $\mathrm{char}(k) = 0$.*

Proof This follows from Proposition 5.1 in view of the Shimada's list [8] of singular fibres and torsion group for elliptic K3 surfaces. $q.e.d.$

In the next sections, we construct some rational points (sections) of infinite order in $E_W(k(t))$ by two different methods: the first one is to make use of correspondence on the curve, while the second depends on the nature of the height formula and symmetric functions (Galois theory).

6. The use of correspondence of the curve

To study the sections on an elliptic K3 surface S, we go back to the canonical isomorphism $E(K) \simeq \mathrm{NS}(S)/T$ where $K = k(t)$, which is the source of the formula (5.4) (cf. [9]). Given any divisor D on S, its class in the Néron-Severi group $\mathrm{NS}(S)$ modulo T determines a point $P \in E(K)$ by the rule: take the restriction of D to the generic fibre E of $f : S \to \mathbf{P}^1$, and then sum up the points on $E(\bar{K})$ to have a point $P \in E(K)$. This method has been used recently to study the Mordell-Weil lattice related to the Kummer surface of a product abelian surface (cf. [10]).

We apply this idea to the following situation. Starting from a correspondence Γ on the curve C, i.e. $\Gamma \subset C \times C$, we take its image under the rational map $C \times C \to J \to S = \mathrm{Km}(J)$, and obtain a rational point $P \in E(K)$ as above. In particular, given an automorphism $\varphi : C \to C$, we take its graph $\Gamma = \Gamma_\varphi$ and obtain a point P_φ.

PROPOSITION 6.1. *In case $\varphi = \mathrm{id}$, the identity automorphism of C, the rational point $P_\varphi = (x, y) \in E_W(k(t))$ is given by the following:*

$$(6.1) \qquad x = \frac{1}{48}(3 - 2c_2 t + (38 + 3c_2^2 - 8c_1 c_3)t^2 - 2c_2 t^3 + 3t^4).$$

Proof In this case, $\Gamma_\varphi = \Delta_C$ is the diagonal in $C \times C$. In the notation of §2, we have then $(x_1, y_1) = (x_2, y_2)$ so that

$$(6.2) \qquad X = 2x_1, t = x_1^2, Y = y_1^2 = f_5(x_1) = x_1(x_1^4 + \cdots + 1).$$

It follows that $x_1 = \pm\sqrt{t}$, and that $X = 2x_1, Y = f_5(x_1)$. Thus we have the two points $P = (X, Y) = (2\sqrt{t}, f_5(\sqrt{t}))$ and its conjugate P' on the curve E defined by (2.5) over $k(t)$. The point P defines a point $(X, Y) = (2\sqrt{t}, (\sqrt{t})^4 + c_1(\sqrt{t})^3 + \cdots + 1)$ on the curve $\mathcal{E}$, (3.1), which are transformed to the point $Q \in \mathcal{E}_W(k(\sqrt{t}))$ defined by (3.2). The sum $P + P' \in E(k(t))$ corresponds to $Q - Q'$, and an explicit computation gives the rational point P_φ ($\varphi = \mathrm{id}$) on $E_W(k(t))$, of the form (x, y) with $\deg(x) = 4, \deg(y) = 6$; explicitly the x-coordinate of this point is given by (6.1). $q.e.d.$

The ring of correspondence on C ([11]) is isomorphic to the ring of endomorphisms of the Jacobian variety $\mathrm{End}(J)$, and the latter is isomorphic to $\mathbf{Z}$ for a general curve C, generated by the identity. The above method can be applied to other correspondence (at least of relatively small degree), but we do not go into this since the structure of $\mathrm{End}(J)$ is not so simple in general.

On the other hand, it should be remarked how the use of correspondence clarifies the occurrence of singular fibres of type I_4 in the special case (ii) or (iii) of Proposition 3.3.

PROPOSITION 6.2. *In the situation of Lemma 3.2, the graph of the automorphism ϕ is mapped to an irreducible component of the singular fibre (of type I_4) over $t = 1$ or $t = -1$.*

Proof Suppose ϕ is the first automorphism defined by (3.4) in case $c_1 = c_3$. The generic point of the graph Γ_ϕ is $(x_1, y_1) \times (x_2, y_2)$ where $x_2 = 1/x_1$. By (2.3), we have

$$(6.3) \qquad X = x_1 + \frac{1}{x_1}, \quad t = x_1 x_2 = 1.$$

Hence Γ_ϕ is mapped into the fibre over $t = 1$. Since the six branch points of the double cover $C \to \mathbf{P}^1$ are stable under ϕ, we may assume for example $d_2 = 1/d_1$. Then we have $d_1 d_2 = 1 = d_3 d_4$ by (2.2). Thus the two I_2-fibres over $t = d_1 d_2$ and $d_3 d_4$ in general has the confluence at $t = 1$ and the resulting fibre is of type I_4 by Proposition 3.3 or 5.1. The other case $c_1 = -c_3$ is similar.   *q.e.d.*

7. An explicit generator of height 1

According to the classical theory recalled in §1, we should have rational points $P \in E_W(k(t))$ of height 1. (The point given by Proposition 6.1 has height 4 in general.) The height formula (4.2) (with $\chi = 2$ for K3)

$$(7.1) \qquad 1 = \langle P, P \rangle = 4 + 2 \cdot 0 - \frac{1}{2} \times 6,$$

suggests a possibility of a section $P = (x, y)$ where $x = \sum_{n=0}^{4} A_n t^n$ is a degree 4 polynomial such that

$$(7.2) \qquad \sum_{n=0}^{4} A_n (d_i d_j)^n = N'(d_i, d_j | d_k, d_l)$$

for all six $i, j (i < j)$. Here the RHS is the x-coordinates of the node of the degenerate Weierstrass cubic (5.1) at $t = d_i d_j$; explicitly we have

$$(7.3) \qquad N'(d_1, d_2 | d_3, d_4) = (d_1 d_2) N(d_1, d_2 | d_3, d_4).$$

Now we can uniquely solve the 6 linear equations (7.2) in the 5 unknown A_n in the same way as in §4, and the result is expressed in terms of the elementary symmetric functions c_n of $d_1, \ldots, d_4$ (see (2.1)) as follows:

$$(7.4) \qquad A_0 = A_4 = \frac{1}{4}, \ A_1 = A_3 = -\frac{c_2}{6}, \ A_2 = \frac{c_1 c_3 + 2}{12}.$$

In this way, we find a beautiful $k(t)$-rational point:

PROPOSITION 7.1. *There is a rational point $P_1 \in E_W(k(t))$ with*

$$(7.5) \qquad x(P_1) = \frac{1}{4} t^4 - \frac{c_2}{6} t^3 + \frac{c_1 c_3 + 2}{12} t^2 - \frac{c_2}{6} t + \frac{1}{4}, \ y(P_1) = \frac{1}{8} \prod_{i<j} (t - d_i d_j).$$

which has height 1. The duplicated point $2P_1$ of height 4 is equal (up to sign) to the point P_{id} in Proposition 6.1 arising from the identity correspondence.

Proof The point (7.5) does not meet the singular point of the cuspidal cubic
(5.1) at $t = 0$ or ∞, but it does meet the node at the six value $t = d_i d_j$ because
it is so arranged by (7.2) above. Hence the height of P_1 is equal to 1 by (7.1).
We can check that this is true even in the confluent cases (ii) and (iii). A direct
computation shows $2P_1 = \pm P_{id}$.   *q.e.d.*

THEOREM 7.2. *The above point P_1 of height 1 is a generator of the Mordell-
Weil group $E_W(k(t))$ modulo the 2-torsion subgroup, for any genus two curve C
with* $\mathrm{End}(J) = \mathbf{Z}$.

Proof If $\mathrm{End}(J) = \mathbf{Z}$, we have $r = \rho(J) = 1$, and we are in the case (i) with
six I_2-fibres. Suppose P is a generator modulo torsion of the rank 1 lattice and
$P_1 = nP$ for some positive integer n. Then the height $\langle P, P \rangle$ of P is equal to $1/n^2$.
On the other hand, it follows from the formula (4.2) that it is an integer or a half
integer. Therefore we have $n = 1$ which implies that $P_1 = P$ is a generator.   *q.e.d.*

COROLLARY 7.3. *If* $\mathrm{End}(J) = \mathbf{Z}$ *(or $\rho(J) = 1$), the Néron-Severi lattice of the
Kummer surface* $\mathrm{NS}(\mathrm{Km}(J))$ *has* $\mathrm{rk} = 17$ *and* $\det = 2^6$.

Proof This is wellknown (cf.[**7**]), but it follows also from the above consideration. In fact, the index I of the narrow MWL $E(K)^0$ in $E(K)$ is equal, in general,
to the index $[N : T \oplus L]$ where $L = T^\perp$ (cf. [**9**]). In our case, $L \simeq E(K)^0$ is
generated by $2P_1$ of height 4. Hence $\det L = 4$ and $I = 2|E(K)_{tor}| = 2^3$. Thus
$\det N$ is equal to $\det(T)\det(L)/I^2 = 2^{10} \cdot 4/2^6 = 2^6$.   *q.e.d.*

Finally let us consider the elliptic K3 surface $\tilde{S}$ which is obtained by the base
change $\mathbf{P}^1_u \to \mathbf{P}^1_t$, $u \mapsto t = u^2$, whose generic fibre $\tilde{E}$ is defined by (3.2) over $k(u)$:

$$(7.6) \qquad\qquad \tilde{E} = \mathcal{E}_W \otimes_{k(t)} k(u) : y^2 = x^3 + a_4(u^2)x + a_6(u^2).$$

THEOREM 7.4. *The elliptic K3 surface $\tilde{S}$ has the semi-stable singular fibres
only: (i) $I_2 \times 12$, (ii) $I_2 \times 8 + I_4 \times 2$, or (iii) $I_2 \times 4 + I_4 \times 4$, according to the three
cases in Lemma 3.1. The Mordell-Weil group $\tilde{E}(k(u))$ contains with finite index
the subgroup generated by $E(k(t))$ and $\mathcal{E}_W(k(t))$. In particular, if $\mathrm{End}(J) = \mathbf{Z}$, it
contains 3 independent rational points $\tilde{P}_1, \tilde{Q}_1, \tilde{R}_1$ of respective height 2,1,1 which
are mutually orthogonal.*

Proof This follows easily from Proposition 3.3, Theorems 4.1 and 7.2, because
the height is multiplied by the degree of the base change (cf.[**9**]).   *q.e.d.*

This result might be of some use in the study of supersingular K3 surfaces in
positive characteristic, since the elliptic fibration is always semi-stable.

References

[1] Cassels, J.W.S.: Lectures on Elliptic Curves, Cambridge Univ. Press (1991).
[2] Hudson, R.W.H.T.: Kummer's Quartic Surfaces, Cambridge Univ. Press 1905; reissued 1990.
[3] Kodaira, K.: On compact analytic surfaces II-III, Ann. of Math. 77, 563-626(1963); 78,
     1-40(1963); Collected Works, III, 1269-1372, Iwanami and Princeton Univ. Press (1975).
[4] Mumford, D.: Abelian Varieties, Oxford Univ. Press (1970).
[5] Nikulin, V.: An analogue of the Torelli theorem for Kummer surfaces of Jacobians, Math.
     USSR Izv. 8, 21–41 (1974).
[6] Oguiso, K., Shioda, T.: The Mordell-Weil lattice of a rational elliptic surface, Comment.
     Math. Univ. St. Pauli 40, 83–99 (1991).

[7] Piateckii-Shapiro, I., Shafarevich, I.R.: A Torelli theorem for algebraic surfaces of type K3, Math. USSR Izv. 5, 547–587 (1971).

[8] Shimada, I.: On elliptic K3 surfaces, Michigan Math. J. 47 (2000), 423–446.

[9] Shioda, T.: On the Mordell-Weil lattices, Comment. Math. Univ. St. Pauli 39, 211– 240 (1990).

[10] — : Correspondence of elliptic curves and Mordell-Weil lattices of certain $K3$ surfaces, Proc. Murre Conf. Leiden, 2004 (in preparation).

[11] Weil, A.: Variétés abéliennes et courbes algébriques, Hermann, Paris (1948/1973).

DEPARTMENT OF MATHEMATICS RIKKYO UNIVERSITY, 3-34-1 NISHI-IKEBUKURO TOSHIMA-KU, TOKYO 171-8501, JAPAN

E-mail address: shioda@rkmath.rikkyo.ac.jp

Contemporary Mathematics
Volume **422**, 2007

Niemeier Lattices and K3 groups

De-Qi Zhang

This paper is dedicated to Professor Dolgachev on the occasion of his Sixtieth Birthday.

ABSTRACT. In this note, we consider algebraic $K3$ surfaces X with an action by the alternating group A_5. We show that if a cyclic extension $A_5.C_n$ acts on X then $n = 1$, 2, or 4. We also determine the A_5-invariant sublattice of the K3 lattice and its discriminant form.

Introduction

We work over the complex numbers field $\mathbf{C}$. A **K3** surface X is a simply connected projective surface with a nowhere vanishing holomorphic 2-form ω_X. In this note, we will consider finite groups in $\operatorname{Aut}(X)$. An element $h \in \operatorname{Aut}(X)$ is **symplectic** if h acts trivially on the 2-form ω_X. A group $G_N \subseteq \operatorname{Aut}(X)$ is **symplectic** if every element of G_N is symplectic.

According to Nikulin [Ni1], Mukai [Mu1] and Xiao [Xi], there are exactly 80 abstract finite groups which can act symplectically on $K3$ surfaces. Among these 80, there are exactly three non-abelain simple groups A_5, $L_2(7)$ and A_6.

To be more precise, as in **(1.0)** below, for every finite group G acting on a $K3$ surface X, the symplectic elements of G (i.e., those h acting trivially on the nonzero 2-form ω_X) form a normal subgroup G_N such that $G/G_N \cong \mu_I$ (the cyclic group of order I in $\mathbf{C}^*$). Namely, we have $G = G_N.\mu_I$ (see **Notation** below). The natural number $I = I(G)$ is determined by G and called the **transcendental value** of G.

It is proved in [OZ3] and [KOZ1, 2] that when G_N is either one of the two bigger simple groups above, the transcendental value $I(G) \neq 3$. As expected or unexpected, the same is true for the smaller (indeed the smallest non-abelian simple group) A_5:

Theorem A. Suppose that $G = A_5.\mu_I$ acts faithfully on a $K3$ surface (assuming that $G_N = A_5$). Then $G = A_5 : \mu_I$ (semi-direct product) and $I = 1$, 2, or 4.

1991 *Mathematics Subject Classification.* Primary 14J28; Secondary 14J50, 14L30.
Key words and phrases. K3 surface, automorphism group, Niemeier lattice.
The first author was supported in part by an Academic Research Fund of NUS..

In general, for a group of the form $G = A_5.C_n$ acting on a $K3$ surface (here G_N might be bigger than A_5; and C_n an abstract cyclic group of order n), we have a similar result:

Theorem B. Suppose that a group of the form $G = A_5.C_n$ acts faithfully on a $K3$ surface. Then $G = A_5 : C_n$ and $I = 1, 2$, or 4. Moreover, $G_N = A_5$ (and hence $C_n = \mu_n$ in the notation above or **(1.0)**) unless $G = G_N = S_5$.

We can determine the A_5-invariant sublattice of the $K3$ lattice in the result below (see Theorem **2.1** for the detailed version), which has application in helping determine the transcendental lattice T_X and hence the surface itself (when rank $T_X = 2$).

Theorem C. Suppose that A_5 acts faithfully on a $K3$ surface X. Then we have:
(1) The A_5-invariant sublattice L^{A_5} of the $K3$ lattice $L = H^2(X, \mathbf{Z})$ has rank 4. The A_5-invariant sublattice $S_X^{A_5}$ of the Neron Severi lattice S_X has rank equal to 1 or 2.
(2) The discriminant group $A_{L^{A_5}} = \mathrm{Hom}(L^{A_5}, \mathbf{Z})/L^{A_5}$ equals one of the following (see Theorem **2.1** for the corresponding intersection forms):

$$\mathbf{Z}/(30) \oplus \mathbf{Z}/(30), \quad \mathbf{Z}/(30) \oplus \mathbf{Z}/(10), \quad \mathbf{Z}/(20) \oplus \mathbf{Z}/(20),$$

$$\mathbf{Z}/(60) \oplus \mathbf{Z}/(20), \quad \mathbf{Z}/(60) \oplus \mathbf{Z}/(20) \oplus \mathbf{Z}/(2) \oplus \mathbf{Z}/(2).$$

Remark D.
(1) Theorem C is applicable in the following situation: Suppose in addition that a non-symplectic involution $g \in \mathrm{Aut}(X)$ commutes with every element in A_5 and that the fixed locus X^g is a union of a genus ≥ 2 curve C and s (≥ 1) smooth rational curves D_i. Then $S_X^{A_5}$ contains $L_0 = \mathbf{Z}[C, \sum_{i=1}^s D_i]$ as a sublattice of finite index d_1. Note that L^{A_5} contains $S_X^{A_5} \oplus T_X$ as a sublattice of finite index d. So $|L_0||T_X| = d_1^2 |S_X^{A_5}||T_X| = d_1^2 d^2 |L^{A_5}|$ while $-|L^{A_5}| = 30^2, 3 \times 10^2, 20^2, 3 \times 20^2$, or 3×40^2 as given in Theorem C. This is a restriction imposed on $|T_X|$. In [Z2], we determined d_1, d, and $|T_X|$ using the existence of the extra μ_4 in (the impossible case:) $A_5.\mu_4$ where T_X then has the intersection form $\mathrm{diag}[2m, 2m]$ for some $m \geq 1$.
(2) The same construction in [OZ3, Appendix] shows that there is a smooth non-isotrivial family of $K3$ surfaces $f : \mathcal{X} \to \mathbf{P}^1$ such that all fibres admit A_6 actions and infinitely many of them are algebraic $K3$ surfaces. So, the symplectic part alone can not determine the surface uniquely, and the study of transcendental value is needed.

In the paper [Z2], we shall prove the following result. Especially, the I in Theorems A and B can only be 1 or 2. The removal of the case $I = 4$ would make the current article lengthy, so it is postponed to [Z2] to be treated in the process of proving other results.

Theorem E (see [Z2]). Suppose that a finite group G acts faithfully on a $K3$ surface. Suppose further that G contains A_5 as a normal subgroup. Then G equals one of the following four groups, each realizable:

$$A_5, \quad S_5, \quad A_5 \times \mu_2, \quad S_5 \times \mu_2.$$

The main tools of the paper are the Lefschetz fixed point formula (both the topo-logical version and vector bundle version due to Atiyah-Segal-Singer [AS2, 3]), the representation theory on the $K3$ lattice and the study on automorphism groups of Niemeier lattices (in connection with Golay binary or ternary codes) where the latter is much inspired by Conway-Sloane [CS], Kondo [Ko1] and Mukai [Mu2].
The reduction to the study of automorphisms of Niemeier lattices was pioneered by Nikulin (see e.g., [Ni3, end of section 1.14]) and further developed by Kondo (see e.g. [Ko1]).

We believe that the way of combining different very powerful machinaries to deduce results as done in the paper should be applicable to the study of other problems. Our humble paper also demonstrates the powerfulness and depth of these algebraic results in the study of geometry.

Notation.
1. S_n is the symmetric group in n letters, A_n ($n \geq 3$) the alternating group in n letters, $\mu_I = \langle \exp(2\pi\sqrt{-1})/I \rangle$ the multiplicative group of order I in $\mathbf{C}^*$ and C_n an abstract cyclic group of order n.
2. For a group G, we write $G = A.B$ if A is normal in G so that $G/A = B$. We write $G = A : B$ if assume further that A is normal in G and B is a subgroup of G such that the composition $B \to G \to G/A = B$ is the identity (we say then that G is a **semi-direct product** of A and B).
3. For groups $H \leq G$ and $x \in G$ we denote by $c_x : H \to G$ ($h \mapsto c_x(h) = x^{-1}hx$) the **conjugation** map.
4. For a $K3$ surface X, we let S_X and T_X be the Neron-Severi and transcendental lattices. So the $K3$ lattice $H^2(X, \mathbf{Z})$ contains $S_X \oplus T_X$ as a sublattice of finite index.

Acknowledgement. This work was done during the author's visit to Hokkaido University, University of Tokyo and Korea Institute for Advanced Study in the summer of 2004. The author would like to thank the institutes and Professors F. Catanese, Alfred Chen, I. Dolgachev, J. Keum, S. Kondo, M. L. Lang, K. Oguiso and I. Shimada for the hospitality and valuable suggestions. He also likes to thank the annonymous referee for valuable suggestions.

§1. Preliminary Results

(1.0). In this section, we will prepare some basic results to be used late. Let X be a $K3$ surface with a non-zero 2-form ω_X and let $G \subseteq \mathrm{Aut}(X)$ be a finite group of automorphisms. For every $h \in G$, we have $h^*\omega_X = \alpha(h)\omega_X$ for some scalar $\alpha(h) \in \mathbf{C}^*$. Clearly, $\alpha : G \to \mathbf{C}^*$ is a homomorphism. A fact in basic group theory says that $\alpha(G)$ is a finite cyclic group, so $\alpha(G) = \mu_I = \langle \exp(2\pi\sqrt{-1}/I) \rangle$ for some $I \geq 1$. This natural number $I = I(G)$ is called the **transcendental** value of G. It is known that $I = I(G)$ for some G if and only if that the Euler function $\varphi(I) \leq 21$ and $I \neq 60$ [MO].
Set $G_N = \mathrm{Ker}(\alpha)$. Then we have the **basic exact sequence** below:

$$1 \longrightarrow G_N \longrightarrow G \xrightarrow{\alpha} \mu_I \longrightarrow 1.$$

For the G in the basic exact sequence, we write $G = G_N.\mu_I$, though there is no guarantee that $G = G_N : \mu_I$ (a semi-direct product).

1.0A. If G is a finite perfect group, i.e., the commutator group $[G, G] = G$ (especially, if G is a non-abelian simple group like A_5), then $G = G_N$.
1.0B. G_N acts trivially on the transcendental lattice T_X (Lefschetz theorem on $(1, 1)$-classes).
1.0C. If a subgroup $H \leq G_N$ fixes a point P, then $H < SL(T_{X,P}) \cong SL_2(\mathbf{C})$ [Mu1, **(1.5)**]. The finite subgroups of $SL_2(\mathbf{C})$ are listed up in [Mu1, **(1.6)**]. These are cyclic C_n, binary dihedral (or quaternion) Q_{4n} ($n \geq 2$), binary tetrahedral T_{24}, binary octahedral O_{48} and binary icosahedral I_{120}.

Lemma 1.1. Suppose that $G := A_5.\mu_I$ acts faithfully on a $K3$ surface X. Then we have:
(1) The Picard number $\rho(X) \geq 19$, and $I = 1, 2, 3, 4, 6$. Moreover, $\rho(X) = 20$ if $I \geq 3$.
(2) We have $G = A_5 : \mu_I$, i.e., a semi-prodcut of a normal subgroup A_5 and a subgroup μ_I of G. Moreover, $G = A_5 \times \mu_I$ if $I = 3$.

Proof. (1) In notation of [Xi, the list], $\rho(X) = \operatorname{rank} S_X \geq c + 1 = 19$. Also the Euler function $\varphi(I)$ divides $\operatorname{rank} T_X = 22 - \rho(X)$ by [Ni1, Theorem 0.1]. So (1) follows.
(2) Let $g \in G$ such that $\alpha(g)$ is a generator of μ_I. Since $\operatorname{Aut}(A_5) = S_5 > A_5$ and the conjugation homomorphism $A_5 \to \operatorname{Aut}(A_5)$ ($x \mapsto c_x$) is an isomorphism onto A_5, the conjugation map c_g equals $c_{(12)a}$ or c_a on A_5 for some $a \in A$. Replacing g by ga^{-1}, we may assume that $c_g = c_{(12)}$ or c_{id}. Thus g^2 commutes with every element in A_5. If $2|I$, then $g^I \in \operatorname{Ker}(\alpha) = A_5$ is in the centre of A_5 (which is trivial) and hence $\operatorname{ord}(g) = I$; thus $G = A_5 : \mu_I$. If $I = 3$, then $\gcd(3, \operatorname{ord}(g)/3) = 1$ as proved in [IOZ] or [Og, Proposition 5.1]; so replacing g by g^ℓ with $\ell = \operatorname{ord}(g)/3$ (or $2\operatorname{ord}(g)/3$), we have $G = A_5 \times \langle g \rangle = A_5 \times \mu_3$.

The second result below [Ni1, §5] is crucial in classifying symplectic groups in [Mu1]. For the first, see [Ni2], [Z1] or [Z2, Lemma 1.2], where the Hodge index theorem is also used here.

Lemma 1.2. (1) Let h be a non-symplectic involution on a $K3$ surface X. Then X^h is a disjoint union of s smooth curves C_i with $0 \leq s \leq 10$. To be precise, X^h (if not empty) is either a disjoint union of a genus ≥ 2 curve C and a few $\mathbf{P}^1$'s, or a disjoint union of a few elliptic curves and $\mathbf{P}^1$'s, or a disjoint union of a few $\mathbf{P}^1$'s.
(2) If δ is a non-trivial symplectic automorphism of finite order on a $K3$ surface X, then $\operatorname{ord}(\delta) \leq 8$ and X^δ is a finite set. To be precise, if $\operatorname{ord}(\delta) = 2, 3, 4, 5, 6, 7, 8$, then $|X^\delta| = 8, 6, 4, 4, 2, 3, 2$, respectively. In particular, if $A_5 \subseteq \operatorname{Aut}(X)$ then $\sum_{\delta \in A_5} \chi_{\mathrm{top}}(X^\delta) = 360$ (see **(1.0A)**).

For an automorphism h on a smooth algebraic surface Y, we split the pointwise fixed locus as the disjoint union of 1-dimensional part and the isolated part: $Y^h = Y^h_{1-\dim} \coprod Y^h_{\mathrm{isol}}$. The proof of (1) below is similar to that for (1) in **(1.2)**.

1.3. (1) $Y^h_{1-\dim}$ (if not empty) is a disjoint union of smooth curves.
(2) The Euler number $\chi_{\text{top}}(Y^h_{1-\dim}) = \sum_C (2 - 2g(C)) = 2n_h$ for some integer n_h, where C runs in the set $Y^h_{1-\dim}$ of curves.
(3) The Euler number $\chi_{\text{top}}(Y^h) = m_h + 2n_h$, where $m_h = |Y^h_{\text{isol}}|$.

The results of [IOZ] below follow from the application of Lefschetz fixed point formula to the trivial vector bundle in Atiyah-Segal-Singer [AS2, AS3, pages 542 and 567]. For a proof, see [OZ1, Lemma 2.3] and [Z2, Proposition 1.4].

Lemma 1.4. Let X be a $K3$ surface and let $h \in \text{Aut}(X)$ be of order I such that $h^*\omega_X = \eta_I \omega_X$ for some primitive I-th root η_I of 1.
(1) Suppose that $I = 3$. Then $m_h = 3 + n_h$ and hence $\chi_{\text{top}}(X^h) = 3(1 + n_h)$. Moreover, $-3 \le n_h \le 6$.
(2) Suppose that $I = 3$. If $\delta \in \text{Aut}(X)$ is symplectic of order 5 and commutes with h. Then $|X^{h\delta}| = 4$.

The following result can be found in [Ni1, Theorem 0.1], [MO, Lemma (1.1)], or [OZ3, Lemma (2.8)].

Lemma 1.5. Suppose that X is a $K3$ surface of Picard number $\rho(X) = 20$ and g an order-3 automorphism such that $g^*\omega_X = \eta_3 \omega_X$ with a primitive 3rd root η_3 of 1. Then we can express the transcendental lattice T_X as $T_X = \mathbf{Z}[t_1, t_2]$ so that $t_2 = g^*(t_1)$, $g^*(t_2) = -(t_1 + t_2)$. In particular, for some $m \ge 1$, the intersection form $(t_i.t_j) = \begin{pmatrix} 2m & -m \\ -m & 2m \end{pmatrix}$.

Now we assume that $G = G_N.\mu_I$ (with $I = I(G)$) acts on a $K3$ surface X. When $G_N = A_5$, we will determine the action of G_N on the Neron Severi lattice S_X of X:

Lemma 1.6. (1) Suppose that A_5 acts on a $K3$ surface X, and rank $S_X = 20$ (this is true if $I \ge 3$ by **(1.1)**). Then we have the irreducible decomposition below (in the notation of Atlas for the characters of A_5), where χ_1 (the trivial character), χ_4 and χ_5 have dimensions 1, 4 and 5, respectively, where χ'_i is a copy of χ_i:

$$S_X \otimes \mathbf{C} = \chi_1 \oplus \chi'_1 \oplus \chi_4 \oplus \chi'_4 \oplus \chi_5 \oplus \chi'_5.$$

(2) For conjugacy class nA (and nB) of order n in A_5 and the characters χ_i of A_5, we have the following **Table 1** from [Atlas], where Z is respectively $1A$, $2A$, $3A$, $5A$ or $5B$:

$$[\chi_1, \chi_2, \chi_3, \chi_4, \chi_5](Z) = [1, 3, 3, 4, 5], \quad [1, -1, -1, 0, 1], \quad [1, 0, 0, 1, -1],$$
$$[1, (1 - \sqrt{5})/2, (1 + \sqrt{5})/2, -1, 0], \quad [1, (1 + \sqrt{5})/2, (1 - \sqrt{5})/2, -1, 0].$$

Proof. The assertion(1) appeared in [Z2]. For the readers' convenience, we re-prove it here. Applying the Lefschetz fixed point formula to the action of A_5 on $H^*(X, \mathbf{Z}) = \oplus_{i=0}^4 H^i(X, \mathbf{Z})$ and noting that $H^2(X, \mathbf{Z})$ contains $S_X \oplus T_X$ as a finite index sublattice, we obtain (see also **(1.0A-B)** and **(1.2)**):

$$2 + \text{rank}\, T_X + \text{rank}(S_X)^{A_5} = \text{rank}\, H^*(X, \mathbf{Z})^{A_5} = \frac{1}{|A_5|} \sum_{a \in A_5} \chi_{\text{top}}(X^a) = 360/60 = 6.$$

Thus rank $S_X^{A_5} = 2$. So the irreducible decomposition is of the following form, where a_i are non-negative integers:

$$S(X) \otimes \mathbf{C} = 2\chi_1 \oplus a_2\chi_2 \oplus a_3\chi_3 \oplus a_4\chi_4 \oplus a_5\chi_5.$$

Using the topological Lefschetz fixed point formula, the fact that $\operatorname{rank} T(X) = 2$ and **(1.0B)**, we have, for $a \in A_5$, that:

$$\chi_{\mathrm{top}}(X^a) = 2 + \operatorname{rank} T_X + \operatorname{Tr}(a^*|S(X))$$

Running a through the five conjugacy classes and calculating both sides, using **(1.2)** and the character Table 1 in (2), we obtain the following system of equations:

$$20 = 2 + 3(a_2 + a_3) + 4a_4 + 5a_5,$$

$$4 = 2 - (a_2 + a_3) + a_5,$$

$$2 = 2 + a_4 - a_5,$$

$$0 = 2 + \frac{1 - \sqrt{5}}{2}a_2 + \frac{1 + \sqrt{5}}{2}a_3 - a_4,$$

$$0 = 2 + \frac{1 + \sqrt{5}}{2}a_2 + \frac{1 - \sqrt{5}}{2}a_3 - a_4.$$

Now, we get the result by solving this system of Diophantine equations.

The two results below are either easy or well known and will be frequently used in the arguments of the subsequent sections.

Lemma 1.7. Let $f : A_5 \to S_r$ $(r \geq 2)$ be a homomorphism.
(1) If $r = 2$, 3, or 4, then f is trivial.
(2) If $\operatorname{Im}(f)$ is a transitive subgroup of the full symmetry group S_r in r letters $\{1, 2, \ldots, r\}$ (whence $r \geq 5$ by (1)), then $r \,||\, |A_5|$ with $|A_5|/r$ equal to the order of the subgroup of A_5 stabilizing the letter 1, so $r \in \{5, 6, 10, 12, 15, 20, 30\}$.

Lemma 1.8. (1) $\operatorname{Aut}(\mathbf{P}^1)$ includes A_5 but not S_5 [Su, Theorem 6.17]. The action of A_5 on $\mathbf{P}^1$ is unique up to isomorphisms.
(2) Every A_5 in $PGL_3(\mathbf{C})$ is the image of an A_5 in $SL_3(\mathbf{C})$.
(3) The conjugation by $(12) \in S_5$ switches the two 3-dimensional characters χ_2 and χ_3 of A_5 [Atlas].
(4) If $\mathrm{id} \neq f \in \operatorname{Aut}(\mathbf{P}^1)$ is an automorphism of finite order, then f fixes exactly two point of $\mathbf{P}^1$ (by the diagonalization of a lifting of f to $GL_2(\mathbf{C})$).
(5) If f_r $(r = 2$ or 3) is an order$-r$ automorphism of an elliptic curve E, then either f_r acts freely on E, or the fix locus satisfies $|X^{f_r}| = 4$ (resp. $= 3$) if $r = 2$ (resp. $r = 3$) (by the Hurwitz formula).

Proof. (1) For the uniqueness of the action of A_5 on $\mathbf{P}^1$, one may assume the representation of $D_{10} = \langle \gamma = (12345), \sigma = (14)(23) \rangle$ is given by $\gamma : z \to \eta z$ with η a primitive 5-th root of 1 and $\sigma : z \to \alpha/z$. Note that $A_5 = \langle \gamma, \varepsilon \rangle$ with $\varepsilon = (12)(34)$. If one lets $\varepsilon : z \to (az + b)/(cz + d)$ be in $\operatorname{Aut}(\mathbf{P}^1)$, then one can check that $d = -a$ because $\operatorname{ord}(\varepsilon) = 2$, and also $b = -c\alpha$ because ε commutes with σ. So $\varepsilon : z \to (z - \alpha e)/(ez - 1)$ with $e = c/a$. The commutativity of $\varepsilon\sigma\gamma^2\varepsilon = (12)(45)$ with $\sigma\gamma^{-1} = (15)(24)$ implies that $e^2\alpha = \eta + \eta^{-1} - 1$. Now let $\rho : z \to e\alpha/z$ be in $\operatorname{Aut}(\mathbf{P}^1)$. Then $\rho^{-1}\gamma\rho : z \to \eta^{-1}z$, $\rho^{-1}\sigma\rho : z \to e^2\alpha/z$ and

$\rho^{-1}\varepsilon\rho : z \to (z - e^2\alpha)/(z - 1)$. Hence the action of A_5 on $\mathbf{P}^1$ is unique modulo isomorphisms.

(2) For an A_5 in $SL_3(\mathbf{C})$, see [Bu, §232]. The inverse $\widetilde{A}_5 \subset SL_3(\mathbf{C})$ of an $A_5 \subset PGL_3(\mathbf{C})$ is of the form $\widetilde{A}_5 = A_5 : \mu_3$ (indeed, a direct product) because the Schur multiplier $M(A_5) = 2$, coprime to 3 [Atlas]. So (2) follows.

§2. Alternating groups actions on the Niemeier lattices

For a $K3$ surface X, denote by $L = H^2(X, \mathbf{Z})$ the **K3 lattice**, $S_X = \operatorname{Pic} X$ (now) the **Neron-Severi lattice** and T_X the **transcendental lattice**. So $T_X = S_X^\perp$ in L and L contains a finite-index sublattice $S_X \oplus T_X$.

(2.0). Suppose that $G_N = A_5$ acts faithfully on X. In this section we shall prove Theorem C which is part of **(2.1)** below. Indeed, by the proof of **(1.6)**, we have $\operatorname{rank} L^{G_N} = \operatorname{rank} T_X + \operatorname{rank} S_X^{G_N} = 4$, so $(\operatorname{rank} T_X, \operatorname{rank} S_X, \operatorname{rank} S_X^{G_N}) = (2, 20, 2)$ or $(3, 19, 1)$.

Denote by $L^{G_N} := \{x \in L \,|\, g^*x = x \text{ for all } g \in G_N\}$ and its orthogonal $L_{G_N} := (L^{G_N})^\perp = \{x \in L \,|\, (x, y) = 0 \text{ for all } y \in L^{G_N}\}$. Then L^{G_N} contains $S_X^{G_N} \oplus T_X$ as a sublattice of finite index by **(1.0A-B)**.
By [Ko1, Lemmas 5 and 6], there are a (non-Leech) Niemeier lattice $N(Rt)$, a primitive embedding $A_1 \oplus L_{G_N} \subset N(Rt)$ and a faithful action of G_N on $N(Rt)$ such that $L_{G_N} = N(Rt)_{G_N}$, and the action of G_N on the summand A_1 is trivial and stabilizes a Weyl chamber (one of whose codimension one faces corresponds to this A_1). Moreover, $G_N \leq S(N(Rt)) := O(N(Rt))/W(N(Rt)) \leq O(Rt)/W(N(Rt))(=: \operatorname{Sym}(Rt))$, where $\operatorname{Sym}(Rt)$ is the full symmetry group of the Coxeter-Dynkin diagram Rt. Note that $\operatorname{rank} N(Rt)^{G_N} = 2 + \operatorname{rank} L^{G_N} = 6$ and the discriminant groups satisfy:

$$(*) \quad A_{L^{G_N}} \cong A_{L_{G_N}}(-1) = A_{N(Rt)_{G_N}}(-1) \cong A_{N(Rt)^{G_N}}.$$

Now $N(Rt)^{G_N}$ is a rank 6 lattice generated by $e_1, \ldots, e_6$ say. Denote by $M = (e_i.e_j)$ the intersection matrix and $M^{-1} = (f_1, \ldots, f_6)$ with f_j column vectors and set $e_i^* = (e_1, \ldots, e_6)f_i$. Then $(N(Rt)^{G_N})^\vee = \operatorname{Hom}(N(Rt)^{G_N}, \mathbf{Z})$ has the dual basis $\{e_1^*, \ldots, e_6^*\}$ with the intersection matrix $(e_i^*.e_j^*)_{1 \leq i,j \leq 6} = M^{-1}$. The discriminant groups satisfy $A_{L^{G_N}} \cong A_{N(Rt)^{G_N}} = \mathbf{Z}[e_1^*, \ldots, e_6^*]/\mathbf{Z}[e_1, \ldots, e_6]$.

In this section, we shall prove the following result (much inspired by [Ko1]), which (and the proof of which) should be useful in studying $\operatorname{Aut}(X)$ from the $K3$ lattice point of view. This result is used in [Z2, Lemma 3.5].

Theorem 2.1. Suppose that $G_N = A_5$ acts faithfully on a $K3$ surface X.
(1) We have $Rt = 24A_1$ or $Rt = 12A_2$. The lattice $N(Rt)^{G_N}$ is of rank 6 and generated by $e_1, \ldots, e_6$ say. Denote by $M = (e_i.e_j)$ the intersection matrix and write $M^{-1} = (f_1, \ldots, f_6)$ with f_j column vectors and set $e_i^* = (e_1, \ldots, e_6)f_i$.

(2) If $Rt = 24A_1$, then the orbit decomposition of the G_N-action on the 24 simple roots is either one of

(i) $[1, 1, 5, 5, 6, 6]$,   (ii) $[1, 1, 1, 5, 6, 10]$,   (iii) $[1, 1, 1, 1, 5, 15]$,   (iv) $[1, 1, 1, 1, 10, 10]$.

If $Rt = 12A_2$, then the orbit decomposition of the G_N-action on the 24 simple roots is either one of

$$(v) \ [1, 1, 1, 1, 10, 10], \quad (vi) \ [1, 1, 5, 5, 6, 6],$$

where in (v) (resp. (vi)) $10A_2$ (resp. $5A_2$, or $6A_2$) is split into two orbits with 10 (resp. 5, or 6) disjoint roots each.

(3) For Case(2i), the intersection matrix $M_1 = (e_i.e_j)$ and its inverse M_1^{-1} are respectively:

$$\begin{pmatrix} -2 & 0 & 0 & -1 & -1 & -1 \\ 0 & -2 & 0 & -1 & -1 & -1 \\ 0 & 0 & -10 & 0 & 0 & -5 \\ -1 & -1 & 0 & -4 & -1 & -1 \\ -1 & -1 & 0 & -1 & -4 & -1 \\ -1 & -1 & -5 & -1 & -1 & -6 \end{pmatrix},$$

$$\begin{pmatrix} -23/30 & -4/15 & -1/10 & 1/6 & 1/6 & 1/5 \\ -4/15 & -23/30 & -1/10 & 1/6 & 1/6 & 1/5 \\ -1/10 & -1/10 & -1/5 & 0 & 0 & 1/5 \\ 1/6 & 1/6 & 0 & -1/3 & 0 & 0 \\ 1/6 & 1/6 & 0 & 0 & -1/3 & 0 \\ 1/5 & 1/5 & 1/5 & 0 & 0 & -2/5 \end{pmatrix}.$$

The discriminant group (cf. (2.0), $A_{L^{G_N}} \cong$) $A_{N(Rt)^{G_N}} =$ $\mathrm{Hom}(N(Rt)^{G_N}, \mathbf{Z})/N(Rt)^{G_N} \cong \mathbf{Z}/(30) \oplus \mathbf{Z}/(30)$ and is generated by cosets $\overline{e}_1^*$ and $\overline{e}_2^* + \overline{e}_3^* + \overline{e}_4^*$ with intersection form:

$$\begin{pmatrix} (\overline{e}_1^*)^2 & \overline{e}_1^*.(\overline{e}_2^* + \overline{e}_3^* + \overline{e}_4^*) \\ \overline{e}_1^*.(\overline{e}_2^* + \overline{e}_3^* + \overline{e}_4^*) & (\overline{e}_2^* + \overline{e}_3^* + \overline{e}_4^*)^2 \end{pmatrix} = \begin{pmatrix} -23/30 & -1/5 \\ -1/5 & -35/30 \end{pmatrix}.$$

(4) For Case(2ii), the intersection matrix $M_2 = (e_i.e_j)$ and its inverse M_2^{-1} are respectively:

$$\begin{pmatrix} -2 & 0 & 0 & -1 & -1 & -1 \\ 0 & -2 & 0 & -1 & -1 & -1 \\ 0 & 0 & -2 & -1 & 0 & 0 \\ -1 & -1 & -1 & -4 & -1 & -1 \\ -1 & -1 & 0 & -1 & -4 & -1 \\ -1 & -1 & 0 & -1 & -1 & -6 \end{pmatrix},$$

$$\begin{pmatrix} -11/15 & -7/30 & -1/10 & 1/5 & 1/6 & 1/10 \\ -7/30 & -11/15 & -1/10 & 1/5 & 1/6 & 1/10 \\ -1/10 & -1/10 & -3/5 & 1/5 & 0 & 0 \\ 1/5 & 1/5 & 1/5 & -2/5 & 0 & 0 \\ 1/6 & 1/6 & 0 & 0 & -1/3 & 0 \\ 1/10 & 1/10 & 0 & 0 & 0 & -1/5 \end{pmatrix}.$$

The discriminant group $A_{N(Rt)^{G_N}}$ is isomorphic to $\mathbf{Z}/(30) \oplus \mathbf{Z}/(10)$ and generated by the cosets $\overline{e}_1^*$ and $\overline{e}_3^*$ with intersection form:

$$\begin{pmatrix} (\overline{e}_1^*)^2 & \overline{e}_1^*.\overline{e}_3^* \\ \overline{e}_1^*.\overline{e}_3^* & (\overline{e}_3^*)^2 \end{pmatrix} = \begin{pmatrix} -11/15 & -1/10 \\ -1/10 & -3/5 \end{pmatrix}.$$

(5) For Case(2iii), the intersection matrix $M_3 = (e_i.e_j)$ and M_3^{-1} are respectively:

$$\begin{pmatrix} -2 & 0 & 0 & 0 & -1 & 0 \\ 0 & -2 & 0 & 0 & -1 & 0 \\ 0 & 0 & -2 & 0 & -1 & 0 \\ 0 & 0 & 0 & -2 & 0 & -1 \\ -1 & -1 & -1 & 0 & -4 & 0 \\ 0 & 0 & 0 & -1 & 0 & -8 \end{pmatrix},$$

$$\begin{pmatrix} -3/5 & -1/10 & -1/10 & 0 & 1/5 & 0 \\ -1/10 & -3/5 & -1/10 & 0 & 1/5 & 0 \\ -1/10 & -1/10 & -3/5 & 0 & 1/5 & 0 \\ 0 & 0 & 0 & -8/15 & 0 & 1/15 \\ 1/5 & 1/5 & 1/5 & 0 & -2/5 & 0 \\ 0 & 0 & 0 & 1/15 & 0 & -2/15 \end{pmatrix}.$$

The discriminant group $A_{N(Rt)^{G_N}}$ is isomorphic to $\mathbf{Z}/(30) \oplus \mathbf{Z}/(10)$ and generated by the cosets $\bar{e}_2^*$ and $\bar{e}_1^* + \bar{e}_4^*$ with intersection form:

$$\begin{pmatrix} (\bar{e}_2^*)^2 & \bar{e}_2^*.(\bar{e}_1^* + \bar{e}_4^*) \\ \bar{e}_2^*.(\bar{e}_1^* + \bar{e}_4^*) & (\bar{e}_1^* + \bar{e}_4^*)^2 \end{pmatrix} = \begin{pmatrix} -3/5 & -1/10 \\ -1/10 & 13/15 \end{pmatrix}.$$

(6) For Case(2iv), the intersection matrix $M_4 = (e_i.e_j)$ and M_4^{-1} are respectively:

$$\begin{pmatrix} -2 & 0 & 0 & 0 & -1 & 0 \\ 0 & -2 & 0 & 0 & -1 & 0 \\ 0 & 0 & -2 & 0 & 0 & -1 \\ 0 & 0 & 0 & -2 & 0 & -1 \\ -1 & -1 & 0 & 0 & -6 & 0 \\ 0 & 0 & -1 & -1 & 0 & -6 \end{pmatrix},$$

$$\begin{pmatrix} -11/20 & -1/20 & 0 & 0 & 1/10 & 0 \\ -1/20 & -11/20 & 0 & 0 & 1/10 & 0] \\ 0 & 0 & -11/20 & -1/20 & 0 & 1/10 \\ 0 & 0 & -1/20 & -11/20 & 0 & 1/10 \\ 1/10 & 1/10 & 0 & 0 & -1/5 & 0 \\ 0 & 0 & 1/10 & 1/10 & 0 & -1/5 \end{pmatrix}.$$

The discriminant group $A_{N(Rt)^{G_N}}$ is isomorphic to $\mathbf{Z}/(20) \oplus \mathbf{Z}/(20)$ and generated by the cosets $\bar{e}_1^*$ and $\bar{e}_3^*$ with intersection form:

$$\begin{pmatrix} (\bar{e}_1^*)^2 & \bar{e}_1^*.\bar{e}_3^* \\ \bar{e}_1^*.\bar{e}_3^* & (\bar{e}_3^*)^2 \end{pmatrix} = \begin{pmatrix} -11/20 & 0 \\ 0 & -11/20 \end{pmatrix}.$$

(7) For Case(2v), the intersection matrix $M_5 = (e_i.e_j)$ and its inverse M_5^{-1} are respectively:

$$\begin{pmatrix} -2 & 1 & 0 & 0 & 0 & 0 \\ 1 & -2 & 0 & 0 & 0 & -1 \\ 0 & 0 & -2 & 1 & 0 & 0 \\ 0 & 0 & 1 & -2 & 0 & -1 \\ 0 & 0 & 0 & 0 & -20 & 0 \\ 0 & -1 & 0 & -1 & 0 & -8 \end{pmatrix},$$

$$\begin{pmatrix}
-41/60 & -11/30 & -1/60 & -1/30 & 0 & 1/20 \\
-11/30 & -11/15 & -1/30 & -1/15 & 0 & 1/10 \\
-1/60 & -1/30 & -41/60 & -11/30 & 0 & 1/20 \\
-1/30 & -1/15 & -11/30 & -11/15 & 0 & 1/10 \\
0 & 0 & 0 & 0 & -1/20 & 0 \\
1/20 & 1/10 & 1/20 & 1/10 & 0 & -3/20
\end{pmatrix}.$$

The discriminant group $A_{N(Rt)^{G_N}}$ is isomorphic to $\mathbf{Z}/(60) \oplus \mathbf{Z}/(20)$ and generated by the cosets $\bar{e}_1^*$ and $\bar{e}_5^*$ with intersection form:

$$\begin{pmatrix} (\bar{e}_1^*)^2 & \bar{e}_1^*.\bar{e}_5^* \\ \bar{e}_1^*.\bar{e}_5^* & (\bar{e}_5^*)^2 \end{pmatrix} = \begin{pmatrix} -41/60 & 0 \\ 0 & -1/20 \end{pmatrix}.$$

(8) For Case(2vi), The intersection matrix $M_6 = (e_i.e_j)$ and M_6^{-1} are respectively:

$$\begin{pmatrix}
-2 & 1 & 0 & 0 & 0 & 0 \\
1 & -2 & 0 & 0 & -1 & 0 \\
0 & 0 & -10 & 0 & 0 & 0 \\
0 & 0 & 0 & -12 & 0 & 0 \\
0 & -1 & 0 & 0 & -4 & 0 \\
0 & 0 & 0 & 0 & 0 & -4
\end{pmatrix},$$

$$\begin{pmatrix}
-7/10 & -2/5 & 0 & 0 & 1/10 & 0 \\
-2/5 & -4/5 & 0 & 0 & 1/5 & 0 \\
0 & 0 & -1/10 & 0 & 0 & 0 \\
0 & 0 & 0 & -1/12 & 0 & 0 \\
1/10 & 1/5 & 0 & 0 & -3/10 & 0 \\
0 & 0 & 0 & 0 & 0 & -1/4
\end{pmatrix}.$$

The discriminant group $A_{N(Rt)^{G_N}} = \mathbf{Z}/(60) \oplus \mathbf{Z}/(20) \oplus \mathbf{Z}/(2) \oplus \mathbf{Z}/(2) = \mathbf{Z}/(10) \oplus \mathbf{Z}/(10) \oplus \mathbf{Z}/(12) \oplus \mathbf{Z}/(4)$ and the latter is generated by the cosets $\bar{e}_j^*$ ($j = 1, 3, 4, 6$).
(9) In both of the cases of M_2 and M_3, the discriminant group $A_{N(Rt)^{G_N}}$ is isomorphic to the group $\langle \bar{t}_1^*, \bar{t}_2^* \rangle \cong \mathbf{Z}/(30) \oplus \mathbf{Z}/(10)$ with the intersection matrix

$$(\bar{t}_i^*.\bar{t}_j^*) = \begin{pmatrix} 1/15 & 1/30 \\ 1/30 & 1/15 \end{pmatrix}.$$

We now prove **(2.1)**. Since rank $N(Rt)^{G_N} = 6$, the G_N-action on the 24 simple roots of Rt has exactly 6 orbits.
We argue as in the proof of [Ko1, Theorem 4]. The fact that $G_N = A_5 < S(N(Rt))$ implies that Rt is one of the following: $24A_1, 12A_2, 6A_4, 6D_4$.
If $Rt = 6A_4$, then $S(N(Rt)) = 2.PGL_2(5)$ ($< 2.S_6$) [CS, Ch 16, §1], where the order 2 element acts as a symmetry of order 2 on each connected component of Dynkin type A_4, and $PGL_2(5)$ acts on the set (identified with $\{0, 1, 2, 3, 4, \infty\}$) of 6 components of Rt as permutations in a natural way. Since A_5 is simple, the composition of homomorphisms below is an injection: $A_5 \subset S(N(Rt)) \to PGL_2(5)$, so we may assume that $A_5 < PGL_2(5)$. Since $G_N = A_5$ fixes one simple root of Rt by the construction, our A_5 is a subgroup of the stabilizer subgroup of $PGL_2(5)$ and this stablizer is of order $|PGL_2(5)|/6 = 20$. This is impossible because $|A_5| = 60 > 20$.
If $Rt = 6D_4$, then $S(N(Rt)) = 3.S_6$, where the order 3 element acts as a symmetry of order 3 on each connected component of Dynkin type D_4, and S_6 acts on the

set of 6 connected components of Rt as permutations. As above, the simplicity of G_N implies that the subgroup G_N of $S(N(Rt))$ is indeed a subgroup of S_6. Since $G_N = A_5$ fixes one simple root of Rt, our group A_5 is a subgroup $(= [S_5, S_5])$ of the stabilizer subgroup S_5 of S_6. So this A_5 acts transitively on the remaining 5 connected components of Rt and hence the G_N-action on the 24 simple roots has exactly 8 orbits, noting that one connected component of Rt is component wise fixed by G_N, a contradiction.

Suppose that $Rt = 12A_2$. Then $S(N(Rt)) = 2.M_{12}$, where the order 2 element acts as a symmetry of order 2 on each connected component of Dynkin type A_2, and the Mathieu group M_{12} acts on the set of 12 connected components of Rt as permutations. Let $r_{2k-1} + r_{2k}$ $(1 \leq k \leq 12)$ be the 12 connected components of Rt with r_j the 24 simple roots. Every non-trivial element of $N(Rt)/Rt$ is of the form $\sum_{i \in H} \pm(r_{2i-1} + 2r_{2i})/3$ where H is an element of the ternary Golay code and $|H| = 6, 9, 12$ [CS, Ch 3, §2.8.5]. Since the group $G_N = A_5$ is simple and fixes one simple root of Rt, this G_N is a subgroup of M_{12} and indeed, a subgroup of the stabilizer subgroup M_{11} of M_{12}. Suppose the G_N-orbit decomposition on the 12 connected components is $1 + a + b$. Then the G_N-orbit decomposition of the 24 simple roots is $[1, 1, a, a, b, b]$ (so $a + b = 11$), where aA_2 (resp. bA_2) is split into two G_N-orbits with a (resp. b) disjoint simple roots each. Thus Case (2v) or (2vi) occurs by **(1.7)**.

Suppose that $Rt = 24A_1$. Then $S(N(Rt)) = M_{24}$. The elements of $N(Rt)/Rt$ form the binary Golay code. Since $G_N = A_5$ fixes one simple root of Rt, our group G_N is the stabilizer subgroup M_{23} of M_{24}. Let the G_N-orbit decomposition of the 24 simple roots be $[1, a, b, c, d, e]$ with $a \leq b \leq c \leq d \leq e$ (so $a + b + c + d + e = 23$). By **(1.7)**, all a, b, c, d, e are in $\{1, 5, 6, 10, 12, 15, 20\}$ and hence Cases (2i) - (2vi) occur.

(4) According to the ordering of $[1, 1, 1, 5, 6, 10]$, we label the orbits as $O_1 = \{r_1\}, O_1' = \{r_2\}, O_1'' = \{r_3\}, O_5 = \{r_4, \ldots, r_8\}, O_6 = \{r_9, \ldots, r_{14}\}, O_{10} = \{r_{15}, \ldots, r_{24}\}$, where r_j are the 24 simple roots. We claim that $O_1 + O_1' + O_1'' + O_5$ (to be precise, after divided by 2) is an octad, and $O_1 + O_1' + O_6$ is also an octad (after relabelling O_1, O_1', O_1''). So

$$e_i = r_i (1 \leq i \leq 3), \ e_4 = \frac{1}{2}(O_1 + O_1' + O_1'' + O_5),$$

$$e_5 = \frac{1}{2}(O_1 + O_1' + O_6), \ e_6 = \frac{1}{2}(O_1 + O_1' + O_{10})$$

form a basis of $N(Rt)^{G_N}$, noting that the last dodecad is the complement of the symmetric sum (a dodecad) of the two octads above and that except for the above-mentioned two octads and two dodecads, there is no any other octad or dodecad which is a union of orbits.

Indeed, let Oct_1 be the unique octad containing O_5. Note that the cycle type in M_{24} of an order-5 element γ in A_5 is (5^4), Appendix B, Table 5.I]. So γ is of type (5^2) (resp. (5)) on O_{10} (resp. on O_5 and O_6). Since $\gamma(Oct_1) \cap Oct_1$ contains O_5, we have $\gamma(Oct_1) = Oct_1$. If Oct_1 contains an element of O_{10} then it contains the five images in O_{10} by the action of $\langle \gamma \rangle$, so $|Oct_1| \geq 10$, absurd. If Oct_1 contains an element r_j in O_6 we may choose γ not fixing r_j (note that the stabilizer subgroup of A_5, regarded as a subgroup of $\text{Sym}(O_6) = S_6$ and fixing an element $(\neq r_j)$ in O_6, has order 10 and hence gives rise to such γ). Then we will get a similar contradiction. Thus $Oct_1 = O_1 + O_1' + O_1'' + O_5$ as claimed.

Let Oct_2 be the unique octad containing the first 5 elements in O_6. Let γ be an order-5 element in A_5 fixing the last element in O_6. Then $\gamma(Oct_2) = Oct_2$. As above, this implies that Oct_2 is disjoint from O_5 and O_{10}. So either $Oct_2 = O_1 + O_1' + O_6$ after relabelling the 1-element orbits, or Oct_2 is the union of the 5 elements in O_6 and the three 1-element orbits (this leads to that the symmetric sum of Oct_1 and Oct_2 is a 10-word Golay code, absurd).

(3) For the orbit decomposition $[1, 1, 5, 5, 6, 6]$, we label the orbits as $O_1 = \{r_1\}$, $O_1' = \{r_2\}$, $O_5 = \{r_3, \dots, r_7\}$, $O_5' = \{r_8, \dots, r_{12}\}$, $O_6 = \{r_{13}, \dots, r_{18}\}$, $O_6' = \{r_{19}, \dots, r_{24}\}$. As in (4), we can prove that both $O_1 + O_1' + O_6$ and $O_1 + O_1' + O_6'$ are octads. Thus $N(Rt)^{G_N}$ has a basis below, noting that except for the two octads, the symmetric sum (a dodecad) of the two octads and the complement (another dodecad) of this dodecad, there is no other octad or dodecad which is the union of orbits:

$$e_i = r_i (i = 1, 2), \ e_3 = O_5,$$

$$e_4 = \frac{1}{2}(O_1 + O_1' + O_6), \ e_5 = \frac{1}{2}(O_1 + O_1' + O_6'), \ e_6 = \frac{1}{2}(O_1 + O_1' + O_5 + O_5')$$

.

(5) For the orbit decomposition $[1, 1, 1, 1, 5, 15]$, we label the orbits as $O_1 = \{r_1\}$, $O_1' = \{r_2\}$, $O_1'' = \{r_3\}$, $O_1''' = \{r_4\}$, $O_5 = \{r_5, \dots, r_9\}$, $O_{15} = \{r_{10}, \dots, r_{24}\}$. As in (4), we may assume that $O_1 + O_1' + O_1'' + O_5$ is an octad after relabelling the 1-element orbits and that there is no any other octad or dodecad which is a union of orbits. Thus $N(Rt)^{G_N}$ has a basis:

$$e_i = r_i (i = 1, 2, 3, 4), \ e_5 = \frac{1}{2}(O_1 + O_1' + O_1'' + O_5), \ e_6 = \frac{1}{2}(O_1''' + O_{15}).$$

(6) For the orbit decomposition $[1, 1, 1, 1, 10, 10]$, we label the orbits as $O_1 = \{r_1\}$, $O_1' = \{r_2\}$, $O_1'' = \{r_3\}$, $O_1''' = \{r_4\}$, $O_{10} = \{r_5, \dots, r_{14}\}$, $O_{10}' = \{r_{15}, \dots, r_{24}\}$. Take an order-5 element γ of A_5. So O_{10} splits into two 5-element subsets on each of which γ acts transitively. Let Oct_j $(j = 1, 2)$ be the unique octad containing the first (resp. second) 5-element subset. As in (4), we can show that each Oct_j is the union of the 5-element subset and three 1-element orbits. The symmetric sum of Oct_1 and Oct_2 is a dodecad which may be assumed to be $O_1 + O_1' + O_{10}$; its complement is also a dodecad. Except for these two dodecads, there is no any other dodecad which is a union of orbits. Thus $N(Rt)^{G_N}$ has a basis:

$$e_i = r_i (i = 1, 2, 3, 4), \ e_5 = \frac{1}{2}(O_1 + O_1' + O_{10}), \ e_6 = \frac{1}{2}(O_1'' + O_1''' + O_{10}').$$

(8) For $Rt = 12A_2$ and the orbit decomposition $[1, 1, 5, 5, 6, 6]$, we label the orbits as $O_1 = \{r_1\}$, $O_1' = \{r_2\}$, $O_5 = \{r_3, r_5, \dots, r_{11}\}$, $O_5' = \{r_4, r_6, \dots, r_{12}\}$, $O_6 = \{r_{13}, r_{15}, \dots, r_{23}\}$, $O_6' = \{r_{14}, r_{16}, \dots, r_{24}\}$, where $r_{2k-1} + r_{2k}$ $(1 \le k \le 12)$ are the 12 connected components of Rt. Every non-trivial element of the group $N(Rt)/Rt$ is represented by some $\gamma_H = \sum_{i \in H} \pm(r_{2i-1} + 2r_{2i})/3$ where H is an element of the ternary Golay code (so $|H| = 6, 9, 12$) which is also the Steiner system $St(5, 6, 12)$ [Atlas]. Let H_i with $i = 1$ (resp. $i = 2$) be the unique element of the ternary Golay code with $|H_i| = 6$ such that $\gamma_{H_1} = \frac{1}{3}\sum_{i=2}^{6} \pm(r_{2i-1} + 2r_{2i}) \pm \frac{1}{3}(r_{2j_1-1} + 2r_{2j_1})$ for some j_1 (resp. $\gamma_{H_2} = \frac{1}{3}\sum_{i=7}^{11} \pm(r_{2i-1} + 2r_{2i}) \pm \frac{1}{3}(r_{2j_2-1} + 2r_{2j_2})$ for some j_2); such Golay code can also be constructed from the binary Golay code = Steiner system $St(5, 8, 24)$ where such H_i is the intersection of a fixed dodecad and an octad. Using

the fact that an order-5 element in A_5 has cycle type (5^2) in M_{12} [EDM], as in the case of $Rt = 24A_1$, we can prove that $N(Rt)^{G_N}$ has a basis:

$$e_i = r_i(i = 1, 2), \quad e_3 = \sum_{k=2}^{6} r_{2k-1}, \quad e_4 = \sum_{k=7}^{12} r_{2k-1},$$

$$e_5 = \frac{1}{3} \sum_{k=1}^{6} (r_{2k-1} + 2r_{2k}), \quad e_6 = \frac{1}{3} \sum_{k=7}^{12} (r_{2k-1} + 2r_{2k}).$$

(7) For $Rt = 12A_2$ and the orbit decomposition $[1, 1, 1, 1, 10, 10]$, we label the orbits as $O_1 = \{r_1\}$, $O_1' = \{r_2\}$, $O_1'' = \{r_3\}$, $O_1''' = \{r_4\}$, $O_{10} = \{r_5, r_7, \ldots, r_{23}\}$, $O_{10}' = \{r_6, r_8, \ldots, r_{24}\}$, where $r_{2k-1} + r_{2k}$ $(1 \le k \le 12)$ are the 12 connected components of Rt. As in (8), we can prove that $N(Rt)^{G_N}$ has a basis:

$$e_i = r_i \, (1 \le i \le 4), \quad e_5 = \sum_{k=3}^{12} r_{2k-1}, \quad e_6 = \frac{1}{3} \sum_{k=1}^{12} (r_{2k-1} + 2r_{2k}).$$

(9) follows from the direct calculation. Indeed, in the case of M_2, the isomorphism $\varphi_2 : \langle \bar{t}_1^*, \bar{t}_2^* \rangle \to A_{N(Rt)^{G_N}}$ is given by $(\varphi_2(\bar{t}_1^*), \varphi_2(\bar{t}_2^*)) = (\bar{e}_1^*, \bar{e}_3^*) \begin{pmatrix} 2 & 7 \\ 1 & 0 \end{pmatrix}$. In the case of M_3, the isomorphism $\varphi_3 : \langle \bar{t}_1^*, \bar{t}_2^* \rangle \to A_{N(Rt)^{G_N}}$ is given as follows, whence **(2.1)** follows:

$$(\varphi_3(\bar{t}_1^*), \varphi_3(\bar{t}_2^*)) = (\bar{e}_2^*, \bar{e}_1^* + \bar{e}_4^*) \begin{pmatrix} 1 & 7 \\ 1 & -4 \end{pmatrix}.$$

§3. The proof of Theorems A and B

In this section, we shall prove Theorems A and B. We prove first the result below which includes Theorem A.

Theorem 3.1.
(1) There is no faithful group action of the form $A_5.\mu_3$ (see **(1.0)**) on a $K3$ surface.
(2) If $G = A_5.\mu_I$ acts faithfully on a $K3$ surface. Then $G = A_5 : \mu_I$ and $I = 1, 2$, or 4. (It is proved in [Z2] that $I = 4$ is impossible.)

(2) is a consequence of (1) and **(1.1)**. Indeed, if $I = 6$, then the subgroup $H = \alpha^{-1}(\mu_3)$ of $G = A_5.\mu_6$ is of the form $H = A_5.\mu_3$ which is impossible by (1). To prove (1), we need the following result first.

Lemma 3.2. Suppose that $G = A_5.\mu_3$ acts on a $K3$ surface X. Let $\zeta_3 = \exp(2\pi\sqrt{-1}/3)$.
(1) We have $G = A_5 \times \mu_3$. Moreover, a generator g of μ_3 can be chosen so that $g^*|S_X \otimes \mathbf{C} = \mathrm{diag}[1, 1, \zeta_3 I_4, \zeta_3^{-1} I_4, \zeta_3 I_5, \zeta_3^{-1} I_5]$, where the decompositoin here is compatible with that in **(1.6)** in the sense that $g^*|\chi_4 \oplus \chi_4' = \mathrm{diag}[\zeta_3 I_4, \zeta_3^{-1} I_4]$ and $g^*|\chi_5 \oplus \chi_5' = \mathrm{diag}[\zeta_3 I_5, \zeta_3^{-1} I_5]$. In particular, $\chi_{\mathrm{top}}(X^g) = -6$.
(2) We have $S_X^G = S_X^g = S_X^{A_5} = H^0(X, \mathbf{Z})^g$. This lattice is of rank 2 (whose $\mathbf{C}$-extension is $\chi_1 \oplus \chi_1'$) and its discriminant group is 3-elementary.
(3) We have $S_X^{A_5} = U = U(1)$, or $U(3)$, where $U(n) = \mathbf{Z}[u_1, u_2]$ is a rank 2 lattice with $u_i^2 = 0$ and $u_1.u_2 = n$.

Proof. (1) The first part is from (**1.1**). For a generator g of μ_3, since $o(g) = 3$ and by the form of the decomposition in (**1.6**), each χ_i ($i = 4, 5$) is g-stable. Since the order-3 element g acts on the rank-2 lattice $S_X^{A_5}$ (which is defined over $\mathbf{Z}$ and whose $\mathbf{C}$-extension is $\chi_1 \oplus \chi_1'$), it has at least one eigenvalue equal to 1 because $G = \langle A_5, g \rangle$ stabilizes an ample line bundle (the pull back of an ample line bundle on X/G). So $g^*|S_X^{A_5} = \mathrm{id}$. The commutativity of g with all elements in A_5 implies that $g^*|\chi_i$ is a scalar multiple, by Schur's lemma.

Thus we can write $g^*|S_X \otimes \mathbf{C} = \mathrm{diag}[1, 1, \zeta_3^b I_4, \zeta_3^c I_4, \zeta_3^d I_5, \zeta_3^e I_5]$, where the ordering is the same as in (**1.6**). Let $a \in A_5$. Then $(ga)^*|T_X = g^*|T_X$ and the latter can be diagonalized as $\mathrm{diag}[\zeta_3, \zeta_3^{-1}]$, noting that $\mathrm{rank}\, T_X = 22 - \mathrm{rank}\, S_X = 2$ [Ni1, Theorem 0.1], (**1.0A-B**). So $\mathrm{Tr}(ga)^*|T_X = -1$. As in the proof of (**1.8**), the topological Lefschetz fixed point formula implies that $\chi(X^{ga}) = 2 + \mathrm{Tr}(ga)^*|T_X + \mathrm{Tr}(ga)^*|S_X = 1 + \mathrm{Tr}(ga)^*|S_X = 3 + \zeta^b\mathrm{Tr}(a^*|\chi_4) + \zeta^c\mathrm{Tr}(a^*|\chi_4') + \zeta^d\mathrm{Tr}(a^*|\chi_5) + \zeta^e\mathrm{Tr}(a^*|\chi_5')$. So for $a = \mathrm{id}, 2A, 3A, 5A$ with nA denoting an element of order n in A_5, we have:

$$\chi_{\mathrm{top}}(X^g) = 3 + 4(\zeta_3^b + \zeta_3^c) + 5(\zeta_3^d + \zeta_3^e),$$

$$\chi_{\mathrm{top}}(X^{g2A}) = 3 + \zeta_3^d + \zeta_3^e,$$

$$\chi_{\mathrm{top}}(X^{g3A}) = 3 + \zeta_3^b + \zeta_3^c - \zeta_3^d - \zeta_3^e,$$

$$\chi_{\mathrm{top}}(X^{g5A}) = 3 - \zeta_3^b - \zeta_3^c.$$

The fact $\chi(X^{g5A}) = 4$ in (**1.4**) implies that $(\zeta_3^b, \zeta_3^c) = (\zeta_3, \zeta_3^{-1})$ after switching χ_4 with χ_4' if necessary. Since $\chi(X^{g3A}) = 0$ is in $\mathbf{R}$ (in $\mathbf{Z}$, indeed), we may assume that $(\zeta_3^d, \zeta_3^e) = (\zeta_3, \zeta_3^{-1})$, or $(1, 1)$. If the former case occurs then the lemma is true. Suppose that the latter case occurs. Then $\chi_{\mathrm{top}}(X^g) = 9$, whence $n_g = 2$ and $|X_{\mathrm{isol}}^g| = m_g = n_g + 3 = 5$ by (**1.4**). Since g commutes with every element in A_5, our A_5 acts on the 5-point set X_{isol}^g. By (**1.7**), A_5 either fixes a point P_1 of the set (and hence $A_5 < SL(T_{X,P_1})$, contradicting (**1.0C**)), or acts transitively as a subgroup ($= [S_5, S_5]$) of S_5, on the set with an order-12 stabilizer (of a point P_1) subgroup $A_4 < A_5$, so $A_4 < SL(T_{X,P_1})$, contradicting (**1.0C**). This proves the assertion (1).

(2) The first part follows from (1), that $g^*|T_X \otimes \mathbf{C} = \mathrm{diag}[\zeta_3, \zeta_3^{-1}]$ w.r.t. to a suitable basis by [Ni1, Theorem 0.1] and that all lattices in (2) are primitive (of the same rank as they turn out to be) in $L := H^2(X, \mathbf{Z})$. We still have to show that the discriminant group $A_{L^g} = \mathrm{Hom}(L^g, \mathbf{Z})/L^g$ of L^g is 3-elementary. Let $L_g = (L^g)^\perp$ be the orthogonal of L^g in L. Then $g^*|L_g$ has only $\zeta_3^{\pm}$ as eigenvalues. Now arguing as in [OZ2, Lemma (1.3)] (for the finite index sublattice $L^g \oplus L_g$ of L, instead of $S_X \oplus T_X$), we can show that A_{L^g} is 3-elementary.

(3) follows from (2). See [CS, Table 15.2a]. $\square$

The fixed locus X^g can be determined:

Lemma 3.3. (1) With the assumption and notation in (**3.1**) and (**3.2**), either $X^g = C \coprod R$ is a disjoint union of a genus-5 curve C and a curve R ($\cong \mathbf{P}^1$) (so $C^2 = 8$, and $S_X^g = U \supset \mathbf{Z}[C, R]$), or X^g equals a sinlge genus-4 curve C (so $C^2 = 6$).

(2) In the former case, $\Phi_{|C|} : X \to \mathbf{P}^5$ is a degree-2 morphism onto either the Veronese-embedded $\mathbf{P}^2$ in $\mathbf{P}^5$ or the normal cone $\overline{\Sigma}_4$ over a rational normal twisted quartic in $\mathbf{P}^4$.

Proof. Since $\chi(X^g) = -6$ by (**3.2**), we have $n_g = -3$ and $m_g = 0$ in notation of (**1.4**). $n(g) < 0$ infers that X^g is a disjoint union of a smooth curve C of genus ≥ 2 and t of $\mathbf{P}^1$'s with $-6 = 2 - 2g(C) + 2t$ (see (**1.2**)). The fact that rank $S_X^g = 2$ in (**3.2**) implies that either $t = 0$ (so $g(C) = 4$), or $t = 1$ (so $g(C) = 5$) so that the two curves in X^g give rise to two linearly independent classes of S_X^g.

If $S_X^g = U(3)$, then $C^2 \equiv 0 \pmod 3$ because C is in S_X^g, whence $C^2 = 6$. This proves the first assertion of the lemma, by virtue of (**3.2**).

Consider the case $X^g = C \coprod R$. By [SD, Theorem 3.1], $|C|$ is base point free and we have a morphism $\varphi := \Phi_{|C|} : X \to \mathbf{P}^5$. Now $8 = C^2 = \deg(\varphi).\deg(\mathrm{Im}\,\varphi)$, where $\deg(\mathrm{Im}\,\varphi) \geq 5 - 1$. Thus either φ is an embedding modulo the curves in $C^\perp$, or φ is a degree-2 map as described in (**3.3**) [SD, Theorem 5.2, Propositions 5.6 and 5.7].

Write $S_X^g = \mathbf{Z}[u_1, u_2]$ with $u_i^2 = 0$ and $u_1.u_2 = 1$. Express $C = a_1 u_1 + a_2 u_2$. Then $8 = 2a_1 a_2$ and we may assume that $(a_1, a_2) = (2, 2)$ or $(4, 1)$ (after replacing u_i by $-u_i$ or switching u_1 with u_2 if necessary). So $C.u_i > 0$ and hence the Riemann-Roch theorem implies that $\dim |u_1| \geq 1$. Write $|u_1| = |M| + F$ with $|M|$ the movable part. Then $0 < C.M \leq C.u_1 = a_2 \leq 2$. If φ is birational then $\varphi(M)$ is a plane conic or a line, whence $M \cong \mathbf{P}^1$, $M^2 = -2$ and $|M|$ is not movable, a contradiction. This proves the lemma.

We now prove (**3.1**) (1). Consider the case in (**3.3**), where $X^g = C \coprod R$ and $\varphi = \Phi_{|C|} : X \to \mathbf{P}^5$ is a degree-2 morphism onto the Veronese-embedded $\mathbf{P}^2$ in $\mathbf{P}^5$. Since C (and hence $|C|$) is G-stable, there is an induced action of G on $\mathbf{P}^5$ (and hence also an action of G on the image $\varphi(X) = \mathbf{P}^2$) so that the map φ is G-equivariant. The $G = A_5 \times \mu_3$ action on the image is also faithful because A_5 is simple and $\deg(\varphi) = 2$ is coprime to 3 $(= |\mu_3|)$. The action of A_5 on the image is via $A_5 \subset SL_3(\mathbf{C}) \subset PGL_3(\mathbf{C})$ and is given in Burnside [Bu, §232, or §266] (**1.8**). In particular, the commutativity of g with the two generators (order 5 and 2) of A_5 in [Bu, §266] shows that g is a scalar and acts trivially on the image $\varphi(X) = \mathbf{P}^2$, a contradiction.

Consider the case in (**3.3**), where $X^g = C \coprod R$ in (**3.3**) and $\varphi = \Phi_{|C|} : X \to \mathbf{P}^5$ is a degree-2 morphism onto the cone $\overline{\Sigma}_4$. Note that the minimal resolution Σ_4 of $\overline{\Sigma}_4$ is the Hirzebruch surface of degree 4. As in the previous case, there is a faithful action of G on $\overline{\Sigma}_4$ such that φ is G-equivariant. Note that the image $\varphi(C)$ is a hyperplane section away from the singularity and with $\varphi(C)^2 = 4$. Let ℓ be a generating line of the cone $\overline{\Sigma}_4$. Then $\varphi(C) \sim 4\ell$ as Weil divisors. This gives rise to a $\mathbf{Z}/(4)$-cover $\pi : Y = Spec \oplus_{i=0}^3 \mathcal{O}_{\overline{\Sigma}_4}(-i\ell) \to \overline{\Sigma}_4$ which is (totally) ramified exactly over $\varphi(C)$. One sees that $Y \cong \mathbf{P}^2$ and $\pi^*\varphi(C) = 4L$ with L a line. Clearly, $A_5 (< G)$ stabilizes the divisorial sheaves $\mathcal{O}(-i\ell)$ and fixes the defining equation of $\varphi(C)$, so there is an induced faithful A_5-action on $Y = \mathbf{P}^2$ so that π is A_5-equivariant (see (**1.7**)). Now L is stabilized by A_5 (because so is $\varphi(C)$). So the defining equation $F_1 = 0$ of L is semi A_5-invariant (and hence A_5-invariant because of the simplicity of the group A_5). But every A_5-invariant form is of even degree by [Bu, §266], noting also that the action of A_5 on $\mathbf{P}^2$ is via $A_5 \subset SL_3(\mathbf{C}) \to PGL_3(\mathbf{C})$ by (**1.8**). We reach a contradiction.

Consider the case $X^g = C$ in (**3.3**). Let $f : X \to Y = X/\langle g \rangle$ be the quotient map. There is an induced faithful action A_5 on Y so that f is A_5-equivariant. Then by the ramification divisor formula, $0 \sim K_X = f^*(K_Y) + 2C$. Pushing down by f_*, one obtains $0 \sim 3K_Y + 2B$ with $B = f_*C = f(C) \cong C$ and $f^*B = 3C$, so $B^2 = 3C^2 = 18$. Solving, one obtains $B = (-3/2)K_Y$ and $K_Y^2 = 8$. Thus the smooth ruled surface Y equals a Hirzebruch surface Σ_d of degree d. The irreducibility of B (being a **Z**-divisor) implies that $d = 0, 2$ [Ha, Ch V, Cor. 2.18].

Suppose that $d = 2$. Then the (-2)-curve M on Y is disjoint from $B = (-3/2)K_Y$ and hence $f^*M = \coprod_{i=1}^{3} M_i$ is a disjoint union of three (-2)-curves not intersecting C. Since M is clearly A_5-stable, the set $\coprod M_i$ is also A_5-stable, whence each M_i is A_5-stable **(1.7)**. An order-5 element $5A$ in A_5 acts on each M_i faithfully by **(1.2)** and has exactly two fixed points by **(1.8)**. But according to **(1.2)**, $4 = |X^{5A}| \geq \sum_{i=1}^{3} |M_i^{5A}| = 6$, a contradiction.

Thus $d = 0$. Clearly, the simple group A_5 stabilizes each ruling and there is an induced action $\rho_i : A_5 \times \mathbf{P}^1 \to \mathbf{P}^1$ with $i = 1, 2$ for the i-th $\mathbf{P}^1$ in $Y = \Sigma_0 = \mathbf{P}^1 \times \mathbf{P}^1$ so that $\rho_1 \times \rho_2$ is the given A_5 action on Y. Changing coordinates suitably we may assume that the A_5 action on Y commutes with the involution ι of Y switching the two $\mathbf{P}^1$'s in Y, i.e., $\rho_1 = \rho_2$ as actions of A_5 on the same $\mathbf{P}^1$ **(1.8)**.

Let $j : Y \to Z = Y/\langle \iota \rangle = \mathbf{P}^2$ be the quotient map. Then there is an induced faithful action of A_5 on $\mathbf{P}^2$ such that j is A_5-equivariant. Now $\iota(B)$ is an irreducible curve with $(\iota(B))^2 = B^2/2 = 9$. It is a cubic curve and A_5-stable because so is $B = f(C)$. The action of A_5 on $Y/\langle \iota \rangle = \mathbf{P}^2$ is via $SL_3(\mathbf{C}) \to PGL_3(\mathbf{C})$ **(1.8)**. The defining equation F_3 of $\iota(B)$ is then a cubic form and semi A_5-invariant (and hence A_5-invariant by the simplicity of the group A_5). However, Burnside [Bu, §266] shows that every A_5-invariant form is of even degree, a contradiction. This completes the proof of **(3.1)** (1) and also of **(3.1)**.

We now prove Theorem B. Suppose that $G = A_5.C_n$ acts faithfully on a $K3$ surface X. By **(1.0A)**, $A_5 \leq G_N$. So $G_N = A_5$, S_5, A_6 or $M_{20} = C_2^{\oplus 4} : A_5$ by [Xi, the list]. In notation of **(1.0)**, for some $m \mid n$, we have $G_N = \text{Ker}(\alpha) = A_5.C_m$ and $G/G_N = \mu_I$, where $n = mI$. By the same proof of **(1.1)**, we have $I = 1, 2, 3, 4$, or 6. Let $h \in G$ such that the coset of h is a generator of $G/A_5 = C_n$. Then $h^*\omega_X = \eta_I \omega_X$ for some primitive I-th root η_I of 1. Note that $n \mid \text{ord}(h)$ and $h^I \in G_N$, whence $\text{ord}(h^I) \leq 8$ by **(1.2)**. Thus $\text{ord}(h) = I \, \text{ord}(h^I)$ and $m \mid \text{ord}(h^I)$. In particular, $|G_N| = m|A_5| \leq 8|A_5|$. Hence $G_N \neq M_{20}$.

If $G_N = A_5.C_m = A_6$, then $m = 6$ and A_6 includes $\langle h^I \rangle \geq C_6$, which is impossible. If $G_N = A_5$, then $C_n = \mu_I$ in notation of **(1.0)**, and Theorem B follows from **(3.1)**. Consider the case $G_N = S_5$. Then $m = 2$ and $n = 2I$. Moreover, $G_N = \langle A_5, h^I \rangle$. So $h^I \in S_5 - A_5$. Since $S_5 \to \text{Aut}(S_5)$ ($x \mapsto c_x$) is an isomorphism, we have $c_h = c_s$ for some $s \in S_5$. Set $g = hs^{-1}$. Then g commutes with every element in S_5 and also $\alpha(g) = \alpha(h)$ is a generator of $\text{Im}(\alpha) = \mu_I$. Now $g^I \in \text{Ker}(\alpha) = G_N = S_5$ is in the centre of S_5 (which is (1)). So $\text{ord}(g) = I$ and $G = G_N \times \langle g \rangle = S_5 \times \mu_I > A_5 \times \mu_I$. By **(3.1)**, we have $I = 1, 2$ or 4. The $S_5 \times \mu_I$ should have an element h such that $h^I \in S_5 - A_5$ (i.e., h^I is not an even permutation). Thus, $I \neq 2$, or 4. Therefore, $I = 1$ and $G = G_N = S_5$. This completes the proof of Theorem B.

References

[1] [AS2] M. F. Atiyah and G. B. Segal, The index of elliptic operators. II. Ann. of Math. 87 (1968), 531–545.

[2] [AS3] M. F. Atiyah and I. M. Singer, The index of elliptic operators. III. Ann. of Math. 87 (1968) 546–604.

[3] [Atlas] J. H. Conway, R. T. Curtis, S. P. Norton, R. A. Parker and R. A. Wilson, Atlas of finite groups. Oxford University Press. Reprinted 2003 (with corrections).

[4] [Bu] W. Burnside, Theory of groups of finite order. Dover Publications, Inc., New York, 1955.

[5] [CS] J. H. Conway and N. J. A. Sloane, Sphere packings, lattices and groups. 3rd ed. Grundlehren der Mathematischen Wissenschaften, 290. Springer-Verlag, New York, 1999.

[6] [EDM] Encyclopedic dictionary of mathematics. Vol. I–IV. Translated from the Japanese. 2nd ed. Edited by Kiyosi It. MIT Press, Cambridge, MA, 1987.

[7] [Ha] R. Hartshorne, Algebraic geometry. Graduate Texts in Mathematics, No. 52. Springer.

[8] [IOZ] A. Ivanov, K. Oguiso and D. -Q. Zhang, The monster and $K3$ surfaces, in preparation.

[9] [KOZ1] J. Keum, K. Oguiso and D. -Q. Zhang, The alternating group of degree 6 in the geometry of the Leech lattice and $K3$ surfaces. Proc. London Math. Soc. (3) 90 (2005), no. 2, 371–394.

[10] [KOZ2] J. Keum, K. Oguiso and D. -Q. Zhang, Extensions of the alternating group of degree 6 in geometry of $K3$ surfaces, European J. Combinatorics: Special issue on Groups and Geometries (in press); also math.AG/0408105.

[11] [Ko1] S. Kondo, Niemeier lattices, Mathieu groups, and finite groups of symplectic automorphisms of $K3$ surfaces. Duke Math. J. 92 (1998), 593–598.

[12] [Ko2] S. Kondo, The maximum order of finite groups of automorphisms of $K3$ surfaces. Amer. J. Math. 121 (1999), 1245–1252.

[13] [MO] N. Machida and K. Oguiso, On $K3$ surfaces admitting finite non-symplectic group actions. J. Math. Sci. Univ. Tokyo 5 (1998), 273–297.

[14] [Mu1] S. Mukai, Finite groups of automorphisms of $K3$ surfaces and the Mathieu group. Invent. Math. 94 (1988), 183–221.

[15] [Mu2] Lattice-theoretic construction of symplectic actions on $K3$ surfaces, Duke Math. J. 92 (1998), 599–603. As the Appendix to [Ko1].

[16] [Ni1] V. V. Nikulin, Finite automorphism groups of Kahler $K3$ surfaces, Trans. Moscow Math. Soc. 38 (1980), 71–135.

[17] [Ni2] V. V. Nikulin, Factor groups of groups of automorphisms of hyperbolic forms with respect to subgroups generated by 2-reflections. Algebrogeometric applications. J. Soviet Math. 22 (1983), 1401–1475.

[18] [Ni3] V. V. Nikulin, Integer symmetric bilinear forms and some of their applications. Math. USSR Izvestija. 14 (1980), 103 – 167.

[19] [Og] K. Oguiso, A characterization of the Fermat quartic $K3$ surface by means of finite symmetries. Compos. Math. 141 (2005), no. 2, 404–424.

[20] [OZ1] K. Oguiso and D. -Q. Zhang, On the most algebraic $K3$ surfaces and the most extremal log Enriques surfaces. Amer. J. Math. 118 (1996), 1277–1297.

[21] [OZ2] K. Oguiso and D. -Q. Zhang, On Vorontsov's theorem on $K3$ surfaces with non-symplectic group actions. Proc. Amer. Math. Soc. 128 (2000), 1571–1580.

[22] [OZ3] K. Oguiso and D. -Q. Zhang, The simple group of order 168 and $K3$ surfaces. Complex geometry (Gottingen, 2000), Collection of papers dedicated to Hans Grauert, 165–184, Springer, Berlin, 2002.

[23] [SD] B. Saint-Donat, Projective models of $K-3$ surfaces. Amer. J. Math. 96 (1974), 602–639.

[24] [Su] M. Suzuki, Group theory. I. Translated from the Japanese by the author. Grundlehren der Mathematischen Wissenschaften 247. Springer-Verlag, Berlin-New York, 1982.

[25] [Xi] G. Xiao, Galois covers between $K3$ surfaces. Ann. Inst. Fourier 46 (1996), 73–88.

[26] [Z1] D. -Q. Zhang, Quotients of $K3$ surfaces modulo involutions. Japan. J. Math. (N.S.) 24 (1998), 335–366.

[27] [Z2] D. -Q. Zhang, The alternating groups and K3 surfaces, J. Pure and Appl. Alg (in press); also math.AG/0506610.

DEPARTMENT OF MATHEMATICS, NATIONAL UNIVERSITY OF SINGAPORE, 2 SCIENCE DRIVE 2, SINGAPORE 117543

Titles in This Series

TITLES IN THIS SERIES

For a complete list of titles in this series, visit the
AMS Bookstore at **www.ams.org/bookstore/**.